Time in the Eternal City

Later Medieval Europe

VOLUME 22

The titles published in this series are listed at *brill.com/lme*

Time in the Eternal City

Perceiving and Controlling Time in Late Medieval and Renaissance Rome

Edited by

Tuomas Heikkilä

BRILL

LEIDEN | BOSTON

Cover illustration: View of Rome, from Schedel, Hartmann. *Buch der Chroniken.* Nürnberg: Anton Koberger, 1493, f. LVIIv. Photo with permission of the National Library of Finland.

The Library of Congress Cataloging-in-Publication Data is available online at http://catalog.loc.gov
LC record available at http://lccn.loc.gov/2020033718

Typeface for the Latin, Greek, and Cyrillic scripts: "Brill". See and download: brill.com/brill-typeface.

ISSN 1872-7875
ISBN 978-90-04-43624-4 (hardback)
ISBN 978-90-04-43625-1 (e-book)

Contents

Illustrations

Tables

Graphs

Notes on Contributors

Marko Halonen
wrote his PhD dissertation at the University of Helsinki on medieval Nordic calendars using digital phylomemetic methods. He has also written several articles concerning medieval calendars, stemmatology and textual criticism. He is currently preparing a new critical edition of Paulus Juusten's (1516–1576) chronicle of Finnish bishops, as well as comparative research about medieval calendars in Europe and the Middle East, particularly from the perspective of climate history. Halonen spent the academic year of 2007–2008 at the University of Edinburgh as a visiting student. During the academic year of 2014–2015 he was the Wihuri-fellow at the Finnish Research Institute in Rome.

Tuomas Heikkilä
is Professor in Church History at the University of Helsinki. He has previously been director of the Finnish Research Institute in Rome (2013–2017); visiting research fellow at the Swedish Collegium for Advanced Study; and visiting scholar at the University of Gothenburg. Originally a historian, Heikkilä has emphasized multidisciplinarity in his scholarship. His fields of expertise include medieval written culture, medieval hagiography, manuscript studies, monastic history, computer-assisted stemmatology, and phylomemetics. Heikkilä has published widely on these topics in several languages. The present volume contains results of a research project he directed at the Finnish Research Institute in Rome.

Holger Kaasik
is a PhD candidate in medieval history at the University of St Andrews, who specializes in medieval calendar manuscripts, especially those originating from 12th to late-15th century central Italy. His research emphasizes the practical and adaptable use of medieval calendars as tools for perpetuating communal time and curating communal memory, and the methodological questions concerning the use of calendars in manuscript localisation in modern scholarship. More broadly, his thematic interests range from calendars and liturgy to sanctity and communal religion, as well as practical questions regarding time, community and the history of memory.

Urpo Kantola
MA, is a PhD candidate in classical philology at the University of Helsinki. His current work concerns Graeco-Roman linguistic and cultural contact through

Roman names in ancient Greek sources. Apart from this, his main research interests lie in the study of Greek and Latin written primary sources, particularly epigraphy. In addition, he is interested in the interaction between Greek and Latin speaking cultures from the Hellenistic period onwards, Graeco-Roman historiography, and Homeric studies.

Jasmin Lukkari
is currently preparing her PhD thesis in ancient history for the Universities of Helsinki and Cologne. She is interested in changes over long periods of time in ancient and medieval Rome and Italy. She has specialized in the study of historiography and historical narratives from the narratological point of view. Her current project explores republicanism and the conceptions of autocracy in ancient Rome.

Saku Pihko
is a PhD candidate in history at Tampere University in Finland, who specializes in medieval inquisition records, especially those originating from 13th and early 14th century Languedoc. His research emphasizes methodological questions concerning the construction of information in premodern legal records and the possibilities and limits of knowledge regarding their content. More broadly, his thematic interests range from heresy and inquisition to communication and lived religion, as well as theoretical questions regarding time, space and the history of experiences.

CHAPTER 1

Time and the Eternal City

Tuomas Heikkilä

1 Studying the Stuff Life Is Made of

The present volume studies time in late medieval and Renaissance Rome. How was time and its passing perceived and understood? How and by whom was it regulated and controlled? How was it present in the lives of Romans? It is with the aid of such basic questions that we attempt to orientate ourselves in the streets of the Eternal City and in the minds of its inhabitants more than half a millennium ago.

Why study time? Because it matters. According to the wise words attributed to Benjamin Franklin (1706–1790), time "is the stuff life is made of." To understand time is to understand life itself. Therefore, time has always been a riddle human beings and societies have tried to solve, be means religion, astrology, or theoretical physics. Time affects everyone's life, and it must be considered a factor in one's activities, regardless of his or her status. Hence, it is no surprise to find out that time has been investigated by scholars and thinkers of all eras, from Aristotle to Albert Einstein, and from Augustine to Thomas Aquinas and Karl Marx. Time has been both covered with religious meanings and cultural contents and studied by the cleverest scientists of all eras.

Time is both personal and universal. Life is a passage through time, and on a personal level a moment squandered is lost forever. Ageing cannot be stopped and today inevitably becomes yesterday. On a more general level, though, time is (at least seemingly) endlessly recurrent: days follow each other in an endless parade, and so do the seasons. The combination of the two ideas results in an understanding of a human lifetime as just a tiny stitch in the huge tapestry of time. In search of understanding time, this tapestry, in turn, has been embroidered with religious ideas.

In a sense, every historical study is a research of and in time. The goal of establishing a common temporal framework spanning, say, ancient pagan Rome and medieval Christian Rome is easy enough to understand from the viewpoint of a historian, since the very nature of the study of history requires a temporal yardstick that can be applied throughout the past. Yet there is much more in time than just chronology. The traditional scholarly approach underlining the importance of chronology fails to convey to us the multitude of

 | DOI:10.1163/9789004436251_002

various ways of keeping and understanding time in the Middle Ages. Elucidating these ways and concepts of understanding provides us with novel tools to better comprehend broad cultural phenomena of societies and communities in the past.

Why Rome? By examining just one city and its inhabitants, we hope to contribute to an important recent trend in scholarship on the cultural history of time: that of building the big picture by adding small pieces of new information to the existing body of knowledge. And Rome, of course, is a special case. While one should not exaggerate the cultural or religious uniformity of Western Christendom, the symbolic capital of which was Rome, the Eternal City enjoyed a special status throughout the Middle Ages and beyond. In terms of time, Rome was present in every parish throughout Western Christianity in three respects. Firstly, the liturgical calendar used was essentially the ancient Roman time-keeper with Christian additions and thus showcased a very concrete continuum from ancient Rome. Secondly, many of the saints present in the calendar and elsewhere in liturgy had a Roman connection. Thirdly, Rome and the pope symbolized the very roots of Western Christianity and its current centre. These three intertwined symbolic aspects – a temporal and ecclesiastical continuum from the past as well as representing religious significance in the present – have been cornerstones of the importance of Rome ever since Late Antiquity but have also had concrete contents in relation to perceiving time in the Middle Ages and the Renaissance.

While the volume at hand deals with a period covering the last centuries of the Middle Ages and those of the Renaissance, it is hard to see any major distinction in the perception and understanding of time in the late medieval and Renaissance periods. Surely, the Renaissance in its various forms drew inspiration from the ancient glorious days of Rome, but so did medieval scholars and artists, who understood their own time as a direct continuum from antiquity.[1]

Medieval Romans stood almost concretely on the shoulders of ancient giants, and a certain air of self-evident continuity from the antiquity was present in their minds, as can be observed in a multitude of time-related sources. The present volume emphasizes the importance of continuity and a long history of inspiration rather than sharp contrasts and swift changes. In the Middle Ages,

1 Ingrid Rowland, "Roma Aeterna," in *The Cambridge Companion to Ancient Rome*, ed. Paul Erdkamp (Cambridge: Cambridge University Press, 2013), 562. In fact, there is evidence suggesting that the difference between the "Middle Ages" and "Renaissance" was not that marked or significant even in the field of art, and many art historians emphasizing the uniqueness of emerging Renaissance art have fallen for the exaggerated propaganda of late medieval rulers, patrons of art, artists and authors. In Roman reality, art and architecture had never truly departed from their classical ideas and examples.

the city of Rome itself enjoyed unique status as a concrete symbol of eternity and continuity from the past. Many of its ancient temples were used as Christian churches, and the age-old civic and imperial rituals and traditions infused the ceremonies of the Western Church. As the city of the pope, Rome was a symbol of apostolic succession from St Peter to the present day and of the transformation of Christianity from a modest, persecuted sect to a religion that dominated most of Europe.

As to Rome being "eternal," the first explicit reference to Rome as *urbs aeterna* and *Roma aeterna* date back to the late 1st century BC. Although present every now and then in the medieval sources, the use of such expressions increased sharply only during the 15th and 16th centuries, as Rome was being recreated, as it were, by the Renaissance popes.[2] The ancient and medieval epithet "eternal" Rome was not to be taken literally – and, thus, was not a temporal expression, per se – but had rather to do with the idea of extreme longevity and continuity from the distant past. It was in this form that eternity can be considered a part of Roman identity within the timescape of this volume.

Despite the extraordinary history of the city, Rome's rulers and inhabitants alike combined past and present in their conceptions of time. The period under scrutiny in the volume at hand brought its own novel additions to this mixture, as the cultural movement of the Renaissance emphasized the rediscovery of classical antiquity. This can be observed, for example, in Urpo Kantola's chapter on temporal expressions in late medieval epigraphy. The changes, however, were more superficial than profound and apparently did not affect the ways of perceiving time.

2 What Is Time?

In a famous passage from his *Confessions*, St Augustine of Hippo (354–430) ponders the essence of time. He asks himself: "What then is time?" and goes on to answer: "If no one asks me, I know; if I want to explain it to a questioner, I do not know."[3] His quote captures well the complex nature of time. On the everyday level, time is something self-evident, but on closer consideration it becomes utterly complex and difficult to comprehend.

2 See, e.g., Elizabeth McCahill, *Reviving the Eternal City: Rome and the Papal Court, 1420–1447* (Cambridge, MA: Harvard University Press, 2013).

3 Augustine, *The Confessions*, XI, 14, ed. Michael P. Foley (Indianapolis: Hackett Publishing Company, 2006), 242.

St Augustine explained the twofold nature of time by arguing that time is simultaneously a phenomenon of human consciousness and a creature of God.[4] In spite of the contributions of later medieval and Renaissance philosophers and theologians to the understanding and explanation of time, most of the protagonists of this volume, medieval Romans, would have found it easy to accept the old Augustinian view. On one hand, time was an everyday consumer goods, but on the other it could be something very solemn and divine. Time, in its various manifestations, was self-evidently present in daily life, but just as obviously it was important when pondering grand existential questions.

Time is tricky. According to the laws of physics, time does not pass with equal speed everywhere. In lower locations all processes are slower, whereas time passes more swiftly the higher one goes. Or, should we stand absolutely still, our time would pass quicker than if we moved.[5] Scientifically, there is not just one objective time, but a plethora of times, all objective in their respective contexts. This is equally true at the level of the everyday experiences of an individual. In the end, this multitude of "objective times" results in all "personal times" being just that: personal and thus subjective. When trying to study and decipher time in medieval and Renaissance Rome, we are dealing with an infinite number of subjective, personal ways of understanding the passing of time and living in it.

There were several commonly recognized and shared temporal frameworks in Rome, from the passing of days and seasons to papal regna and to ecclesiastical calendar(s). In fact, it was not the personally felt and uniquely lived time but the strive towards following a commonly shared organization of time that created coherence, brought order to chaos and subjected people to this order.

Studying time has its caveats. In terms of time perception, the past is a foreign country with familiar traits. It is tempting to think that we thoroughly understand Augustine's feelings from 1,600 years ago. We share many of the cultural features important to him – but at the same time his exact cultural context inevitably escapes us. On one hand, the past influences the present time through direct continuums that may be hard to pinpoint due to their

4 William Alexander Hernandez, "St. Augustine on Time," in *International Journal of Humanities and Social Science* 6:6 (2016), 37–40; Wesley Stevens, "A Present Set of Things Past: *Quid est enim tempus?*," in *Time and Eternity: The Medieval Discourse*, ed. by Gerhard Jaritz & Gerson Moreno-Riaño, International Medieval Research 9 (Turnhout: Brepols 2003), 9–14; cf. John L. Morrison, "Augustine's Two Theories of Time," *New Scholasticism* 45 (1971), 600–610.

5 For a recommendable introduction to time in physics, see Carlo Rovelli, *The Order of Time* (London: Allen Lane, 2018).

self-evident nature, and on the other it is as an endless treasure trove of material that can be moulded to serve various needs.

Albeit time and its passing were omnipresent in the lives of medieval men and women, it escaped an exact definition, just as it does if we try to define it today. In fact, even St Augustine had to confess in the end: "And I confess to thee, O Lord, that I am still ignorant as to what time is."[6]

3 Controlling Time, Controlling People's Lives

The interest of humankind in measuring, organizing and thus controlling time is itself timeless. Even animals organize their activities according to the predictable regularities of nature, like the darkness of the night and the brightness of daylight, or the varying seasons. As to humans, our endeavour to measure and thus master time are probably as old as humanity itself. There are already Palaeolithic finds that probably record a system of keeping track of time,[7] and for instance, the huge Neolithic stone circle in Stonehenge served according to popular theory as a lunisolar calendar.

Time is an abstraction and easily escapes precise definitions. It can only be made visible by measuring it by a common agreement. All kinds of timekeepers – sundials, mechanical clocks, calendars, etc. – serve to find and partly create regularities and thus order in a seemingly chaotic world. Even the creation stories of various cultures tell their own eloquent story of making sense of things that are senseless. They are one way of understanding the distant past and explaining the present.

The importance of time is already stressed in the very first extant literary works of mankind. In the 8th century BC, Hesiod dedicated a whole book, *Works and Days* (*Ἔργα καὶ Ἡμέραι*), to the topic,[8] and after him there is a wealth of time-related literature in antiquity. Many of these texts have been immensely influential in many ways in the Western sphere of culture throughout later centuries, but none of them have been as important as the Hebrew texts of the Old Testament, describing time and the past as a salvation history. It was against the background of this Hebrew *Heilsgeschichte* that the teachings of Jesus of Nazareth took place. It was also this centuries-old story of salvation

6 Augustine, *Confessions*, XI, 25.

7 J.T. Fraser, *Of Time, Passion, and Knowledge* (Princeton: Princeton University Press, 1990), 48.

8 Hesiod, *Theogony, Works and Days, Testimonia*, ed. Glenn W. Most, Loeb Classical Library (Harvard University Press, 2006).

that later Christian theology moulded into a multi-phased plan by God, in which the phases were separated by Christ's life and resurrection as well as his second coming. Such developments paved the way for a shared understanding of time that prevailed in medieval Rome, the physical context of this book.

The context of the articles of this volume is Christian, and we find ourselves in the very centre of Western Christianity, Rome. The ways and aims of controlling time for a medieval Roman were Christian on the surface, but underneath their Christian camouflage of saints' feasts and other aspects of *interpretatio Christiana* lies a more general human interest in locating oneself within the maelstrom of time.

Time is intricately linked with power. The one who controls people's time, be it through interpretations of the past, calendars or other timekeepers, controls people. This is why both temporal and religious rulers have tried to control, make use of, and make sense of time as long as we can see into the darkness of the past. It is not a coincidence that our present calendar is a result of inventions, innovations and reforms of ancient Roman rulers and the church. In fact, practically all new regimes have tried to influence the life of the people through controlling time: from Julius Caesar to Augustus, from the counter-Reformation Catholic Church to Fascist Italy, or from the French Revolution to the Communist Soviet Union. Social time has been and continues to be embedded in the social relations of power.

The medieval power aspect of controlling time was dependant on the fact that time was shared. Whereas time was experienced individually, it was lived commonly. The ways of keeping track of time were forms of social organization and contributed to constructing a social time.[9]

In medieval and Renaissance Rome, time was a topic much dominated – albeit not totally, as we shall see in Saku Pihko's and Marko Halonen's articles – by the views of the Christian faith. The Church has understood to use shared ideas of time as a tool since its earliest centuries. On one hand, the division between the past, the present and the future is quite central to the Christian salvation story. On the other hand, time-reckoning and calendars have been used to regulate the lives and to strengthen the faith of Christians, both on the level of daily routines (e.g., the regulations of monastic rules, and canonical hours) as well as the longer cycles of ecclesiastical calendars.

9 The idea was coined by Émile Durkheim (1858–1917). On his aftermath, see, e.g., William Watts Miller, "Durkheimian Time," in *Time & Society* 9:1 (2000), 5–20. On time as a social construct, see, e.g., Norbert Elias, *Time: An Essay*, trans. by E. Jephcott (Oxford: Blackwell, 1992), 46.

4 Calendars as Keys to Understanding Shared Time

There is a myriad of calendars from various epochs and from all parts of the world. Practically every culture has tried to make sense of the passing of time by creating a calendar.[10] A calendar is a device to regulate and control time *par excellence,* and it plays an important role in this study as a source genre through which we can reach aspects of time in the late Middle Ages and the Renaissance.

Calendars are carriers of ideas and influences, and they combine very long traditions with novel ideas and innovations. On one hand, it is easy to see calendars as something very traditional and resistant to change. The kings of ancient Egypt had to swear that they would not change the calendar; any change to the old Jewish calendar required the meeting of the Great Sanhedrin; the Julian calendar reform required quasi-imperial command and the Gregorian calendar reform needed papal authority.[11] Practical examples of well-known long calendrical traditions include the names of the weekdays and months, which still derive from ancient Rome in most Western languages. On the other hand, calendars are extremely flexible. As the whole point of their existence is to reflect the basic values, ideas, mythopoesis and most important historical personalities of the community upholding them, they were prone to change.

A calendar is, self-evidently, a tool to map time and to visualize its invisible passing. Even more importantly, a calendar is a shared cultural memory-aid for a community that upholds it. The *status quo* of a calendar is not a coincidence, but the result of important and noteworthy things, persons and events accumulating on various dates. Consequently, an ecclesiastical calendar in, say, medieval Rome was not just a list of 365 units that were important to keep track of time, for the dates carried significance *per se*. In a way, liturgical calendars brought the Christian past to the present day and thus allowed an interaction with the past as important moments in the history of their own community were being relived. The same phenomenon of bringing the past close to the audience has, in fact, been observed in medieval literature. As Mary Carruthers has suggested, while the past is the time of narrative, the present is the tense of narrative.[12]

10 Good introductions to various calendars and calendar systems are, e.g., Bonnie J. Blackburn & Leofranc Holford-Strevens, *The Oxford Companion to the Year* (Oxford: Oxford University Press, 1999); E.G. Richards, *Mapping Time: The Calendar and its History* (Oxford: Oxford University Press, 1998).

11 See Richards, *Mapping Time*, 110.

12 Mary Carruthers, "Meditations on the 'Historical Present' and 'Collective Memory' in Chaucer and *Sir Gawain and the Green Knight,"* in *Time in the Medieval World,* ed. by Chris Humphrey and W.M. Ormrod (York: Boydell & Brewer, 2001), 137–155.

In late medieval Rome, ecclesiastical calendars were used in every church, and much of their contents was familiar not only to the clergy using them but also to the laymen and -women whose lives were influenced by them. They provided many of the organizing principles of social life and dictated the rhythm of work and rest. The calendars were not only ubiquitous and necessary for the liturgy but also important in both building and manifesting the self-understanding of the community. In fact, the choice of contents of a calendar often mirrors the values and cultural contacts of the community maintaining the calendar. The real significance of calendars for both the contemporary community and a much later historian is that they reveal a carefully composed and updated list of values, ideas, and historical or quasi-historical personalities that the community considered important to remember. Thus, the study of calendars can bring invaluable insights into understanding medieval communities.

The idea of just one universal calendar to be used in the whole of Christendom was alien to the Middle Ages. Yes, there was a common basic framework of time-keeping applied throughout Western Christianity, but there was not one single calendar with a universal authority over the other local calendars of which there was a plethora at any given moment. Unlike modern national calendars, medieval calendars did not follow an entirely uniform model but rather showed a great deal of variation within certain limits. The contents of the local calendar were influenced by the calendars of other communities. Hence, the multitude of local variation within the source genre of calendars enables, in fact, the study of underlying cultural contacts, transmission of ideas, values, and the cults of saints, for instance.[13]

5 Towards a Better Knowledge of Time

In his seminal article from 1960 about the time of the Church and the time of the merchant, Jacques Le Goff wrote about the future possibilities of research on medieval time perception. He stated as his hope that "an exhaustive investigation will someday be made with the intention of showing in a particular historical society the interaction between objective structures and mental frameworks, between collective adventures and individual destinies, and between the various times within Time."[14]

13 For a recent approach resulting in new tools to study cultural influence through calendars, see Tuomas Heikkilä & Teemu Roos, "Quantitative Methods for the Analysis of Medieval Calendars," *Digital Scholarship in the Humanities* 33:4 (2018), 766–787.

14 Jacques Le Goff, "Au Moyen Âge: temps de l'Église et temps du marchand," *Annales. Economies, sociétés, civilisations* 15:3 (1960), 417–433. (Citation from the English translation:

Much ground has been covered by scholars of various disciplines since Le Goff's days. His hypothesis of binary time, often misunderstood to be meant as a watertight conclusion, has been contested and shown to be too simplistic.[15] Possibly the most important new finding has been the multitude of attitudes towards and ways of using time. The medieval concept of time was much more complex and fluid than anticipated by Le Goff and his generation. E.g., the church's time and the merchant's time were not mutually exclusive but were applied by the same individuals – alongside of other temporal frameworks that coexisted without confrontation. In all, it has turned out to be exceedingly difficult to capture the versatility and flexibility of time perception in just one study by one scholar.[16] While acknowledging the lack of exhaustivity, the present volume is hopefully taking a step in this direction within the context of late medieval and Renaissance Rome.

Time is omnipresent and has to do with every field of life. Thus, there is a myriad of possible and equally relevant approaches towards time in history, including, e.g., existential questions, memory, divination and future-telling, philosophy, development of time measurement, cosmologies, apocalyptic ideas, ageing, temporality and afterlife, etc. Thus, it is not the aim of the present volume to be exhaustive – that task would be impossible – but our goal is to study and represent some important and interlinking aspects to showcase various ways of controlling, understanding, and living time. As a whole, this book offers an insight into a variety of ways in which medieval people were engaged in the challenging yet necessary task of controlling and understanding time. It attests to the fact that the medieval experience of time was both varied and complex. The Romans presented in this book lived within a "timescape,"[17] a

---

Jacques Le Goff, *Time, Work, & Culture in the Middle Ages*, 38.) I thank Saku Pihko for pointing out Le Goff's vision.

15 See, e.g., Jonathan Martineau, *Time, Capitalism and Alienation: A Socio-Historical Inquiry into the Making of Modern Time* (Leiden & Boston: Brill, 2015), 68–75.

16 For meritorious contributions to the field, see, e.g., Peter Laslett, *The World We Have Lost Further Explored*, 3. edn (London: Routledge, 2005); Peter Burke, "Reflections on the Cultural History of Time," *Viator* 35 (2004), 617–626; Gerhard Jaritz & Gerson Moreno-Riaño, eds, *Time and Eternity: The Medieval Discourse*, International Medieval Research 9 (Turnhout: Brepols, 2003); Chris Humphrey, & W.M. Ormrod, eds, *Time in the Medieval World* (York: Boydell & Brewer, 2001); Gerhard Dohrn-van Rossum, *History of the Hour: Clocks and Modern Temporal Orders* (Chicago and London: University of Chicago Press, 1996), 1–15; Johannes Fabian, *Time and the Other* (New York: Columbia University Press, 1983), 1–35. More critically, Jeffrey Cohen, *Medieval Identity Machines* (Minneapolis and London: University of Minnesota Press, 2003), 3.

17 The concept was coined by Barbara Adam and it has become a standard in time studies. E.g., Barbara Adam, *Time* (Cambridge: Polity Press, 2004); Barbara Adam, *Timewatch: The Social Analysis of Time* (Cambridge: Polity Press, 1998); cf. Martineau, *Time, Capitalism and Alienation*, 43–46.

multitude of coexisting temporalities and time frames, only some of which can be touched upon in the following articles.

The articles of the volume at hand approach time from the viewpoint of the original sources, as they offer us the best insight into how something as self-evident yet complex and abstract as the passing of time was experienced and explained in Rome. The approach is mostly a practical one: it is not our intention to delve into the philosophical finesses shared by a handful of exceptional individuals but rather to trace practices and everyday ideas shared by many.[18]

The seven articles are divided into two parts, roughly corresponding to two categories of the community's ways of dealing with time. The first three articles deal with examples of "communicative and present time" and are concerned with manifestations of living in time in late medieval and Renaissance Rome. It contains a number of examples of practical, often informal everyday ways of understanding time, using and communicating various temporal frameworks in the present to one's contemporaries. The second part of the book – the four last articles – has to do with "cultural and historical time" that draws material from the past for the use of the present. Its articles have to with the complex practical and intellectual construct of the calendar and highlight a more formal view of structuring time in the past and a more ceremonial way of using it.

In such a categorization – "communicative and present time" vs "cultural and historical time" – we have drawn inspiration from the conceptual framework developed by Jan Assman for the study of premodern societies. According to him, societies were (and are) bound together by various shared manifestations of their cultural identification and joint memory, such as rituals, ceremonies and beliefs.[19] It is easy enough to add perceptions of time to this list of ideas building communities, as shared ideas of the shared past building the shared present were cornerstones of late medieval societies.

Admittedly, the division of the articles into two categories is by no means watertight. Past and present were mixed in all manifestations of understanding and keeping track of time, since there could be no present without the past in the Christian context of Rome. The versatility of the topics of individual articles of this volume lets the reader experience something of the multiformity of medieval ways of dealing with something as indefinite yet self-evident as

18 For a very concise introduction to the varying philosophical views on time, see, e.g., Hans Lenz, *Universalgeschichte der Zeit*, 2., überarbeitete Auflage (Wiesbaden: marixverlag, 2013), 13–25.

19 Jan Assman, *Religion und kulturelles Gedächtnis. Zehn Studien* (München: C.H. Beck, 2000); Jan Assman, *Das kulturelle Gedächtnis. Schrift, Erinnerung und politische Identität in frühen Hochkulturen* (München: C.H. Beck, 1992). See also Roy A. Rappaport, *Ritual and Religion in the Making of Humanity* (Cambridge: Cambridge University Press, 1999).

time. Although there are several points of connection between individual contributions in the volume, they also serve to show the fragmented state that the historical study of time finds itself in, ranging from studies of everyday life to philosophical ponderings, and from liturgy to *computus* or historiography.

6 Time Perception and Time Control: Seven Approaches

The first chapter takes us straight to the streets of medieval Rome. Saku Pihko's article offers an interesting insight into how individuals' notions of time were translated into more general conceptions when writing down the testimonies of witnesses in the canonization process of St Francesca Romana. It eloquently shows the interplay between practical, often fluctuating, individual ways of perceiving time and a more formal and uniform approach of keeping track of time. Through a careful analysis of the canonization acts of St Francesca Romana, Pihko showcases how time was understood and categorized by late medieval Romans. Interestingly, Pihko's article also includes a contribution to the old debate about the influence of the arrival of mechanical clocks. In Rome, however, mechanical clocks certainly did not replace celestial clockwork during the Middle Ages or Renaissance.

Many ways of controlling time, e.g. ecclesiastical calendars, martyrologies, chapter books of monastic communities, are built on the past. This past was, naturally, not objective and directly inherent in the events, but was always a construction, even a creation. This is beautifully highlighted in Jasmin Lukkari's contribution on the first Christian jubilee in 1300. Every care was taken to anchor the newly constructed jubilee tradition to a more distant past and a *mos maiorum,* as it were, although the event was in reality unprecedented. Linking the first Roman jubilee to an assumed previous tradition was instrumental for the success of the holy year of 1300. In fact, this masterful demonstration of papal power proved to be so successful that the jubilee tradition still continues.

One aspect of social time was the numerous temporal references visible to everyone throughout the Eternal City. Urpo Kantola focuses on temporal references in the non-funerary epigraphy of Rome from 1000 to 1527. Although most of these references were literally set in stone, it turns out that their ideals were not. Kantola shows important changes occurring with the emergence of the Renaissance or humanist style epigraphy during the second half of the 15th century.

Moving on to the latter part of the volume, attention is focused on ecclesiastical calendars and their use as a window to more formal ways to control time. Firstly, Holger Kaasik gives – for the first time – a comprehensive overview of the rich corpus of medieval calendars of S. Pietro in Vaticano and S. Maria Maggiore in Rome. The calendars are an invaluable source for studying the conceptions of time and the trans-generational curation of communal memory in medieval societies, as well as the diffusion of cults and cultural influence between communities. As medieval calendars are commonly used to localise and date their host manuscripts, this study has implications for a wide range of fields.

Kaasik then goes on to navigate the various cycles of time in one 13th century Vatican calendar. Through close reading of just one calendar, he succeeds in reconstructing its detailed use. Moreover, Kaasik ponders the one exception in the otherwise rather flexible system of medieval calendar markings: the long-standing use of the cumbersome way of ancient Romans to indicate the dates. It has remained unchanged since antiquity and thus proves the long-enduring authority of tradition. On the other hand, the primary aim of liturgical calendars was not so much to date new events than to recall traditional ones, and tradition and conservatism easily went hand in hand.

While ecclesiastical calendars were mainly used as the key to liturgy and hence as a very practical way of controlling time, the choice of their varying contents yields information about the time-related self-understanding of the religious communities managing them. In another article, Holger Kaasik studies all known extant medieval calendars of the churches of S. Pietro in Vaticano and S. Maria Maggiore. In both cases, the calendars tying together the heroic Christian past and the present were results of selecting material and thus constructing time. The most important liturgical feasts were the major events that urban population was likely to witness, and they were important in expressing the ecclesiastical and cultural agendas of the churches.[20] As an important by-product of the article, the material presented by Kaasik constitutes a substantial expansion of the knowledge on the calendars used in the main Roman basilicas of the Middle Ages.

The last chapter of a volume opens up broader views. Although Christian religion and its various presentations were ubiquitous in Rome, there was much room for pagan customs and superstition – even on the leaves of the very same calendars that were in daily use in every church. Based on very broad source material, Marko Halonen's article concentrates on the non-liturgical aspects of calendars. After all, medieval and Renaissance Romans used and followed the ancient pre-Christian calendar only superficially modified through *interpretatio Christiana*. Halonen's study shows interestingly that Christian and pagan continuums went hand in hand. Ecclesiastical calendars were used

20 Cf. McCahill, *Reviving the Eternal City*, 148.

by a small learned and orthodox minority of medieval and Renaissance Romans, but they certainly reveal even non-orthodox (or pagan) practices of understanding and trying to control time, patterns that reveal ideas of the majority of the Roman community. Similarly, calendars with their calculations regarding astronomy were simultaneously end products of generations of very high-class scholarship and contained down-to-earth popular beliefs.

Bibliography

Sources

Augustine. *The Confessions*, Ed. Michael P. Foley. Indianapolis: Hackett Publishing Company, 2006.

Hesiod. *Theogony, Works and Days, Testimonia*. Transl. Glenn W. Most. Loeb Classical Library 57. Harvard University Press, 2006.

Tibullus. *Elegies*. In *Catullus. Tibullus. Pervigilium Veneris*. Transl. F.W. Cornish, J.P. Postgate, J.W. Mackail. Revised by G.P. Goold. Loeb Classical Library 6. Cambridge, MA: Harvard University Press, 1913.

Virgil, *Aeneid*. 1. Ed. Randall T. Ganiban. Indianapolis: Hackett Publishing, 2008.

Literature

Adam, Barbara. *Time*. Cambridge: Polity Press, 2004.

Adam, Barbara. *Timewatch: The Social Analysis of Time*. Cambridge: Polity Press, 1998.

Assman, Jan. *Das kulturelle Gedächtnis. Schrift, Erinnerung und politische Identität in frühen Hochkulturen*. München: C.H. Beck, 1992.

Assman, Jan. *Religion und kulturelles Gedächtnis. Zehn Studien*. München: C.H. Beck, 2000.

Blackburn, Bonnie J. and Leofranc Holford-Strevens. *The Oxford Companion to the Year*. Oxford: Oxford University Press, 1999.

Burke, Peter. "Reflections on the Cultural History of Time." *Viator* 35 (2004), 617–626.

Carruthers, Mary. "Meditations on the 'Historical Present' and 'Collective Memory' in Chaucer and *Sir Gawain and the Green Knight*." In *Time in the Medieval World*, edited by Chris Humphrey and W.M. Ormrod, 137–155. York: Boydell & Brewer 2001.

Cohen, Jeffrey. *Medieval Identity Machines*. Minneapolis and London: University of Minnesota Press, 2003.

Dohrn-van Rossum, Gerhard. *History of the Hour: Clocks and Modern Temporal Orders*. Chicago and London: University of Chicago Press, 1996.

Elias, Norbert. *Time: An Essay*, transl. by E. Jephcott. Oxford: Blackwell, 1992.

Fabian, Johannes. *Time and the Other*. New York: Columbia University Press, 1983.

Fraser, J.T. *Of Time, Passion, and Knowledge.* Princeton: Princeton University Press, 1990.

Gerhard Jaritz and Gerson Moreno-Riaño, eds, *Time and Eternity: The Medieval Discourse.* International Medieval Research 9. Turnhout: Brepols, 2003.

Heikkilä, Tuomas and Teemu Roos. "Quantitative Methods for the Analysis of Medieval Calendars." *Digital Scholarship in the Humanities* 33:4 (2018), 766–787.

Hernandez, William Alexander. "St. Augustine on Time." *International Journal of Humanities and Social Science* 6:6 (2016), 37–40.

Humphrey, Chris and W.M. Ormrod, eds, *Time in the Medieval World.* York: York Medieval Press, 2001.

Laslett, Peter. *The World We Have Lost Further Explored.* 3. ed. London: Routledge, 2005.

Le Goff, Jacques. "Au Moyen Âge: temps de l'Église et temps du marchand." *Annales. Economies, sociétés, civilisations* 15:3 (1960), 417–433.

Lenz, Hans. *Universalgeschichte der Zeit.* 2., überarbeitete Auflage. Wiesbaden: marixverlag, 2013.

Martineau, Jonathan. *Time, Capitalism and Alienation: A Socio-Historical Inquiry into the Making of Modern Time.* Leiden & Boston: Brill, 2015.

McCahill, Elizabeth. *Reviving the Eternal City: Rome and the Papal Court, 1420–1447.* Cambridge, MA: Harvard University Press, 2013.

Morrison, John L. "Augustine's Two Theories of Time." *New Scholasticism* 45 (1971), 600–610.

Pratt, Kenneth J. "Rome as Eternal." *Journal of the History of Ideas* 26:1 (1965), 25–44.

Rappaport, Roy A. *Ritual and Religion in the Making of Humanity.* Cambridge: Cambridge University Press, 1999.

Richards, E.G. *Mapping Time: The Calendar and its History.* Oxford: Oxford University Press, 1998.

Rovelli, Carlo. *The Order of Time.* London: Allen Lane, 2018.

Rowland, Ingrid "Roma Aeterna." In *The Cambridge Companion to Ancient Rome,* ed. Paul Erdkamp, 558–574. Cambridge: Cambridge University Press, 2013.

Stevens, Wesley. "A Present Set of Things Past: *Quid est enim tempus?*" In *Time and Eternity: The Medieval Discourse,* ed. by Gerhard Jaritz & Gerson Moreno-Riaño, 9–28. International Medieval Research 9. Turnhout: Brepols, 2003.

Watts Miller, William. "Durkheimian Time." In *Time & Society* 9:1 (2000), 5–20.

CHAPTER 2

Temporal Expressions in Canonisation Processes and *Diari*, and the Perception of Time in Late Medieval Rome

Saku Pihko

The ways in which time was perceived in the Middle Ages has been the topic of much scholarship and discussion,[1] but examining it is no easy task, as historians do not have access to the ways people thought about time, only to the expressions they used to denote it.[2] The objective of this article is to examine the variety of temporal expressions used in canonisation processes and *diari* from 15th-century Rome. Which temporal schemes did people use when placing events in time or estimating its passing? What can an inquiry of this sort tell us about the overall nature of late medieval time perception?

The existence of separate normative time schemes in a given context does not equal the human experience of time. Instead, there is a complicated connection between temporal infrastructure,[3] and how people adhere to it in their lives.[4] My aim is to reach an understanding of this connection, an understanding that respects the complexities and nuances involved, and looks at the inhabitants of late medieval Rome as individuals. The analysis builds upon the

1 Karen Elaine Smyth, "Changing Times in the Cultural Discourse of Late Medieval England," *Viator* 35 (2004), 435. For an overview of the trajectories and developments in the historiography regarding cultural perceptions of time, see Peter Burke, "Reflections on the Cultural History of Time," *Viator* 35 (2004), 617–26.

2 Jacques Paul, "Expression et perception du temps d'après l'enquête sur les miracles de Louis d'Anjou," in *Actes des congrès de la Société des historiens médiévistes de l'enseignement supérieur public, 13e congrès, Aix-en-Provence, 1982, Temps, mémoire, tradition au Moyen-Âge* (1982), 21.

3 Paul Glennie and Nigel Thrift, *Shaping the Day. A History of Timekeeping in England and Wales 1300–1800* (Oxford: Oxford University Press, 2009), 102 have coined the concept *temporal infrastructure* to mean the general availability of temporal information in a given place, as well as the impact of both man-made and natural time signals and cues on this availability.

4 For an overview of the philosophical problems related to the idea of expressing and perceiving time, see, e.g., J.T. Fraser, *Of Time, Passion, and Knowledge* (Princeton: Princeton University Press, 1990 (1975)), 72–94; Robin LePoidevin, "The Experience and Perception of Time," *The Stanford Encyclopedia of Philosophy*, https://plato.stanford.edu/entries/time-experience/ (accessed 21 September 2019).

 | DOI:10.1163/9789004436251_003

idea of time as a lived experience, an inseparable, although often unconscious backdrop, in front of which daily life unfolds.[5]

Information regarding the quotidian existence of medieval people is fleeting, and sources pertaining to it are rare. Records of juridical depositions are one of the most fruitful options at our disposal.[6] As a result of the increasing textualisation of the legal sphere, in full drive by the 13th century,[7] more and more records of witness depositions were drawn up in both secular and ecclesiastical courts. One example of such proceedings are canonisation processes, juridical trials aimed at establishing whether certain miracle-working individuals of pious reputation were indeed saints worthy of papal canonisation.[8] During these investigations, papally appointed clerical officials, aided by notaries, gathered testimony from selected witnesses who had known the putative saint or experienced alleged miracles.[9] The records are often formulaic, and

5 Cf. Chris Humphrey, "Time and Urban Culture in Late Medieval England," in *Time in the Medieval World*, ed. Chris Humphrey and W.M. Ormrod (York: York Medieval Press, 2001), 105–06.

6 Monique Bourin, "Quel jour, en quelle année? A l'origine de la « révolution calendaire » dans le Midi de la France," in *Le Temps, sa mesure et sa perception au Moyen Âge. Actes du Colloque Orléans 12–13 avril 1991*, ed. Bernard Ribémont (Caen: Paradigme, 1992), 37–38 notes them to be among the best sources for the study the medieval time perception.

7 For an overview of these developments, see e.g. Paul Bertrand, *Les écritures ordinaires. Sociologie d'un temps de revolution documentaire (entre royaume de France et empire, 1250–1350)* (Paris: Publications de la Sorbonne, 2015); M.T. Clanchy, *From Memory to Written Record. England 1066–1307*,. 3rd Edition (Sussex: Wiley-Blackwell, 2013).

8 After ground-breaking studies, such as Ronald Finucane, *Miracles and Pilgrims. Popular Beliefs in Medieval England* (London: J.M. Dent & Sons Ltd., 1977) and André Vauchez, *La sainteté en occident aux derniers siècles du Moyen Âge. D'après les procès de canonisation et les documents hagiographiques* (Rome: École française de Rome, 1988) canonisation processes have been used widely as sources for the history of religion and every-day life in the medieval West. Emphasis has gradually shifted from quantitative analysis of large sets of records to qualitative close reading of selected processes, which the study at hand also represents. For a historiographical overview of research using canonisation processes, see Sari Katajala-Peltomaa, "Recent Trends in the Study of Medieval Canonizations," *History Compass* 8/9 (2010), 1083–092.

9 For overviews of how canonisation processes typically played out, see e.g. Didier Lett, *Un procès de canonisation au Moyen Âge. Essai d'histoire sociale* (Paris: Presses universitaires de France, 2008); Sari Katajala-Peltomaa, *Gender, Miracles, and Daily Life. The Evidence of Fourteenth-Century Canonization Processes* (Turnhout: Brepols, 2009), 23–70. On the history and development of canonisation in general, see Vauchez, *La sainteté en occident*, 15–67. On the legal details of canonisation processes, see Thomas Wetzstein, *Heilige vor Gericht. Das Kanonisationsverfahren in europäischen Spätmittelalter* (Köln, Weimar, Wien: Böhlau Verlag, 2004). On miracles as proof in canonisation processes in general, see Michael Goodich, "Reason or revelation? The criteria for the proof and credibility of miracles in canonisation processes," in *Procès de canonisation au Moyen Âge: aspects juridiques et religieux = Medieval*

emphasis was placed on the accumulation of relevant information.[10] This also included a wealth of details about time,[11] but in spite of this, canonisation processes have rarely been used as evidence of medieval time perception.[12]

---

canonisation processes: legal and religious aspects, ed. Gábor Klaniczay (Rome: École française de Rome, 2004), 181–97.

10 Gábor Klaniczay, "Proving sanctity in the canonization processes (Saint Elizabeth and Saint Margaret of Hungary)" in *Procès de canonisation au Moyen Âge: aspects juridiques et religieux = Medieval canonization processes: legal and religious aspects*, ed. Gábor Klaniczay (Rome: École française de Rome, 2004), 128.

11 For discussion on the applicability of canonisation records for examining time perception, see Paul, "Expression et perception du temps," 21. Lett, *Un procès de canonisation au Moyen Âge*, 246 states that the documentation at hand strongly defines the references to time available for the researcher, and concludes – somewhat pessimistically – that "*si les paroles de témoins peuvent nous livrer quelques renseignements généraux sur la perception du temps, nous observons surtout une mémoire du temps construite par et pour la procédure.*" With this warning in mind, emphasis will be placed on working out the limits within which temporal expressions in the records can be said to relate to those possibly used in lived reality.

12 Previously, the only systematic large-scale treatment of temporal information and its relation to medieval time perception in canonisation records has been Paul, "Expression et perception du temps" which uses the 1308 canonisation process of St Louis of Toulouse. Lett, *Un procès de canonisation au Moyen Âge*, 237–46 gives a brief yet insightful overview of the same topic in the 1325 canonisation process of St Nicholas of Tolentino. To list some other publications in the rather sporadic historiography relating to the medieval experience of time, see e.g. Bourin, "Quel jour, en quelle année ?," which examines references to time in medieval inquest material from the French Midi; Paul Brand, "Lawyers' Time in England in the later Middle Ages," in *Time in the Medieval World*, ed. Chris Humphrey and W.M. Ormrod (York: York Medieval Press, 2001), 73–104 examines time as a tool in the legal profession in late medieval England; Pierre Dubuis, "Le temps des paysans alpins au moyen âge: tour d'horizon et inventaire des problèmes," *Traverse: Zeitschrift für Geschichte = Revue d'histoire* 4 (1997), 63–72 is focused on the experience of time in the medieval Alpine rurality; Steven A. Epstein, "Business Cycles and the Sense of Time in Medieval Genoa," *The Business History Review* Vol. 62, No. 2 (1988), 238–60 examines references to time in business documents from 13th-century Genoa; Arnold Esch, "Gli interrogatori di testi come fonte storica. Senso del tempo e vita sociale esplorati dall'interno," *Bullettino dell'Istituto Storico Italiano per il Medio Evo* 105 (2003), 253–59 gives an overview of the possibilities offered by witness depositions for the study of time perception; Claude Gauvard, *"De Grace especial". Crime, État et société en France à la fin du Moyen Âge* (Paris: Publications de la Sorbonne, 1991), 480–91 examines temporal expressions in remission letters from late medieval France; Humphrey, "Time and Urban Culture in Late Medieval England," examines time and urban culture in late medieval York; Jacques Le Goff, *Time, Work, & Culture in the Middle Ages*, trans. Arthur Goldhammer (Chicago & London: The University of Chicago Press, 1980), 29–42 contains the classic formulation of the division between the time of the church and the time of the merchant; Emmanuel Le Roy Ladurie, *Montaillou. Cathars and Catholics in a French village 1294–1324*, trans. Barbara Bray (London: Penguin Books, 1990 (1978)), 277–82 takes a brief look at temporal information in 14th-century inquisition records; Alexander Murray, "Time and Money," in *The Work of*

In the context of 15th-century Rome, a rich and often used source is the documentation related to the attempts to canonise a local holy woman Francesca Bussa dei Ponziani (1384–1440), later known as S. Francesca Romana. She was a woman of noble birth, well known in Rome for her ascetic lifestyle, pious works, mystical visions, and miracles that allegedly happened due to her intercession.[13] The proceedings for her canonisation were opened after Francesca's death in 1440, and after an initial failure, they were resumed in 1443 and 1451, with some additional testimony collected in 1453.[14] For reasons unknown, the medieval attempts to grant Francesca the official status of a saint failed, but she was later canonised in 1608.[15] As a positive consequence of this delay and the multiple consecutive sets of records, there is obvious overlap in the depositions, which often affords us many depictions of the same events given by the same deponents at different times, which has enabled a more nuanced reading.

The level of detail in canonisation records can be seductive, but it must be understood how epistemologically problematic these seemingly lively texts really are. The depositional event was not a venue of free discussion about the

---

Jacques Le Goff and the Challenges of Medieval History, ed. Miri Rubin (Woodbridge: The Boydell Press, 1997), 3–25 pays homage to Jacques Le Goff through a brief overview of temporal information in some 13th- and 14th-century inquest material; Smyth, "Changing Times," looks at time in the cultural discourse of late medieval England. See also the articles collected in Pascale Bourgain & Jean-Yves Tilliette, ed., *Le Sens du Temps. The Sense of Time. Actes du VIIe Congrès du Comité International de Latin Médiéval. Proceedings of the 7th Congress of the International Medieval Latin Committee (Lyon, 10–13.09.2014)* (Genève: Librairie Droz, 2017).

13 For a detailed overview of Francesca's life, see Alessandra Bartolomei Romagnoli, S. *Francesca Romana. Edizione critica dei trattati latini di Giovanni Mattiotti* (Città del Vaticano: Libreria Editrice Vaticana, 1994), 59–138.

14 The documentation of Francesca's 15th-century canonisation processes has been edited in Placido Lugano, ed., *I processi inediti per Francesca Bussa dei Ponziani (santa Francesca Romana) 1440–1453*, Studi e Testi 120 (Città del Vaticano, 1945), henceforth cited as *Processi*. For information about these processes, see Bartolomei Romagnoli, *Santa Francesca Romana*, 141–47; Arnold Esch, "I processi medioevali per la canonizzazione di Santa Francesca Romana (1440–1451)," in *La canonizzazione di Santa Francesca Romana. Santità, cultura e istituzioni a Roma tra medioevo ed età moderna. Atti del Convegno internazionale Roma, 19–21 novembre 2009*, ed. Alessandra Bartolomei Romagnoli and Giorgio Picasso (Firenze: Edizioni del Galluzzo, 2013), 39–51; Placido Lugano, "Introduzione," in *I processi inediti per Francesca Bussa dei Ponziani (santa Francesca Romana) 1440–1453*, ed. Placido Lugano (Città del Vaticano: Studi e Testi 120, 1945), xiii–xxix.

15 Letizia Pellegrini, "La sainteté au XVe siècle entre procès et droit canonique: avant et après Bernardin de Sienne," in *Procès de canonisation au Moyen Âge: aspects juridiques et religieux = Medieval canonization processes: legal and religious aspects*, ed. Gábor Klaniczay (Rome: École française de Rome, 2004), 318 n. 21.

putative saint, but rather a technical examination of evidence aimed at a juridical decision.[16] Although the records do, to a degree that varies from process to process, reflect the personal and collective memories recounted by the deponents,[17] in many cases they are not transcripts of the once spoken dialogue.[18] This caveat is especially pertinent to the Francesca processes. The people who deposed about her life and miracles gave oral testimony pertaining to a predetermined list of questions in Italian, after which the appropriated information was truncated to fit the needs of the officials and translated into Latin. The extant records are strongly shaped by later reorganisation of the information, due to which the original interrogatory articles have been supplemented with information given by the deponents.

There are also limits to the social representativeness of canonisation records. Despite the delightful variety of people giving testimony for Francesca's canonisation processes, the recorded information does not offer us a complete cross-section of the urban society in late medieval Rome. The deponents form a rather homogeneous group of people from the higher social strata of the Trastevere and Campidoglio areas.[19] With these caveats in mind, we must be

16 Aviad Kleinberg, "Canonization without a canon," in *Procès de canonisation au Moyen Âge: aspects juridiques et religieux = Medieval canonization processes: legal and religious aspects*, ed. Gábor Klaniczay (Rome: École française de Rome, 2004), 10. Although there was uniformity of procedure among different medieval canonisation processes, there was also a lot of variance in the details of these proceedings. In more detail, see e.g. Laura Ackerman Smoller, "Northern and southern sanctity in the canonization of Vincent Ferrer: The effects of procedural differences on the image of the saint," in *Procès de canonisation au Moyen Âge: aspects juridiques et religieux = Medieval canonization processes: legal and religious aspects*, ed. Gábor Klaniczay (Rome: École française de Rome, 2004), 289–90.

17 Sari Katajala-Peltomaa and Christian Krötzl, "Approaching Twelfth- to Fifteenth-Century Miracles: Miracle Registers, Collections, and Canonization Processes as Source Material," in *Miracles in Medieval Canonization Processes. Structures, Functions, and Methodologies*, ed. Christian Krötzl and Sari Katajala-Peltomaa (Turnhout: Brepols, 2019), 9 have emphasised canonisation records as the "*outcome of the negotiations between the interrogators, notaries and witnesses.*" For discussion on the effects of memory on the formation of canonisation records, see e.g. Jussi Hanska, "The hanging of William Cragh: anatomy of a miracle," *Journal of Medieval History* 27 (2001), 121–38; Katajala-Peltomaa, *Gender, Miracles, and Daily Life*, 247–87; Laura A. Smoller, "Miracle, Memory, and Meaning in the Canonization of Vincent Ferrer, 1453–1454," *Speculum* Vol. 73, No. 2 (1998), 429–54.

18 For an overview of the problems arising from the passage of information from oral to written, and from vernacular to Latin, which defines the information found in canonisation records, see Lett, *Un procès de canonisation au Moyen Âge*, 257–70.

19 Arnold Esch, "Santa Francesca Romana e la società romana del suo tempo," in *Francesca Romana. La santa, il monastero e la città alla fine del medioevo*, ed. Alessandra Bartolomei Romagnoli (Firenze: Edizioni del Galluzzo, 2009), 11–12. According to Paolo Golinelli,

sure to take nothing in the records at face value. It is important to rigorously evaluate how the information at hand came to be, and what the limits are within which it can be said to relate to language and experiences in lived reality beyond the formulaic records.

I will supplement the records of Francesca's canonisation processes with two *diari* written by Roman ecclesiastics, Antonio di Pietro[20] and Stefano Caffari,[21] which depict public and private events in 15th-century Rome, often in enchanting detail.[22] These texts are supplied with a lot of temporal information, making them a fitting source for our pursuits. The entries in Di Pietro's *diario* run from 1404 to 1417, while the more varied collection of writings by Caffari ranging from, for example, documentation of business transactions to chronicles of family events, were written between 1423 and 1477. While canonisation records were the result of a complicated interaction between multiple parties in which information was given in retrospect, *diari* were formed by single individuals and written, at least usually, very soon after the depicted events

"Social aspects in some Italian canonization trials. The choice of witnesses," in *Procès de canonisation au Moyen Âge: aspects juridiques et religieux = Medieval canonization processes: legal and religious aspects*, ed. Gábor Klaniczay (Rome: École française de Rome, 2004), 166 Italian canonisation trials had, by the 14th century, developed towards a model that privileged the testimony of churchmen and members of the upper echelons of society. Gauvard, "*De Grace especial*", 486–88 has noted the lack of influence social standing or profession seems to bare on expressions medieval people used to denote time, but Lett, *Un procès de canonisation au Moyen Âge*, 239–40 suggests some possible correlation between social standing, profession, and preferences regarding ways of expressing time.

20 Antonio di Pietro was a canon of the Vatican Basilica. His *diario* has been edited in Francesco Isoldi, ed., *Il Diario Romano di Antonio di Pietro dello Schiavo. Dal 19 ottobre 1404 al 25 settembre 1417*, Rerum Italicarum Scriptores, Tomo XXIV – Parte V (Città di Castello, 1917), henceforth cited as *Diario Romano*.

21 Stefano Caffari was a man of noble birth, who pursued an ecclesiastical career and kept extensive written records of his public and private life. This collection of documents has been edited in Alba Ingletto and Stefania Santi, eds, *Stefano Caffari. Memorie di una famiglia della Roma del Quattrocento* (Roma: Società Romana Storia Patria, 2009), 71–275, henceforth cited as *Stefano Caffari*. For more details on Caffari's person, see the introduction to the volume (41–60).

22 On the genre of *diario* writing, prominent from the 14th century onwards throughout Italy, see e.g Luca Robertini, "Tagebuch, I. Italien," in *Lexikon des Mittelalters VIII. Stadt (Byzantinisches Reich) bis Werl* (Darmstadt: WBG), 423–25. In addition to those of Di Pietro and Caffari, there are also many other *diario*-type texts from 15th-century Rome. However, I have specifically chosen these two *diari* because they would seem to represent the style of writing pertinent to the genre, especially on the level of temporal information, and the possibilities offered by the inherent similarities and differences between them. The content of these texts has also contributed to them being chosen: Di Pietro is the only Roman diarist who wrote about the construction of Rome's first public clock, and Caffari's writings overlap with some of the information in Francesca's canonisation processes.

had transpired. Both types of sources come with their own sets of possibilities, but also with their own sets of problems. The parallel use of these different types of sources has allowed insights and conclusions, which would have been impossible using only one type of material.[23]

I will proceed by conducting an extensive examination of the different kinds of temporal expressions that can be found in the selected sources, moving from larger and longer timeframes, such as years, towards shorter ones, such as the internal divisions of a single day. Based on this, I will conclude with a discussion on what the use of multiple different time schemes can reveal about the overall nature of time perception in Quattrocento Rome.

1 From Years to Days

First, we must assess what kinds of temporal schemes did the people found in our sources adhere to and use in the 15th century when they wanted to pinpoint events in time or estimate its passing. We will begin by examining references to calendar dates.

1.1 *Calendar Dates*

A standard part of medieval legal documents was the *datatio*, and canonisation processes are no exception, as they were first and foremost legal trials that strove for juridical formality. The records of Francesca's canonisation processes begin with intricate dating apparatuses, which state the year of the Lord, the current pope, the indiction and the running calendar date.[24] This was standard practice, clearly visible over a wide array of medieval documents. However, while the official *datatio* of the documentation is uniform and standard, there are clear differences in other dating practices visible within the records. For example, with a few exceptions,[25] the individual depositions are not dated at all in the first two processes, but in the later 1451 process, the date of the deposition is often noted. Perhaps there was an aspiration for more legal precision and calendrical exactitude in 1451, when the attempt to canonise Francesca had already failed twice,[26] and this reflected on the temporal information in the records.

23 Smyth, "Changing Times," 435, 453 has called for a combination of unconnected perspectives to evaluate medieval time perception.

24 *Processi*, 3, 209.

25 *Processi*, 8, 9, 10.

26 Gábor Klaniczay, "The Power of the Saints and the Authority of the Popes. The History of Sainthood and Late Medieval Canonization Processes," in *Church and Belief in the Middle*

Canonisation processes were mainly focused around the pious life and alleged miracles of the putative saint. Especially the conditions of the miracles, their time and place, were of emphasised legal importance.[27] For example, article XVI of the canonisation documents states that sister Allegra,[28] a Dominican tertiary, had such severe pains in her chest that she could not walk nor lie down, but had to sit day and night with her knees to her chest, and often vomited blood. The record indicates that she was miraculously healed with Francesca's help in 1440, on the 24 February.[29] Similar exact calendar dates for miracles or other events can only be found in the records of the 1451 process.[30] In

---

Ages. Popes, Saints, and Crusaders, ed. Kirsi Salonen and Sari Katajala-Peltomaa (Amsterdam: Amsterdam University Press, 2016), 133 notes that only around half of the cases that were proposed proceeded into the inquest phase, and less than half of the actual canonisation processes resulted in canonisation.

27 Dávid Falvay, "Memory and Hagiography: The Formation of the Memory of Three Thirteenth-Century Female Saints," in *The Making of Memory in the Middle Ages*, ed. Lucie Dolezalová (Leiden & Boston: Brill, 2010), 354; Lett, *Un procès de canonisation au Moyen Âge*, 129–30. Placing events in time was just as important in other legal spheres in the Middle Ages, in more detail, see e.g. Brand, "Lawyers' Time," 88–98; Gauvard, "*De Grace especial*," 480. Paul, "Expression et perception du temps," 21 notes how there is only a single case in his material in which temporal information is non-existent.

28 When citing by name the people who appear in the records, I use the Latin names that can be found in the records and do not try to guess what the Italian name might have been. Didier Lett, "Les noms des hommes, des filles et des épouses dans les Marches d'après le procès de canonisation de Nicolas de Tolentino (1325)," *Mélanges de l'École française de Rome – Moyen Âge* 119/2 (2007), 402–03 has criticised the practice of trying to translate Latinised names in canonisation records, by emphasising the effect of notarial conventions of translation, and stating that "[*l*]*e passage de l'oral à l'écrit, du vulgaire au latin, affecte les forms antroponymiques et surtoit accentue leurs fluctuations.*"

29 *Processi*, 271: "*anno Domini* M°CCCC°XL°, *die vero* XXIIII *mensis februarii, quedam domna nomine soror Allegra* [...] *tercii Ordinis sancti Dominici, paciebatur quamdam gravissimam infirmitatem in pectore, propter quam nec ambulare poterat nec iacere, nec se movere de loco, sed die ac nocte supersedebat tenens continue genua iuncta cum pectore, et sepe evomebat maximam quantitatem sanguinis per os suum; et cum non posset iuvari aliquo remedio medicorum, visitavit eam beata Francisca* [...] *ei compaciens posuit manum suam super caput ipsius infirme et dixit ei: 'Confortare et habe bonam spem in Domino et confide in ipsius pietate quia cito liberaberis ab infirmitate ipsa.' Et his dictis statim infirma sanata fuit.*" For discussion on the miracles attributed to Francesca, see Bartolomei Romagnoli, *Santa Francesca Romana*, 80–92.

30 For miracles that are given a year, a month, and a date, see *Processi*, 270, 271, 281, 284, 291, 295, 297, 300, 308, 315, 318; for miracles that are given a year and a month, see *Processi*, 261, 265, 266, 267, 268, 274, 275, 278, 279, 288, 290, 292, 299, 302, 304, 312, 319; for miracles that are only given a year in the 1451 process, see *Processi*, 261, 262, 263, 264, 265, 267, 268, 272, 274, 276, 277, 279, 282, 299, 300, 313. It should be noted that most of these more or less detailed dates are given in the article part of the record, not the deposition. However, cf. *Processi*, 320.

the earlier processes, even references to years,[31] or months of a certain year, are rare.[32] An exception to this are the records that deal with the events surrounding Francesca's death, burial, and subsequent exhumation, in which exact calendar dates are used.[33]

A similar tendency to date events according to the calendar, often with great precision, can be found throughout the *diari*. The text of a typical *diario* was organised according to the running year, and entries often began with a running calendar date and the day of the week. For example, in 1404, Antonio di Pietro wrote that Ladislao, king of Naples, entered Rome through the gate of S. Giovanni in Laterano on Sunday the 19 October.[34] He also dates many other events, such as military movements[35] and papal processions,[36] in similar fashion. The authors of the Roman *diari* not only wrote about matters related to high politics or spiritual intrigue, but also about other events of personal or familial interest. Stefano Caffari noted many births and deaths, including that of his horse,[37] all supplied with detailed dating information. He also noted Francesca's passing, writing that in the abovementioned year, that is 1440, the 9 March, the venerable lady Ceccholella, as Francesca was sometimes called, died. Her body was on display for three days in the church of S. Maria Nova, and the whole city honoured her funerary rites.[38] The skill to express time correctly had become a part of literary eloquence during the Middle Ages,[39] and the painstaking lengths these authors often went to in writing the long dating litanies by hand would suggest this was often done for literary effect. Perhaps it gave the texts an air of legitimacy, comparable to the aspirations towards exact dating visible in the juridical records. Even so, it seems clear that at least for the literate part of society in late medieval Rome, giving dates according to the calendar was a widely used practice.

31 *Processi*, 51.
32 See *Processi*, 51 for a reference to the year 1431, and *Processi*, 85–86 for a reference to October of 1434.
33 *Processi*, 140–47.
34 *Diario Romano*, 3: "*Item die dominicho* XIX *supradicti mensis, intravit dominus rex Ladislaus rex Neapolitanus per portam Sancti Iohannis de Laterano.*"
35 *Diario Romano*, 7.
36 *Diario Romano*, 18.
37 *Stefano Caffari*, 145: "*Die* III *mensis ianuarii. Mortuus fuit equus.*"
38 *Stefano Caffari*, 132: "*Anno ut supra et die* IX *mensis martii. Obiit venerabilis domina Ceccholella* [...]. *Que domina tribus diebus mortua stetit super terram in ecclesia Sancte Marie Nove* [...] *tota civitas honoravit suum funus.*" Ingletto and Santi, eds, *Stefano Caffari*, 37 note that Stefano Caffari was the only Roman diarist to reference Francesca's passing.
39 Smyth, "Changing Times," 447.

While the sources consulted here can – at least to an extent – attest to the widespread use of the calendar scheme in 15th-century Rome, closer inspection of the canonisation material reveals epistemological challenges. For example, according to article XCVII of the 1440 Francesca process, a four-year-old girl called Camilla, who lived in the Campidoglio area, had been mute since birth. Her mother took her to Francesca, who healed Camilla by touching her tongue with her finger. The article states that this took place in 1438. The case seems to have been of high interest to the officials in charge of the proceedings, as the article is followed by extensive testimony. However, none of the deponents are recorded to have explicitly stated the year of the events. A year is referenced only in additional testimony, given by Agnes Pauli Lelli in 1453, who had been present when Francesca touched the girl's tongue. Interestingly enough, she claims that this happened in 1439.[40] It is left frustratingly unclear who actually stated the date or year of the events in question. This serves to remind us that certain temporal schemes may have been emphasised in the records, and that the relationship between the thoughts and memories of the deponents and the extant textual record is full of subtle complexities, which work effectively to obscure our conclusions.[41]

In some cases, especially in those which deal with events that took place during the short three-day period when Francesca's body was on display in S. Maria Nova, it is possible to see more clearly how the authorities may have translated vaguer temporal references given by the deponents into more exact dates. A telling example of these subtle processes of standardisation is article XLIIII and related witness testimony from the 1451 process. According to her deposition, Ceccha Nucii had suffered from a fever, and on the day after Francesca's death it came to her in a dream that she should visit the church that housed her body. She did this and was healed.[42] Although no apparent mention of the date is made, the article states this to have happened in the year of the Lord 1440, on the 10 March.[43] Perhaps it was filled in by the authorities for

40 *Processi*, 156–58. There is also another case in the 1440 process, in which a year is mentioned in the article but not the depositions (*Processi*, 97–98), and one in which a year is given in the article, but only in an additional deposition made in 1453 (*Processi*, 85–86).

41 Modern psychologists have noted that exact dates of recalled events are seldom remembered, but rather inferred from contextual information. In more detail, see e.g. Martin A. Conway, *Autobiographical Memory. An Introduction* (Philadelphia: Open University Press, 1990), 60, 172.

42 *Processi*, 311: "*cum graviter infirmaretur ex febribus quartanis* [...] *die sequenti post obitum b. F^e in sompnis habuit quod ipsam visitare deberet, prout fecit et subito* [...] *sanata est.*"

43 *Processi*, 310: "*anno Domini M°CCCC°XL°, die vero decimo mensis martii.*" For another similar case, see the previous article XLIII (*Processi*, 310).

juridical purposes based on Ceccha's reference to Francesca's death, which was probably the most prominent temporal marker in her personal memory of the events.[44] Similar underlying standardisation may have also affected some of the other entries in the records, although the information supplied is often so heavily edited that it is impossible to know for sure.

It is especially difficult to differentiate between the individuals accountable for giving certain bits of temporal information in the records of the first Francesca process. This is due to the way the notaries often combined the information related to a single article given by the deponents in 1440 into a single text, at least if the content of the depositions was uniform enough.[45] Article XXIII exemplifies the manner in which this causes problems for analysing the temporal perceptions and preferred methods of expressing time. According to the related testimony, Francesca was in the church of S. Maria in Trastevere in 1431 when she encountered the Devil masquerading as an angel. She did not believe the Devil who was trying to trick her, and steadfastly yelled out to the Lord for help. There are three witnesses, and the deposition of two of them who had been present, the priest Johannes and Rita Covelli, are combined into one. Either one of them, both, or neither mentioned the year to the authorities, but we can never know for sure. Based on later records, neither one of the two witnesses mentioned the year of the events in depositions they gave in 1443.[46] The documentation of the depositions that witnesses gave in 1451 and 1453 entail a clearer differentiation between individual witnesses compared to the earlier records, which alleviates some of the problems regarding their interpretation.

---

44 It is also possible that written records of miracles that happened in S. Maria Nova, where Francesca's body was on display, may have been made, and these records could have been used during the canonisation proceedings to supplement e.g. temporal information given by the deponents. This seems especially plausible considering the fact that there are many more exact dates in the records of the third 1451 process, which took place over a decade after many of the depicted events. While there is no explicit information about the consultation of other texts in the records, there is reason to note that this was done during the contemporaneous Italian canonisation process of Bernardin of Siena. In more detail, see Letizia Pellegrini, "Testifying to Miracles: A Report on the Canonization Process of Bernardin of Siena," in *Miracles in Canonization Processes: Structures, Functions, and Methodologies*, ed. Christian Krötzl and Sari Katajala-Peltomaa (Turnhout: Brepols, 2018), 105–27.

45 Esch, "I processi medioevali," 45 notes that although it was suggested during Francesca's final and successful canonisation process in the 17th century that deponents were not interviewed separately, it is "*invece probabile che fossero stati interrogati individualmente, ma che ne venisse poi trascritta una risposta collettiva, una deposizione di gruppo.*"

46 *Processi*, 50–52.

1.2 *The Passing of Years and Months*

Special significance was placed on alleged miracles that healed an especially long and arduous illness, and therefore canonisation records are full of estimates of duration, often made in reference to years and months. For example, according to article CXXI of the 1440 process, Perna Jacobi had for 25 years endured an ailment that had made it impossible for her to urinate without great difficulties and great pains. The estimate is agreed upon in the testimony portion of the record, and the same estimate is made in later testimony from 1443.[47] Similar references to estimates of duration in years are numerous in the Francesca processes,[48] and estimates of shorter timeframes were often given in reference to a number of months.[49] Compared to Francesca's canonisation processes, the Roman *diari* consulted in this study contain barely any estimates of temporal duration. This emphasises the fact that the nature and original purpose of our sources dictates the possibilities of examining medieval perceptions of time.[50]

Apart from the long illnesses or other ailments, another context in which deponents were asked to estimate and express long durations of time was when they gave information about how long they had known Francesca. Those witnesses who had known the putative saint for the longest were regarded as especially trustworthy, and therefore the records of Francesca's canonisation processes are full of references to people stating they had known her for many years, sometimes decades. For example, the priest Johannes told officials in 1440 that he had been Francesca's spiritual father for 11 years,[51] while Anastasia Thome Clarelli claimed she had known Francesca for 40 years.[52]

This practice seems just as straightforward as many of the estimates of long illnesses. However, at a closer look, the more nuanced information provided in some cases raises questions, which warrant closer examination in relation to

47 *Processi*, 186: "*domna Perna passa fuit quandam infirmitatem, nec poterat emittere urinam per annos vigintiquinque absque maxima difficultate et maximis doloribus.*"

48 *Processi*, 14, 15, 23, 25, 35, 42, 43, 112, 126, 132, 134, 159, 166, 195, 200, 202, 203, 220, 221, 222, 223, 224, 226, 227, 228, 230, 232, 234, 253, 263, 265, 267, 275, 284, 294, 307, 330. Denoting the passage of time in reference to *anno Domini* is only done once in the records, see *Processi*, 251–52.

49 *Processi*, 121, 122, 127, 134, 138, 142, 161, 165, 169, 172, 190, 192, 197, 202, 271, 276, 281, 289, 293, 300, 312, 316, 330, 331. See *Processi*, 120 for a case in which the deposition contains an estimate in months, but it is not mentioned in the article.

50 For a few rare estimates of temporal durations made by Di Pietro and Caffari, see *Diario Romano*, 15, 36, 57; *Stefano Caffari*, 115.

51 *Processi*, 11: "*ipse fuisset pater spiritualis dicte b. F^ae^ spacio undecim annorum continuorum.*"

52 *Processi*, 15: "*XL annorum quibus continue cum b. F^a^ conversabatur.*"

the extent the recorded information can be said to portray temporal schemes used in quotidian thought. The information given by one of the women who was close to Francesca during her lifetime, Rita Covelli, is especially telling. According to the record, she told the officials in 1443 that she had stayed in Francesca's house in Trastevere for 16 years.[53] Later on in the documentation, Rita commented about similar things in vaguer terms on three further occasions. She had been in contact with Francesca for a long time,[54] she had payed witness to Francesca's humble life until her death,[55] and lived with Francesca in the house of her husband for a long time.[56] Perhaps, from the viewpoint of the authorities, it was sufficient that Rita initially expressed the duration of her contact with Francesca according to the scheme of calendar years, but after this information was recorded once, the officials chose to record what were vaguer and perhaps more common expressions used by Rita to talk about her memories.[57] It is possible that in this case we can catch a glimpse of the ways the authorities' evident desire for exactitude affected the recorded temporal information. Noteworthy in this regard is the record of the deposition of Laurentius Benedicti Obitionis from 1451, according to which he referred to the entire time he had known Francesca, that is 25 years before her death.[58] While we can seldom know for sure whose choices the recorded information regarding the period of time reflects, the records of the deposition of both Rita and Laurentius serve as sobering reminders of the subtle complexities defining the information at our disposal.

Similar suggestions can be made based on those parts of the records that have lacunae in places where some kind of temporal information, such as the

53 *Processi*, 11: "*stetisset in domo habiationis b. F^ae^ sita in reg. Transtiberim spacio sedecim annorum continuorum.*" For similar cases in which the record states deponents had known Francesca for a certain amount of years, see e.g. *Processi*, 15, 25, 35, 42, 99.

54 *Processi*, 17: "*testis per magna tempore huius beate Dei famule fuerat et erat domestica secreta familiaris et filia in Christo.*"

55 *Processi*, 24: "*usque ad eius b. F^e^ felicem obitum vidit ut supra.*"

56 *Processi*, 56: "*ipsa testis maneret cum predicta beata in domo sui viri per multa tempora.*"

57 It should also be noted that even with these seemingly explicit preferences of the authorities, some deponents simply stated that they had known Francesca for many years, or referred to the whole time they had known her, which again points to the fluidity regarding the choice of these temporal expressions. See *Processi*, 14, 244, 253.

58 *Processi*, 222: "*toto tempore quo beatam F^am^ cognovit, scilicet 25 annis ante ipsius obitum.*" See also *Processi*, 224. Similar speculations can be made based on two other entries, according to which deponents had been asked to further specify the time they had known Francesca, based on the fact that the length of this time in years is added to the record in parenthesis, perhaps to supplement the vaguer terms the deponents had initially used. See *Processi*, 48, 227.

duration of an illness, would typically go.[59] Apparently these spaces were left into the records for later supplementation, probably due to the importance of temporal information for the officials. However, for some reason this never happened, at least not in these cases. This also raises the question about the extent to which other temporal information in the records may have been sought for after the initial testimony, which yet again problematizes the relationship of the oral deposition and the written record.

1.3 *Weeks and Days*

While there is no doubt that the weekly cycle was an important part of life in late medieval Rome, it is referred to very rarely in the material consulted here.[60] Weekdays are also rather absent from the records of Francesca's canonisation processes.[61] However, they are omnipresent in the elaborate dating apparatuses used by both di Pietro and Caffari, and therefore there is no reason to doubt that they were a natural temporal marker of choice used by late medieval Romans. Even so, we can only find them in the Francesca processes in those parts of the records that deal with Francesca's revelation of her impending death. For example, Franciscus Sclavi told the authorities in 1440 that when he spoke with Francesca before her death, she told him the end of her life would be on the coming Wednesday or Thursday.[62] The record of his 1451 deposition tells us that Francesca spoke to him about her impending death on Sunday, 3 March.[63] The visibility of weekdays is a good example of the way the nature of the sources dictates the available temporal information.

Compared to the almost non-existent references to weeks as estimates of duration,[64] deponents frequently mentioned a period of eight days, which can be understood as a synonym for a week.[65] For example, Julianus de Janua, who

59 *Processi*, 124, 128, 135, 160, 169. In the edition, these extra spaces are simply marked by "…", with a footnote stating "*lacuna nel testo,*" or with only the footnote. Despite numerous attempts to contact the monastery of Tor de'Specchi, where the manuscripts of the Francesca processes are housed, it was not possible to consult them during the writing of this article. In addition to taking a closer look at these lacunae indicated by the editor, it would have been extremely illuminating to see whether other temporal information in the records would seem to have been filled into similar lacunae, possibly left into the records at a previous stage of textual production.

60 For a single reference in the Francesca processes, see *Processi*, 151–53.

61 For rare references, see *Processi*, 197, 233, 234.

62 *Processi*, 101: "*die mercurii vel die iovis erit finis vite mee.*"

63 *Processi*, 254: "*et hec verba b. Fa die dominica, tercia marcii, testi dixit.*"

64 For a single reference, see *Processi*, 306–07. Paul, "Expression et perception du temps," 36 also notes a similar lack of the number of weeks used to denote the duration of events.

65 Cf. the commonplace use of the concept of the *octave* to denote a week's interval in the liturgical sphere. Based on the old Roman practice of inclusive counting, it is ubiquitous in Roman liturgical material from the 15th century.

had struck himself in the foot with an axe while gathering wood, was beyond medical help for five months, but after Francesca visited him, Julianus healed in eight days. This timeframe is mentioned in both the article CVIII of the 1440 Francesca process, and testimony given about the case by Rita Covelli, both in 1440 and 1443.[66] Furthermore, there are numerous other cases in Francesca's canonisation processes in which the records state certain events took eight days.[67]

Certain cases make the estimate of eight days problematic and point to a vaguer use. For example, according to article LXXXVI of the 1440 Francesca process, Jacobella Bartholomei from Trastevere had endured pains and open sores in her breasts for a year, and doctors had been unable to cure them. When she heard Francesca's body was on display in S. Maria Nova, she went there and placed Francesca's hands over her breasts, after which the pain vanished, and the sores healed in eight days. In the initial testimony portion of the text, Jacobella only affirms the contents of the article, but in additional testimony she gave in 1443, it is stated that the sores in her breasts were healed after a moderate amount of time.[68] This begs the question, whether Jacobella had forgotten the exact timeframe of her recovery in the three years that passed, or whether she perhaps also initially used eight days to denote a vaguer time of about one week. While we can never know for sure, other cases also point to a vaguer use, which should make us wary of the other references as well, even though they seem to give us no explicit reason to doubt their exactitude. In the very next article of the 1440 Francesca process, it is stated that seven-year-old Jacobus, also from Trastevere, was healed from a two-year throat sickness in eight days after visiting Francesca's body. In the initial testimony given by the boy's mother in 1440, the same eight-day timeframe is mentioned, but in an additional testimony given in 1443, the temporal estimate suddenly doubled into 15 days.[69] It would seem much more plausible to argue that an initially vague estimate of around a week had mutated into a vague two weeks or thereabouts in the memories recounted by the mother for a second time, rather than an exact eight days to have almost doubled in length. What mattered to the deponents was the miracle, not its timeframe.

Like the estimate of eight days, an estimate of three days was also very common in the canonisation processes. For example, once when Francesca was gravely ill, her confessor allowed her to drink wine, but as a result of this she

66 *Processi*, 172–73.

67 *Processi*, 117, 138, 139, 194, 197, 294, 311, 315. In some of the cases, the estimate mentioned by the deponent did not find its way into the article part of the record. The timeframe of eight days is also used in *Diario Romano*, 8; *Stefano Caffari*, 72, 113.

68 *Processi*, 138–39.

69 *Processi*, 139–40.

vomited for three days.[70] It seems to have been common enough to denote the duration of events counting the number of days, and there is a lot of variation to the lengths of the timespans denoted in this way.[71] Still, due to sheer quantity, the exactitude of both eight and three-day estimates fall suspect. There is of course no reason to suggest that none of the ailments or other durations of events referred to by in these terms had not taken exactly as long as the deponents claimed, but even so it is much more likely that we are witnessing the use of much vaguer terms, seemingly common when recounting memories, and hence probably also usual in everyday life.[72] Based on these observations, it seems clear that years, months, and weeks were commonly used temporal units that the inhabitants of late medieval Rome used to structure their memories of the past, but a rigid standard of exactitude was not necessarily a part of their mentality regarding time.

1.4 *Resorting to Approximations*

It was common for deponents in the canonisation material to use approximate temporal expressions, for example stating that events had taken place in a certain year or a little bit after.[73] It seems plausible to suspect that similar incertitude may have lurked beneath the surface of other dates, which were given by deponents but written down in a more laconic fashion. Some events are not dated at all, and they are simply referred to have taken place often,[74] or, for example, one day,[75] one evening,[76] or one night,[77] without any calendrical dating information. Sometimes the lack of a year or a specific date is due to the fact that the deponent had claimed not to remember it,[78] but in many cases no reason for the lack of temporal information is recorded.

Although there may have been efforts to try to obtain exact temporal estimates from the deponents, timespans of different lengths are also referred to

70 *Processi*, 27. Similar three-day estimates, see *Processi*, 137, 159, 162, 178, 283, 317; *Diario Romano*, 15, 107.

71 *Processi*, 109, 110, 136, 164, 179, 181, 185, 189, 249, 296, 297, 309, 313, 316, 320, 326, 327, 330. See also *Diario Romano*, 7, 13, 36, 38, 39, 72, 74, 75, 79, 81, 88, 98, 106, 112; *Stefano Caffari*, 188.

72 Paul, "Expression et perception du temps," 34–35 also suggests similar conclusions.

73 *Processi*, 45, 280.

74 *Processi*, 92–93.

75 *Processi*, 22, 29, 38, 43, 61, 77, 87, 95, 96, 151, 180, 182, 262, 263, 288, 296, 329.

76 *Processi*, 248.

77 *Processi*, 148, 248, 272, 301.

78 *Processi*, 152–53, 232. Hanska, "The hanging of William Cragh," 137 notes that *"it is only natural that sometimes the witnesses were betrayed by their own memory. They did not know that one day they would be called to testify and thus they had no specific reason to memorise the details particularly well."*

with very vague and ambiguous vocabulary. For example, the records state that events, often an illness, had lasted for many years[79] or simply a long time.[80] In similar fashion, shorter timespans are often noted to have stretched for a few days or many days,[81] and the *diario* material contains similar expressions.[82] We can find an especially telling case in the first Francesca process from 1440. According to article CVIIII, Paulus Jacobi de Portu was severely wounded, but after Francesca visited him, he was healed after a few days. According to the text that was based on Paulus' testimony, he told the officials he had been beyond the remedies of doctors for a long time, but after seeing Francesca he had been healed in a few days. Rita Covelli, who also gave testimony about the same case, seems to have stated that she had heard from Paulus how doctors had been unable to help him for a long time. She also told the officials that Paulus was healed after a few days, but it is unclear whether she had heard this from Paulus, or witnessed it herself and happened to give the same approximate estimate of the length of the healing process.[83] Although more exact terms might have been preferred, all of these vague temporal estimates seem to have been good enough for the officials, as they found their way into the records.

Sometimes, by chance, we get to peer behind these vague expressions. For example, according to article LV and subsequent testimony in the 1440 process, a carpenter by the name of Antonius Petri Johannis had been without confessing his sins for a very long time. However, moved by devotion, he changed his mind after visiting Francesca's body after her death and confessed his sins. Alexius Pauli Mentabona and Vannotia Sancte Crucis also gave testimony about the case, and although their depositions have been combined into one text which obscures personal choices regarding temporal expressions, it can be seen that Antonius had told them about his long absence from confession, although apparently not specifying the length of it any further. This case would seem rather similar as the previous one we looked at, but when we examine the later testimony given by Antonius in 1443, we can see that he had been without confessing for a period of three years.[84] This case is a fitting reminder of the fact that the choices related to temporal expressions of both the

79 *Processi*, 58, 59, 227, 233, 244, 261, 268, 274, 294.

80 *Processi*, 10, 18, 20, 21, 110, 158, 185, 230, 285, 295, 301, 317, 320, 325, 329, 330.

81 *Processi*, 33, 35, 175, 180, 187, 248, 273, 288, 293, 296, 319, 328.

82 See e.g. *Diario Romano*, 23, 28, 89.

83 *Processi*, 174.

84 *Processi*, 106–07. The next article (LVI) pertains to an almost identical case, although in this context we are not given any further estimation about how long the "*magna tempora sine confessione*" had actually been. In more detail, see *Processi*, 108.

deponents and the notaries could vary situationally even on an individual level, and we can usually only get a glimpse of one alternative out of many different possibilities.

When the sources happen to provide a glimpse behind the vaguer temporal terms used by the deponents, it becomes clear that the expressions are not commensurate when compared to each other. The case of the confession-dodging carpenter is not the only one. For example, Agnes Pauli Lelli, giving testimony about the illness of a certain Francisca de Verulis in 1443, apparently stated the woman had been gravely ill for a long time, that is, from the first advent up until the feast of St Gregory in March,[85] which is around four months. Then again, according to article LVIII, to which no testimony is connected in the record, a certain Dominicus from the Parione region had problems with his right arm because he had once fallen on it, and he had endured this infirmity likewise for a long time. However, in this case the long time is stated to refer to 12 years.[86]

The vagueness of expression did not seem to matter to the deponents, for whom this was probably a common way to remember the temporal aspects of certain events, as exactitude concerning time may not have been as important in daily life and thought as it was in the legal context. If prompted to do so, many could give more exact estimates, as we have seen, but it is questionable how primary and natural this way of thought was, even if the records seem to emphasise it. Past events and phases were probably remembered as events and phases, rather than in terms of calculable numerical time. Furthermore, the readiness on the part of the officials to accept these kinds of vague, non-commensurate estimates does raise the question whether it even mattered to them whether, for example, an illness had lasted for a few months or multiple decades, whether it was persistent enough and beyond the help of doctors, hence making the cure potentially miraculous. The miracle was remembered, temporal information was secondary.

1.5 *Liturgical Feast Days*

Although it seems that the annual cycle divided into months comprised of a running series of dates was a common scheme to refer to when dating events and estimating the passage of time, there were also alternative temporal

85 *Processi*, 109: "*Francisca de Verulis* [...] *per magnum tempus graviter infirmaretur, a prima die scilicet adventus usque ad festum sancti Gregorii de mense marcii.*"

86 *Processi*, 110: "*Quidam Dominicus* [...] *de regione Parionis, diu fuerat passus infirmitatem antiquissimam in bracchio dextro; ceciderat enim super illud bracchium et fregerat bracchium predictum* [...] *quam infirmitatem passus fuit per annos duodecim.*" For similar cases, see *Processi*, 158, 284.

schemes that people made regular reference to. A common and traditional way was to mention a liturgical feast day.[87] For example, article XXXIII of the first Francesca process states that on the feast of St Stephen of the year of the Lord 1440, Francesca entered the church of S. Giovanni in Laterano with two of her sisters-in-faith, and fell into ecstasy.[88] Two deponents, Agnes and Vannotia, whose testimonies have been combined, seem to have claimed that the events took place one day, on the feast of St Stephen of the current year.[89] The reference to the ongoing year has probably been supplanted into *anno Domini*, perhaps for future reference, but the feast day in question seems to have served as an unproblematic way to state the date, which met the requirements of the juridical proceedings. The synonymous overlap of these schemes is rather clear based on this example, because had the officials seen the need, they could have easily replaced the feast day with a calendar date. The simultaneous use of these two schemes becomes especially evident when looking at the writings of Antonio di Pietro and Stefano Caffari, who often use both calendar dates and feast days.[90]

Our view of feast days as temporal markers is complicated by the fact that in the canonisation processes consulted here, there are several accounts of people stating events took place on a certain feast day, but not implying the year in question. For example, article XXXI of the first Francesca process depicts how she endured a long ecstasy and holy visions around the feast of the Nativity of Christ. Numerous deponents gave extensive witness statements related to the events, and about what they heard Francesca had experienced, which would point to the fact that they were all talking about a very specific case among Francesca's numerous ecstasies. Most witnesses, in 1440 and in later depositions, do refer to the feast, but no one mentioned the year to the authorities.[91] This kind of use makes the feast days mentioned not so much references to actual points in time but rather narrative contexts in which the events were

87 Lett, *Un procès de canonisation au Moyen Âge*, 242; Paul, "Expression et perception du temps," 29 note the abundance of liturgical feast days used as temporal markers in the 14th-century canonisation processes they have examined.

88 *Processi*, 73: "*in die sancti Stephani de anno domini Millesimo CCCC°XL. Cum predicta domina Francisca causa devotionis intraret sociata duabus sororibus ecclesiam venerabilem sancti Johannis lateranensis simili modo et forma rapta fuit in estasi.*" For more entries in the records that give a feast day and a year, see *Processi*, 18, 20, 97, 122, 124, 140–41, 269, 277, 309.

89 *Processi*, 73: "*una die, in festo beati Stephani anni presentis.*"

90 See e.g. *Diario Romano*, 7, 13, 15, 20, 29, 36, 47, 50, 54, 64, 70, 82, 90, 108; *Stefano Caffari*, 161, 163, 202, 209, 212, 227, 246, 268.

91 *Processi*, 66–70. For similar cases, in which a feast day is mentioned without a year, see *Processi*, 64, 65, 66, 70, 134, 286, 320.

remembered and later recounted, reminding us that time is not only linear chronology, but the venue in which human life has always enfolded.

The interval between feast days is also used to denote the duration of time. For example, according to article LXXXX of the 1440 process, lady Margarita, a Franciscan tertiary from Trastevere, had for some time endured grave haemophilia, but was healed after visiting Francesca's body in S. Maria Nova. At the end of the article it is stated that the illness had lasted from the time of the indulgence of St Lawrence in August until the octave of the Assumption in the said month.[92] As it is so uncharacteristic – compared to the other records – to estimate the passage of time in reference to the liturgical calendar, it seems plausible that it was the deponent herself who resorted to this choice, even though the deposition part of the text contains no temporal references.[93] It is also telling of the variation and overall fluidity of temporal perception in late medieval Rome that this was recorded as such, although if needed, it could have easily been translated to, e.g. a certain number of days, a common expression in many other cases.

In some rare cases, feast days and running calendar dates were combined to estimate a duration. For example, according to article LXXVI of the 1440 process, a three-year-old boy had problems with his legs that made it impossible for him to walk from the previous All Saints' Day up until the 9 March. It is of course noteworthy that the 9 March was the day of Francesca's death, and therefore we cannot be completely sure if the deponent actually used this date, or if it was supplemented into the article by the authorities after the woman giving testimony told them she took the boy to Francesca's body in S. Maria Nova after hearing about her passing.[94] Even so, it is interesting to witness the interplay of the different calendrical schemes.[95] While it perhaps cannot be

92 *Processi*, 147: "*Quedam domna Margarita, de regione Transtiberim, et tertii Ordinis sancti Francisci de observantia* [...] *cum aliquod temporis pateretur gravem infirmitatem fluxus sanguinis* [...] *visitavit corpus istius humillime ancille Christi in ecclesia Sancte Marie Nove* [...]. *Existens vero ibidem in oratione se liberam ab illa infirmitate divina favente gratia sentiit* [...]*; que passa fuerat dictam infirmitatem a tempore indulgentie sancti Laurentii de mense augusti usque ad octavam sante Marie Assumptionis beate virginis de dicto mense.*"

93 Cf. *Processi*, 109 where a temporal estimate of a certain illness is given in reference to feast days in initial testimony from 1440, and additional testimony from 1443, but it is not recorded in the article part of the text.

94 *Processi*, 130: "*puer tibiam et pedes habebat inhabiles ad ambulandum, et a festo omnium Sanctorum proxime preterito usque ad nonas martii nisi baiularetur ambulare non poterat.*" It should be noted that according to the old Roman calendar, the Nones fell on 7 March. However, the reference to "*nonas martii*" must be a mistake in the original manuscript, or in Lugano's edition, as the deposition is clearly related to a time when Francesca was already dead, most likely 9 March 1440.

95 For another example of a mixture of running calendar dates and liturgical feasts, see e.g. *Processi*, 143–44.

denied that days of the year baptised with the names of saints were easier to remember, compared to abstract running numerals,[96] this notion hardly captures the complex relationships between intermingling dating schemes towards the end of the Middle Ages. Different temporal schemes were interchangeable with each other, and probably not understood as separate, as they perhaps are in modern times. An individual could choose either, or both, based on situational context and personal preference.

1.6 *Past Events as Temporal Markers*

Like feast days, incidents from recent history were also used as temporal reference points, when people attempted to pinpoint events in time.[97] Popes and their activities were often used in this way. For example, article I of the 1440 process deals with Francesca's pious conduct in her childhood. To get eyewitness information about these distant events, the officials heard testimony by an old Roman woman Johanna, from the Parione. It seems to have been important to prove her high age to legitimise her statement. Therefore, the record states that Johanna said, and from her aspect it was demonstrated, that she is the age of 70 years and more, because she said that she had been pledged to her late husband when Pope Urban V came to Rome.[98] Arnold Esch has pointed out how interesting it is that the first – although not permanent – return of the Avignon pope to Rome held temporal significance so long after the fact.[99] The

---

96 Cf. Giovanni Cherubini, "Stagioni, cicli, lavoro: il tempo tardomedievale," in *Spazi, tempi, misure e percorsi nell'Europa del Bassomedioevo. Atti del XXXII Convengo storico internazionale. Todi, 8–11 ottobre 1995* (Spoleto: Centro italiano di studi sull'Alto Medioevo, 1996), 51.

97 Lett, *Un procès de canonisation au Moyen Âge*, 237–38 notes an almost complete lack of historical events used as temporal markers in the canonisation process of St Nicholas of Tolentino. Cf. also Esch, "Gli interrogatori di testi come fonte storica," 255–59.

98 *Processi*, 9: "*domna Johanna que ut ipsa dixit et, ex sui aspectu demonstrat, est etatis LXX annorum et ultra, quoniam asserit quod quando Urbanus papa quintus venit Romam, ipsa erat a dicto quondam suo viro subarrata.*" Urbanus V returned to Rome for the first time in 1367, but it seems justified to emphasise that "*quando Urbanus papa quintus venit Romam*" is a temporal marker and a specific time in its own right, not only a mere substitute for the AD in question. Esch, "I processi medioevali," 42–43 has also pointed out incongruence pertaining to this case, as a septuagenarian living in 1440 would have been born around 1370. Taking this into regard, it is even more interesting that the entry has found its way into the official record. For more references to popes as temporal markers, see *Processi*, 91, 290–91. For discussion on old age and canonisation records, see Christian Krötzl, "*Sexaginta vel circa*. Zur Wahrnehmung von Alter in hagiographischen Quellen des Spätmittelalters," in *Alterskulturen des Mittelalters und der frühen Neuzeit. Internationaler Kongress, Krems an der Donau, 16. bis 18. Oktober 2006* (Wien: Verlag der Österreichischen Akademie der Wissenschaften, 2008), 109–15.

99 Esch, "I processi medioevali," 43. Also discussed in Esch, "Gli interrogatori di testi come fonte storica," 258–59.

diario writers, especially Antonio di Pietro, wrote a lot about what the popes were doing in Rome, so perhaps it is not that surprising that at least some of these events caught the attention of others too, and served as temporal landmarks[100] in their memories.[101]

While the choices made by the deponents of the canonisation processes to make temporal reference to certain events seem random at times, mainly because no reason for these choices can be deduced from the records, in other cases the choice is clearly more contextual. Article XXXV of the 1440 process states that before Pope Eugenius left the city, Francesca had predicted that many people would have to flee the city.[102] Eugenius IV was among those fleeing Rome due to a social uprising in 1434,[103] so temporal reference to him seems logical. This timeframe is given by the priest Johannes, first in his initial deposition in 1440,[104] and later with more exactitude by him in 1443, when he told the officials that Francesca had told him about the revelation about a year before the flight of the pope.[105] The authorities seem to have been fine with this vaguer contextual time reference, although it could probably have been supplanted with a reference to the chronological scheme. Perhaps they agreed with Johannes, who claimed the events were not only known in Rome, but in the whole world,[106] and hence did not deem it necessary to standardise the temporal information supplied by the deponent.

Francesca's death was also an event that served as a temporal landmark in its own right, at least in the context of the canonisation documentation. In the 1440 process, undergone right after Francesca's death, the events surrounding

100 On the psychological aspects of dating past events in relation to landmark events, see e.g. Conway, *Autobiographical Memory*, 53–55.

101 Popes were not the only ones whose activities facilitated temporal pinpointing of events recounted from memory. For example, the momentary reign of the Neapolitan king Ladislao in Rome is used as a dating device twice in the 1440 process (*Processi*, 28, 170), and twice in additional depositions given in 1453 (*Processi*, 31).

102 *Processi*, 75: "*Antequam dominus noster Eugenius recederet de Urbe, ipsa predicta domina Francisca predixit* [...] *quod plures gentes aufugere debebant de Urbe.*" See also *Processi*, 88.

103 For more details about this period of political turmoil in Rome, see e.g. Alan Ryder, "The papal states and the kingdom of Naples," in *The New Cambridge Medieval History Volume VII c.1415–c.1500*, ed. Christopher Allmand (Cambridge: Cambridge University Press, 1998), 575.

104 *Processi*, 75: "*presbiter Johannes deposuit quod ante recessum ab Urbe dicti pape Eugenii, ipsa domina predixerat de huiusmodi recessu et gentibus ab Urbe fugientibus.*"

105 *Processi*, 77: "*dicta domina* [...] *eidem dixit quod ante recessum ipsius pape de Urbe per annum vel quasi, quod ipse dominus noster papa et plures gentes in Urbe existentes metu maximo ducti debebant de Urbe recedere.*"

106 *Processi*, 77: "*notum est non solum in Urbe sed per totum mundum.*"

her passing are dated only in general terms, referencing the time after her death,[107] or the time when her body was on display at S. Maria Nova.[108] We could speculate that this was due to the recent nature of the events, but looking at the additional testimony given years later, in 1443 and 1453, it seems that the deponents still referred to events that took place around her death without specific dates.[109] Time was also counted in relation to Francesca's passing. For example, Anastasia Thome Clarelli claimed she did not remember what day and year a certain miracle, in which a small amount of leftover bread fed all the women living in Francesca's pious community, had happened, but she said it was several months before the death of Francesca.[110] Significant events seem to have mixed quite freely with calendar dates when people attempted to classify their personal memories into temporally coherent periods, but the context of these recollections probably influenced these choices quite strongly.[111]

Other events are also casually used as temporal markers. At least in some cases, these are not supplanted by calendar dates, even though doing so would probably have been possible, for example, when deponents referred to the time of Francesca's wedding,[112] or to a certain famine, during which Francesca distributed grain to the poor.[113] In other cases, the quotidian temporal expressions found in the records seem to have been more contextual, and served as narrative devices when deponents recounted events to the officials. Telling is the way Caterina Guidolini from Trastevere, giving testimony in 1443, dated a miracle in which her son, who had fallen into the Tiber and drowned, was revived due to Francesca's intervention, to one of the times when the Tiber was flooding.[114] The Tiber flooded so often that it would be impossible to relate this expression to a more specific date in retrospect.

107 *Processi*, 114, 115, 119, 121, 130, 132, 134–35, 136, 223, 229, 252, 310. There is also a single reference to the day preceding her death (*Processi*, 77–78), the time of her funeral (*Processi*, 320), and a few references to the night of her death (*Processi*, 122, 137).

108 *Processi*, 106, 117, 118, 120, 122, 123, 126, 127, 130, 133, 140.

109 See e.g. *Processi*, 106, 107, 113, 119, 127.

110 *Processi*, 152–53: "*Non recordatur quibus die et annis, sed dixit ea plures menses ante obitum dicte beate anime fuisse facta.*" For similar entries stating that an event took place a certain number of years before Francesca's death, see *Processi*, 45, 99, 222, 227, 232, 234, 238, 240, 241, 244, 253, 277, 301.

111 Cf. Lett, *Un procès de canonisation au Moyen Âge*, 240–41 who notes the abundant use of the death of St Nicholas of Tolentino as a temporal marker in his canonisation process, but argues that it did not have a lot of relevance as a temporal marker outside the proceedings.

112 *Processi*, 26: "*tempore nuptiarum.*"

113 *Processi*, 33: "*tempore penurie.*" See also *Processi*, 26, 35.

114 *Processi*, 192–93: "*quadam vice flumine Tiberis crescente.*"

1.7 *Natural Rhythms*

The cycles of nature by no means lost their relevance as temporal markers during the Middle Ages,[115] but references to the passing of the seasons are extremely rare in the records I have consulted.[116] Perhaps there are more subtle processes of standardisation at play in the records than what is visible on the surface, and references to nature no longer had room in the juridical discourse, although nature has always had its part in the ways people think about time. However, some temporal references to nature do appear in very specific cases in which the time of the year had direct bearing on the legitimacy of the miracle itself, and hence on the overall agenda of the canonisation process.[117] For example, Augustina Angeli, according to the record of her deposition in 1453, stated she was often present when Francesca took holy communion in the church of S. Maria in Trastevere, and even in wintertime the sweet scent of roses and lilies filled the air.[118] The idea of pleasant odours was often associated with sainthood, and the reference to the season reinforced the notion that the sweet scent was due not to chance or natural causes, but to Francesca's holiness.

Very interesting is the way in which similar reference to the season is made more implicitly through the use of other temporal schemes. For example, in article CXI of the first process, it is noted that in the month of April two figs fell on top of Francesca and her relative Vannotia, who were standing outside talking to each other.[119] In later depositions, both the priest Johannes[120] and Agnes Pauli Lelli,[121] noted how miraculous this was because of the season in question. There is also another telling case in which Francesca and her companions,

115 Cherubini, "Stagioni, cicli, lavoro," 45, 49; Le Goff, *Time, Work, & Culture in the Middle Ages*, 34–35, 49. Dubuis, "Le temps des paysans alpins au moyen âge," 65 notes that especially the people of rural areas in the Middle Ages lived under immense temporal constraints placed upon them by nature.

116 For a rare reference to summertime, see *Processi*, 33. The lack of temporal references to natural cycles, e.g. harvests and seasons, has also been noted in Gauvard, "*De Grace especial*", 481; Lett, *Un procès de canonisation au Moyen Âge*, 241–42; Paul, "Expression et perception du temps," 29.

117 Cf. Katajala-Peltomaa, *Gender, Miracles, and Daily Life*, 261, who notes that "[*t*]*he importance of the date of a miracle undoubtedly depended on the occasion.*"

118 *Processi*, 167: "*sepius fuit cum b. F^{a} in ecclesia S. Marie in Transtiberim tempore communionis Corporis Christi, et suavissimum odorem rosarum et liliorum sensiit eciam tempore hyemali.*"

119 *Processi*, 176: "*cum sic starent loquentes et esset de mense aprilis duo coctana ceciderunt super eas.*"

120 *Processi*, 176: "*miraculose videbantur venisse dicta duo coctana presertim in tali mense.*"

121 *Processi*, 176: "*de pomo coctano comederunt in mense aprilis: tali tempore anni similia poma in ortis et arboribus non reperiuntur.*"

while gathering wood outside Porta Portuensis, grew thirsty, but were able to quench their thirst as they found grapes growing from a tree, even though this was after the feast of the Nativity of Christ, in January, as one of the deponents specified in her 1453 deposition.[122] The point of placing the events into a specific time is yet again to argue for the legitimacy of the miracle due to it taking place in wintertime. However, the stage of the annual cycle of seasons is sometimes referenced in a straightforward manner, sometimes using the calendar, and sometimes using a well-known feast day, which was known to take place annually in wintertime.

2 The Day and the Night

Compared to dating events, canonisation processes did not place as much emphasis on denoting the exact moment when the events in question had transpired, unless they had special significance,[123] and references to different times of the day or the night[124] are more numerous in the *diario* material. Even so, both sources offer us information about the different temporal schemes people used to divide the day and the night. Chris Humphrey has posed the question regarding how rapid the developments in time perception that came with the emergence of mechanical clocks were, and what happened to the traditional ways of time-keeping that perhaps began to be supplanted.[125] Examining the situation in 15th-century Rome offers interesting insight into these processes.

2.1 *Cycles of Light and Darkness*

If nature was referenced rarely in our sources when denoting the time of the year, it still had its place within the division of the day, as references to the different phases of daylight and darkness are used to denote specific moments

122 *Processi*, 153–56.

123 Paul, "Expression et perception du temps," 31. However, Brand, "Lawyers' Time," 89–90 notes that the specific hour of a crime was important in the legal context of medieval England.

124 Lett, *Un procès de canonisation au Moyen Âge*, 245 notes that "[*l*]*e jour n'est pas, au Moyen Âge, une durée de vingt-quatre heures comme nous le concevons. Il revêt un sens astronomique. Il s'agit d'un point du temps, un moment, un* locus, *et non une plage de temps. La nuit n'est donc pas, comme pour nous aujourd'hui, une sorte de complément oblige du jour pour comptabiliser une journée, mais un autre* locus."

125 Humphrey, "Time and Urban Culture in Late Medieval England," 105.

in the material.[126] For example, sunrise is referenced as a temporal marker in its own right, without supplementation with other temporal information in the Francesca processes.[127] However, it is noticeable that these kinds of traditional references to daylight are rather rare in the canonisation processes. They are much more numerous in the writings of Antonio di Pietro and Stefano Caffari. For example, di Pietro wrote how he went to dinner with a few other men in the hour right after sundown.[128] Caffari and di Pietro also used morning,[129] midday,[130] evening,[131] night,[132] and midnight,[133] just like sundown, to pinpoint events to a certain time of the day. It is therefore evident that nature, the grand timekeeper, had not disappeared from the scene, not even in the face of novelties.

2.2 *Canonical Hours*

As we have already seen, different time schemes were not separable from each other and they often mixed together rather freely. Due to this, references to daylight were often used to supplement and sharpen other temporal schemes, such as the canonical hours.[134] For example, Antonio di Pietro wrote how a

---

126 Psychologists have noted that memories regarding the time of the day when a specific event took place are surprisingly accurate even after long intervals of time. One possible explanation for this is the abundance of natural time cues, like the level of sunlight, embedded in the visual memory of the event. In more detail, see Steen F. Larsen, Charles P. Thompson and Tia Hansen, "Time in autobiographical memory," in *Remembering our past. Studies in autobiographical memory*, ed. David C. Rubin (Cambridge: Cambridge University Press, 1996), 148–49.

127 *Processi*, 76, 91, 295, 296. There are also some references to evening (*Processi*, 312, 327), morning (*Processi*, 10, 298, 325), night (*Processi*, 10, 142, 312), and midnight (*Processi*, 301).

128 *Diario Romano*, 36: "*hora statim post hocasum solis* [...] *ivimus ad cenandum*." For more references to sundown, see *Diario Romano*, 15, 25, 26, 28, 61, 69, 82, 83, 98, 102.

129 *Diario Romano*, 6, 8, 16, 17, 18, 20, 21, 22, 24, 25, 27, 29, 35, 38, 40, 42, 46, 47, 48, 55, 57, 61, 65, 74, 84, 90, 95, 96, 111; *Stefano Caffari*, 116, 167, 187, 188, 191, 200, 260, 263, 266.

130 *Diario Romano*, 25. Paul, "Expression et perception du temps," 32 notes the lack of references to midday in his material.

131 *Diario Romano*, 4, 18, 24, 40, 44, 52, 62, 66, 73, 96; *Stefano Caffari*, 112, 188, 260, 266.

132 *Diario Romano*, 5, 6, 9, 11, 12, 20, 33, 34, 45, 53, 58, 64, 69, 70, 74, 79, 85, 86, 89, 91, 97; *Stefano Caffari*, 167, 235.

133 *Diario Romano*, 17, 20, 24, 54; *Stefano Caffari*, 72, 74.

134 The *horae canonicae*, or canonical hours, are the seven daily times of prayer that structure liturgy in the Catholic Church in remembrance of the Passion. Matins, Prime, Terce, Sext, None, Vespers, and Compline are based on the old Roman division of the day, and they correspond with different phases of daylight. For example, Matins was at sunrise, Sext at midday, and Vespers at sunset. On canonical hours and their development in more detail, see e.g. Gerhard Dohrn-van Rossum, *History of the Hour. Clocks and Modern Temporal Orders*, trans. Thomas Dunlap (Chicago: The University of Chicago Press, 1996 (1992)),

certain prior died after Compline, around sundown.[135] Naturally, the canonical hours chimed by church bells that rang out across Rome,[136] are referenced as temporal markers also without other supplemental information. They are very prominent in the writings of both di Pietro and Caffari. For example, the latter made note of a deal pertaining to a future matrimony, supplied with detailed dating information, and a reference to how the deal was struck at Vespers.[137] Di Pietro makes references to canonical hours in similar fashion, temporally pinpointing, for example, a death[138] and a baptism[139] according to this traditional time scheme.[140]

Perhaps it is not surprising that these writers denoted such solemn events according to this daily cycle related to church activities. However, canonical hours were also used to refer to much more mundane events.[141] It seems, yet again, that in some cases the time scheme that people used to reflect on past events was contextual as in the previous examples, and in other cases at least seemingly coincidental. For example, Antonio di Pietro writes about two

29–33; David S. Landes, *Revolution in Time. Clocks and the Making of the Modern World* (Cambdridge, Ma.: Harvard University Press, 1983), 59–63.

135 *Diario Romano*, 26: "*eodem die post completorium Sancti Petri, quaxi in ocasum solis, defunctus est dominus Iohannis Nutii prior et canonicus basilice Sanctorum apostolorum de Urbe.*" For a similar temporal reference combining a canonical hour and sundown, see *Diario Romano*, 15.

136 After an upsurge of bell tower building in Rome at the end of the 11th and the 12th century, bell towers were omnipresent throughout the city. On some occasions, the bells could even ring for the entire day. In more detail, see John H. Arnold and Caroline Goodson, "Resounding Community: The History and Meaning of Medieval Church Bells," *Viator* Vol. 43, Issue 1 (2012), 99–130; Jean-Claude Maire Vigueur, *The Forgotten Story: Rome in the Communal Period*, trans. David Fairservice (Roma: Viella, 2016 (2010)), 271. In this regard A.J. Gurevich, *Categories of Medieval Culture*, trans. G.L. Campbell (London: Routledge, 1985 (1972)), 106 has aptly noted that medieval people were very used to hearing time.

137 *Stefano Caffari*, 195.

138 *Diario Romano*, 8: "*obiit Paulus Stati post hora vesperorum.*" Di Pietro's readiness to freely use different temporal schemes in his writings is, again, aptly demonstrated by the fact that immediately after the note on the death of Paulus Stati he writes about another death, that of a cardinal, giving it a date and stating it took place "*ante diem.*"

139 *Diario Romano*, 14: "*eodem die post hora vesperorum fuit batizata filia dicti Paulis de Ursini.*"

140 The canonical hours are not the only temporal scheme related to church services present in our sources. For example, di Pietro also makes many references to "*hora missarum,*" temporally pinpointing events in relation to the different masses of S. Pietro. See *Diario Romano*, 8, 16, 17, 20, 22, 30, 34, 35, 38, 39, 40, 41, 45, 47, 48, 50, 54, 56, 57, 59, 61, 64, 65, 66, 69, 73, 75, 78, 82, 103, 104, 111. See also *Processi*, 69; *Stefano Caffari*, 200, 222, 233.

141 Jean-Claude Schmitt, *Les rythmes au Moyen Âge* (Paris: Gallimard, 2016), 312 notes that canonical hours are not only an abstract measurement, but an entire dimension of time that was sensible to clerical and lay-folk alike.

decapitations that he pinpoints using canonical hours.[142] He also wrote how in the year of the Lord 1408, 1st indiction, on Saturday 7 January, in the morning, at Matins, that is to say when the bells of S. Pietro were rung, armed men rode from Rome towards Monterotondo.[143] Perhaps in this case the memory of the events taking place simultaneously with the bells ringing incited di Pietro to use Matins as a temporal marker.[144] While we can only speculate on the reasons behind his choice, it is again evident that he went to extreme lengths in temporally pinpointing the events, and made reference to many different time schemes in the same passage.

While it is clear that canonical hours still played their part in the temporal mindscape of late medieval Romans, we cannot always know how definite these references were.[145] For example, in the previous case, Matins was probably used in an exact manner, as reference is made to the actual moment of the bell stroke, but in other cases we simply cannot know. It seems justifiable to speculate on the possibility that in the minds of late medieval Romans, the canonical hours created vague timespans around the actual moments of church service, which were then referred to and thought about in a more approximate manner, nevertheless exact enough for contemporary needs. Furthermore, while there is no direct evidence of this, the overrepresentation of Vespers as a temporal reference in our sources could point towards it being simply a more solemn way to refer to evening.[146]

In addition to noting the time of the day, canonical hours were also used as an estimate of temporal duration. For example, the prior of S. Maria Nova,

---

142 *Diario Romano*, 7, 15.

143 *Diario Romano*, 23: "*anno Domini* MCCCCVIII *indictione I mensis ianuarii die sabbati* VII *dicti mensis de mane hora matutinali, videlicet quando pulzabant campane Sancti Petri, hequitavit Beccharinus cum multis aliis de gentibus armorum de Paulis Ursinis per pontem Molli versus Montem Rotundum.*" See also *Diario Romano*, 46, 80.

144 For more references to the ringing of church bells used as temporal markers, see *Diario Romano*, 15, 16, 22, 23, 24, 25, 52, 77, 97, 99, 108; *Stefano Caffari*, 74, 112.

145 In rare cases, incertitude is explicit in the text, see *Diario Romano*, 21, 23, 88, 99.

146 For references to Vespers made by di Pietro, see *Diario Romano*, 5, 6, 7, 8, 11, 14, 15, 17, 18, 21, 22, 23, 24, 27, 28, 31, 32, 36, 37, 39, 43, 45, 47, 48, 50, 53, 55, 57, 58, 59, 60, 61, 65, 66, 71, 72, 74, 75, 76, 77, 78, 79, 83, 85, 87, 88, 90, 92, 95, 98, 99, 100, 102, 105, 109, 111, 112. Di Pietro makes roughly half as many references to Terce, and much fewer references to the other canonical hours. Caffari, who uses canonical hours much less, makes seven references to Vespers (*Stefano Caffari*, 73, 113, 117, 172, 195, 200, 207), but only a few references to the other canonical hours. The statistical overrepresentation of Vespers as a temporal marker has also been noted in Gauvard, "*De Grace especial*", 481; Paul, "Expression et perception du temps," 32. It should also be noted that church bells typically rang for longer at Matins and Vespers, compared to the moment of the other canonical hours, in more detail, see e.g. Schmitt, *Les rythmes au Moyen Âge*, 315–16.

friar Ypolitus, told the authorities that he had for many years witnessed Francesca and her companions visit the church, stay there for the entirety of the holy office from Prime to Sext, hearing mass, confessing, and consuming the body of Christ with great devotion.[147] This choice of temporal scheme is obviously contextual, as the events took place inside a church, and were related to the church service, which was temporally structured according to the canonical hours. It was the most logical time scheme of choice when the deponent talked about the events to the authorities, but it also fits in with the overall picture of the fluidity of these choices of expression, which could all be used synonymously. Adherence to a specific scheme was situational, and here Ypolitus seems to have resorted to the most useful and available one, which also served the authorities looking to record the temporal dimensions of the deponents' recollections.

The ever-present possibility for variation in the use of temporal expressions is portrayed by the fact that canonical hours, at least in some cases, mixed freely with references to sunlight also in the context of denoting duration. For example, according to article VII of the 1440 process, which dealt with Francesca's humility, she sat with poor foreign pilgrims in S. Paolo fuori le Mura begging for alms from morning until Vespers.[148] Rita Covelli also gave testimony of the same event in 1443, stating that Francesca sat with the poor begging until Vespers,[149] but not referring to the amount of time she spent with them. In similar fashion, Francesca was also said to have been in ecstasy at the same church from morning until Compline.[150] While these examples are inarguably interesting, they also beg the question posed above, of whether these were exact estimates, or rather a more ceremonial way to express that the events had taken the entire day, from morning to evening or sunrise to sunset.

147 *Processi*, 227: "*sepius vidit* [...] *b. F*am *cum domna Vannotzia et eciam cum domna Rita* [...] *prefatam ecclesiam visitare: et in eadem per totum divinum officium a Prima usque ad Sextam permanere et missas audire, confiteri et Corpus Christi sumere magna cum devocione.*" For a similar entry denoting duration with canonical hours, see *Processi*, 96.

148 *Processi*, 22: "*quadam die cum esset indulgentia in ecclesia sancti Pauli* [...] *et populi magna copia esset in dicta ecclesia, ista anima Deo devota posuit se in una trabe ad sedendum inter certas alienigenas pauperculas ad petendum elemosinam et ibi sedit, a mane usque ad vesperos.*"

149 *Processi*, 24: "*illa se posuit ad sedendum inter alias pauperculas et alienigenas in una trabe sedentes ad elemosinam petendam; erat enim ibi magna populi multitudo seditque ibi usque ad horam vespertinam.*"

150 *Processi*, 72: "*ipsa coram altari maiori rapta in extasi, sic a mane illius diei usque ad horam completorii continuo permansit.*"

2.3 *Temporal or Equal Hours?*

Alongside the ways in which late medieval Romans divided the day with references to natural phenomena and the traditional canonical hours, we can find a lot of references to *horae*. But what kind of hour scheme do these entries refer to?

In the Francesca processes, the word *hora* is sometimes used simply in reference to an ambiguous moment, rather than an hour of the clock.[151] For example, the priest Franciscus Sclavi, giving testimony in 1443, told the officials how people visited Francesca's unburied body in S. Maria Nova until a late hour.[152] Alongside these rather ambiguous references, we can also find cases in which a seemingly specific moment is noted with a reference to an hour. For example, according to article XXIII of the 1451 process, in December of 1437 Ceccha de Clarellis had given birth prematurely, but because all seemed fine, she did not have her newborn daughter baptised straight away. However, during the following night around the second hour, Francesca came and wanted the baby to be baptised without delay. After this was agreed upon and the baptism performed, the baby fell ill and died, but not unbaptised.[153] Only one of the three deponents, Brigida Clarelli, explicitly mentions Francesca arriving on the second hour of the night.[154] Agnes Pauli Lelli, who was present during the events, gave no temporal marker, at least according to the surviving documentation, and Ceccha, the mother, simply claimed she believed Francesca to have come due to divine knowledge, as she came at a non-customary hour, at night.[155]

Similar entries denoting an hour are very rare in the canonisation material,[156] and they all refer to night-time hours. This unfortunate combination of sparsity and nocturnal nature of the evidence leaves it entirely unclear as to what type of hour scheme these entries are in reference to. By the 15th century, there were two potential ways to divide the day and the night: the traditional system of temporal hours and the more novel system of equal hours, which gained

151 *Processi*, 33, 75, 100, 104, 125, 127, 129, 256, 275, 295, 305.

152 *Processi*, 105: "*usque ad horam tardam.*" For an identically worded entry, see *Processi*, 88.

153 *Processi*, 275–76: "*anno Domini M°CCCC°XXXVII° de mense decembris. Cum praefata domna Ceccha* [...] *filiam peperisset ante debitum nature tempus, quia tamen apparebat sana et incolumis baptizari eam subito non fecit. Nocte vero sequenti circa horam secundam supervenit beata Francisca* [...] *et voluit quod statim puella absque mora baptizaretur* [...]. *Quo facto supervenit subito infirmitas et mortua est. Quod ex divina revelatione creditur factum, ne dicta puella sine baptismo moreretur.*"

154 *Processi*, 276: "*b. Fa subito hora secunda noctis veniens.*"

155 *Processi*, 276: "*Credit divinitus cognovisse et quia preter horam consuetam noctis tempore eam visitavit.*"

156 *Processi*, 280, 298, 324.

prominence when mechanical clocks began to be adopted into use.[157] Mechanical clocks that struck the hours began to be erected at central locations of medieval towns and cities in the 14th century,[158] but Rome seems to have been a late bloomer in this sense, as the first clock was built only in 1412 onto the façade of S. Maria in Aracoeli on the Campidoglio.[159] The Italian form of reckoning these new equal hours was a 24-hour system, in which counting the hours started over again each evening, half an hour after sundown, from the first hour and continued to the 24th hour, which was the last hour of daylight of the next day.[160] With this in mind, it becomes evident that the second hour of the night, apparently mentioned by Brigida in her deposition, could be a reference to either of the two, and because no daytime hours are mentioned in the records, it seems too risky to speculate in favour of either one of the possible hour schemes. Perhaps this was not of great relevance to most contemporaries, or even the authorities in charge of the canonisation records.

Due to the apparent strive towards temporal detail in the *diari*, the writings of Antonio di Pietro and Stefano Caffari offer us more insight into the question regarding the choice between temporal and equal hours in 15th-century Rome. Both of the *diario*-writers mention specific hours quite often. There are numerous references to ambiguous night-time hours,[161] but there is also an abundance of references to hours of the day. For example, Caffari wrote about a

157 Until the 14th century, daily division of time was organised according to the so-called temporal hours, a system that divided the day and the night into 12 parts, the length of which changed according to the season. The new so-called equal hours, told by the new mechanical clocks, which divided the span of one day and one night into 24 segments of constant length, replaced temporal hours during the later Middle Ages. In more detail, see e.g. Humphrey, "Time and Urban Culture in Late Medieval England," 106.

158 The details regarding the invention of the mechanical clock are not entirely known, but they became sought after status symbols for urban centres during the 14th century, especially in Italy, although Rome does seem to have been an exception in this development. While clocks eventually became common, it has to be noted that even in the 15th century they were very expensive to build and maintain, and also very unreliable as time reckoning devices, as most clocks either lost or gained time even within the span of a day, and had to be regularly reset according to a sundial. For more information about the development and diffusion of clocks in the Middle Ages, see e.g. Carlo M. Cipolla, *Clocks and Culture 1300–1700* (New York & London: W.W. Norton & Co., 2003 (1967)); Dohrn-van Rossum, *History of the Hour*; Landes, *Revolution in Time*.

159 Antonio di Pietro noted in his *diario* that the new clock began to ring out the hours on 27 December 1412, on the feast day of St John the Evangelist. For more details, see *Diario Romano*, 76. See also E. Zinner, "Wurde die Räderuhr in Deutschland oder in Italien erfunden?," *Die Himmelswelt* 53 (1943), 20–21.

160 Dohrn-van Rossum, *History of the Hour*, 114.

161 *Diario Romano*, 6, 57, 89, 93, 96, 97, 100, 101, 102, 108; *Stefano Caffari*, 72, 74, 112, 171, 201, 209, 263, 264.

solar eclipse, stating that it happened around the 21st hour, and that it became so dark that it was like the 1st hour of the night.[162] All of the day-time hours mentioned in the *diari* are clear and explicit references to the Italian system of equal hour reckoning,[163] which would imply that the night-time hours are also quite probably references to equal hours. Even so, this seemingly systematic use of the equal hour scheme should perhaps not be overemphasised, nor should it be taken as unproblematic evidence of the sudden dominance of the mechanical clock. The use of equal hours is not something that instantly overrode the more traditional ways of time reckoning, but had rather settled in among them, as yet another extension to the list of various ways in which individuals could express time.[164] It seems to have been a welcome addition to the intricate dating apparatuses of the *diari*, but equal hours did not supplant the other schemes; they were simply added to the mix. While cheaper technology later canonised clocks as primary timekeepers, this was not something that was inevitable, nor most likely even expected by medieval people.[165] Everything was still open, and this openness was the normal state of affairs in the 15th century.[166]

In addition to telling the time of the day or the night, hours were used to express duration. For example, a clear area of interest for the officials running the canonisation proceedings were Francesca's recurrent holy ecstasies, and hours were sometimes used to note their duration. One of the entries reads that Francesca's confessor Johannes told the authorities how he had witnessed the ecstatic Francesca stay immobile for about an hour and more, and after this remained in a similar state, although moving around, talking about divine

162 *Stefano Caffari*, 107: "*circha horam XXI*^{am}*, luna opposuit se soli et sol obscuratus fuit, quasi si esset prima hora noctis.*"

163 *Diario Romano*, 41, 88; *Stefano Caffari*, 72, 95, 110, 118, 120, 147, 162, 186, 187, 188, 191, 235, 245, 252.

164 Cf. Gauvard, "*De Grace especial*," 484 who notes that clock hours did not override natural time signals, and Humphrey, "Time and Urban Culture in Late Medieval England," 116 who argues that clocks were initially not thought of as a complete self-standing system of time reckoning.

165 Nancy Mason Bradbury and Carolyn P. Collette, "Changing Times: The Mechanical Clock in Late Medieval Literature," *The Chaucer Review* Vol. 43, No. 4 (2009), 351–52 note that one of Chaucer's texts "*reveals both a receptive interest in the clock's proto-scientific potential to produce more exact quantifications of time, and, at the same time, a spirited and mocking resistance to the idea that the new* quantitative *time offered by clocks can possibly displace the richly* qualitative *time that had always shaped human lives in innumerable and profound ways.*"

166 For similar observations in the context of late medieval York, see Humphrey, "Time and Urban Culture in Late Medieval England," 113.

things and singing melodically for two or three more hours.[167] This is an exceptional case, because it gives us a rare glimpse of comparison between different numbers of hours. Although Johannes' estimates were probably approximations, it is still interesting to see hours used in such a seemingly systematic way. This apparent exactitude also reminds us that based on the more sparingly worded entries referencing hours as duration, we cannot know if they were meant as commensurate units or as a vaguer estimate. It seems much more likely that the latter was often the case, as people reflected upon events based on their memories, sometimes even years after the fact. Telling of these kinds of underlying problems is the deposition made by Rita Covelli in 1443, pertaining to Francesca being immersed into ecstasy for two hours or thereabouts, as she had often seen happen.[168] This would imply that the hour estimate Rita gave was not so much an estimate of the actual event in question, but a vaguer expression deduced from memories related to many different occasions. Both of the *diario*-writers also use hours to denote duration,[169] although rather rarely. Hours were clearly one way to express the temporal length of an event in 15th-century Rome, but it must be noted again that in all of these cases, both in the canonisation material and in the *diari*, we simply cannot know what kind of hours these references signify.

While hours as an estimate of duration seem common enough in the consulted sources, it is sometimes unclear from whom these references to hours originate in the canonisation context. For example, according to article XL of the 1451 Francesca process, Paulus Guidolini had endured pains in his chest for five years, and when he also caught a grave fever, a piece of Francesca's clothing was placed on his chest, and within one hour he rose from his bed completely free from both ailments.[170] However, when we look at the witness statements related to the article, no one makes explicit reference to the hour mentioned in the article. Was it simply not recorded, or did the authorities supplement vaguer temporal information given by the deponents? This is yet another reminder of the possibility that, in many cases, the way time is expressed in the text may not be the way it was talked or thought about in daily life. The way many of

167 *Processi*, 62. For more references to estimates of duration given in hours, see *Processi*, 12, 13, 14, 130, 184, 188, 191, 193, 194, 230, 231, 232, 247, 294, 326, 327.

168 *Processi*, 52: "*vidit ipsam beatam raptam in extasim per duas horas vel quasi, sicut sepe sepius videbat.*"

169 *Diario Romano*, 8, 24, 80, 88, 97, 100; *Stefano Caffari*, 188.

170 *Processi*, 308: "*passus fuisset per annos quinque maximum dolorem in pectore, et cum eciam pateretur gravissimas febres* [...] *et cum pannus ipsius beate positus esset supra pectus infirmi, infra unius hore spatium, statim de lecto surrexit et liber totus ab utraque infirmitate.*"

Francesca's ecstasies are denoted to have taken a certain time[171] or a long time[172] would also seem to point to certain vagueness being the accepted standard beyond the record, but also within it.

Dividing the hours into smaller units seems to have been rather rare, as there are only very few references to it in the material.[173] To give a few examples, one and a half hours is given once as an estimate of duration in a deposition of the 1443 Francesca process,[174] and Antonio di Pietro wrote about a severe storm that lasted for a quarter of an hour.[175] Likewise, Stefano Caffari divides hours very rarely.[176] It would seem that dividing the hours was a part of the late medieval Roman mindset, but the references to it are so rare that it is plausible to argue it was perhaps not much used in daily life. Hours seem to have been exact enough for contemporary needs.[177]

Instead of minutes or divided hours, it seems that short spans of time were sometimes expressed in vaguer everyday terms.[178] A good example of this can be found in a deposition made by friar Ypolitus, according to which there was discord among the friars in his monastery about whether or not they should accept a certain Frenchman into their ranks. The man was taken to Francesca, who made an estimate of his intentions after looking at him for the duration of one Ave Maria, or thereabouts, after which he was denied the right to join the friars of S. Maria Nova.[179] The same estimate can also be found in the article (XXVIII) part of the register.[180] It seems that Ypolitus' quotidian expression of choice served the authorities just as well as it had served the deponent. The commonness of referring to the duration of a prayer to denote a short span of time has also been noted in previous research,[181] but this is the only such entry in the documentation consulted here. It is impossible to tell whether the use of

171 *Processi*, 63, 65, 68, 69, 230, 231.

172 *Processi*, 185, 230.

173 Dohrn-van Rossum, *History of the Hour*, 20 also notes that dividing hours into smaller parts was rare in the Middle Ages.

174 *Processi*, 193: "*steterat in flumine per horam cum dimidia.*"

175 *Diario Romano*, 101: "*fuit maxima tempestas* [...] *et duravit per quartam partem unius hore.*"

176 *Stefano Caffari*, 74, 188, 191.

177 Cipolla, *Clocks and Culture 1300–1700*, 40 notes that 15th-century clocks often had no minute hand due to low requirements of precision.

178 Paul, "Expression et perception du temps," 33 has also noted that in his material, short durations are expressed "*par un vocabulaire très varié,*" and Smyth, "Changing Times," 443–44 has reminded us not to forget how common subjective time measurements are even today.

179 *Processi*, 60.

180 *Processi*, 59: "*per spatium unius avemarie vel circha.*"

181 Le Roy Ladurie, *Montaillou*, 277; Paul, "Expression et perception du temps," 34.

such references had diminished by the 15th century, or whether the rarity is due to chance.

In many cases the passage of short timespans is noted in even vaguer terms, which give no opportunity for comparative retrospect estimates. For example, in relation to article XXV of the 1440 process, the testimony of three women is combined into a single text, which states that one day they went to Francesca's chamber, and found her there unconscious, after which they stood there stupefied and did not dare wake Francesca, and after a certain while, Francesca came to her senses and they asked her what had happened. She replied that while she was praying and reading the scriptures, she was harassed and attacked by demons.[182] Similar vague estimates of short time spans can be found in abundance in the records.[183] It is left unclear in these cases whether the authorities had attempted to get the deponents to give more exact estimates. However, based on the overall picture, it seems justified to suspect that if they had supplied this information, it would have found its way into the extant record, which yet again points to the overall variability related to ways people temporally construed their memories and expressed time when recounting past events in their depositions.

3 Conclusions: Living in Time

Antonio di Pietro wrote in his *diario* that in the year of the Lord 1408, 1st indiction, on Friday 20th of January, which was the feast day of the holy martyrs Fabianus and Sebastianus, in the morning after the ringing of the bells, that is for Matins, it began to snow, and continued to do so well for an hour.[184] His contemporaries seldom used all of these temporal schemes together, but this telling example sums up the argument I have been attempting to construct

182 *Processi*, 55–56: "*una dierum ipse testes ad dictam domum accedentes et applicuerunt ad dictam cellam et intraverunt eam, et invenerunt ipsam dominam* [...] *et ipsa domina erat ibidem quasi insensibilis, et ipse videntes predicta stabant stupefacte nec audebant loqui et eam vocare. Et sic stantibus per certum spatium, et revertente ipsa domina Francisca in eius sensibus interrogabant, qualiter et quomodo talia evenerant, que domina eis respondebat* [...] *dum essem in orationibus et scripturas legerem demones me insultarunt et acriter percusserunt.*"

183 See e.g. *Processi*, 14, 28, 33, 51, 63, 65, 68, 75, 82, 84, 88, 123, 139, 170, 172, 194, 230, 231, 232, 286, 291, 315, 325.

184 *Diario Romano*, 24: "*anno Domini* MCCCCVIII *indictione I mensis ianuarii die veneris* XX *dicti mensis, que fuit festum sanctorum martirum Fabiani et Sebestiani de mane post pulzationem campanarum, etcetera, videlicet ad matutinum, incepit cecidere nix, et duravit per bene per unam horam.*"

throughout this article. It seems clear that there existed multiple interchangeable ways in which people in late medieval Rome could express time, whether to place events in time, or to estimate its passing. A strong situational variation seems to have been typical. Sometimes the choice of the temporal scheme was contextual, sometimes it seems random. The nuances are telling. It is also evident that different practices blended together seamlessly in both juridical depositions and in the writings of single individuals. The sheer number of schemes available to choose from when reflecting on their personal past, and how situational their use seems to have been, speaks volumes about the complexity of the pool of distinct cultural resources people could draw from when thinking and talking about time.

While the nature of the sources at our disposal may overemphasise intricate temporal information, it also offers us glimpses of the lived realities of time perception and the organic need to anchor events in time. Especially telling in this regard is a case from Francesca's 1440 canonisation process and additional related testimony that was gathered in 1443 and 1453. According to the records, Francesca and a group of women had visited the basilica of S. Paolo fuori le mura, where Francesca had fallen into ecstasy, which endured for the remainder of the day. It became late and they left. Francesca walked towards the city with the other women, still ecstatic with her hands crossed and her eyes turned to the sky. Catherina Guidolini and Jacobella de Brunomonte apparently both recalled in their depositions that Francesca had suddenly come to her natural senses and demanded to know where she was and what hour it was. According to the records, they had replied to her that they were coming from S. Paolo, they were near the gate of the city, and it was night.[185]

This textual depiction of Francesca snapping out of her ecstasy and wanting to know the time is one of the most important singular pieces of evidence in our quest to circumnavigate the subtle problems of information in our sources, because in this case the reactionary everyday need for temporal information is clearly visible, and cannot be reduced to mere juridical demand or literary convention. Diminished temporality is often part of ecstatic experiences, and the sense of time is one of the psychological safeguards that protects the integrity of the human ego.[186] Our evidence reinforces the universality of this

185 *Processi*, 70–72.

186 Fraser, *Of Time, Passion, and Knowledge*, 292, 308–11. Cf. also Peter Berger and Thomas Luckmann, *The Social Construction of Reality. A Treatise in the Sociology of Knowledge* (London: Penguin Books, 1991), 42, who note that "[*o*]*nly within this temporal structure does everyday life retain for me its accent of reality. Thus in cases where I may be 'disoriented' for one reason or another* [...], *I feel an almost instinctive urge to 'reorient' myself within the temporal structure of everyday life.*"

phenomenon. As a natural reaction to the apparent anxiety brought on by the suppression of her spatial and temporal senses, Francesca looked to reorientate herself. She wanted to place herself in time. According to the record, she asked for the hour, but as the group was on the road outside the walls of Rome, the only temporal scheme her companions could use in the situation was a reference to daylight, the most traditional and omnipresent of all temporal cues.

Traditional temporal schemes were still a vital part of everyday life and textual representation in 15th-century Rome. Novelties, like the scheme of equal hours told by the mechanical clock, did not come to replace or perhaps even compete with the old ways people were used to thinking about time. Rather, they were accepted alongside the old methods. The old and the new coexisted nonhierarchically, without conflict, as integral parts of the shapeless whole.[187] This holistic nature of the late medieval temporal discourse is telling of the open-ended attitude towards time. The realm of time perception was open to innovation and in constant flux, but this gradual development was neither linear nor fast.

Despite the influential dichotomies that have defined some of the earlier scholarship regarding medieval perceptions of time,[188] it must be emphasised that lived reality is seldom reducible to clear divisions between different groups. People in a city such as Rome were largely parts of the same cultural pool, which seems especially pertinent in regard to temporal expressions and time perception. Telling of this is how fluidly and situationally temporal expressions were mixed together and used, and how these choices seem to have been ultimately down to individual preference, rather than a collective mode of thought. Because of this, we must be sensitive towards medieval people and their prerogative to a personal sense of time. They must be treated as individuals who used the available pool of cultural models selectively when thinking about and denoting time.[189] The variability of these choices does not seem to have affected comprehensibility for their contemporaries, as different temporal schemes could be thought of as largely synonymous.

---

187 Cf. Humphrey, "Time and Urban Culture in Late Medieval England," 106–07 and Smyth, "Changing Times," 452–53 for somewhat similar conclusions based on very dissimilar sources in the context of late medieval England.

188 For an overview of these influential historiographical juxtapositions, the most famous being Jacques Le Goff's division of time in the Middle Ages into that of the Church and that of the merchant, see Burke, "Reflections on the Cultural History of Time."

189 Cf. Larsen, Thompson & Hansen, "Time in autobiographical memory," 137 who note that "[*t*]*he structure of a person's temporal schemata originates from a mixture of natural, cultural-societal, and personal regularities.*"

Canonisation processes proved to be a privileged vantage point from which to engage with these questions. The textual records of the oral depositions were created through the interaction of both parties of the depositional event. Although a position of agency is to a certain extent implied for the deponents as contributors to the content of the records, the information accumulated while assessing Francesca's potential sainthood was controlled by the officials and the notaries running the proceedings. While there does seem to have existed a will for temporal exactitude on their part, we can find no single systematic scheme into which the temporal information in the records was moulded, even with the vast possibilities of later correction, translation, and standardisation. The officials and the deponents are not visible as representatives of two different groups perceiving or expressing time in contrasting ways, as the multifaceted and situational mixture of different time schemes in the records makes it impossible to argue for a clear divide between them. Looking at the ways both churchmen and lay folk perceived and expressed time has made it possible to draw conclusions about similarities and connectedness, rather than contrasts. Aron Gurevich has asked what degree of unity can be found between the categories different groups used when thinking about time.[190] I propose that unity can be found in the fluid interplay of different temporal schemes, so fundamental to the general sense of time and its expression in the Middle Ages, shared by all social groups, clerical and lay, noble and lowly. The *diario* material complements this view, as the seamless combination of different temporal schemes by Antonio di Pietro and Stefano Caffari echoes the practices represented in the depositional records.

According to Chris Humphrey, the Middle Ages occupy an ambiguous place in the history of time, because the period is simultaneously "recognisably premodern, and yet also arguably the birthplace of a modern temporal sensibility."[191] Perhaps it is the lure of these similarities that has obscured the fluid nature of medieval ways of relating to time in previous research, and made it difficult to look at the emergence of seemingly modern concepts, such as the equal hours of the clock, and understand them to have been changeable parts in the non-hierarchical system of fluctuating ways to think of and express time. This must be kept in mind when we read our sources. For example, the records of a canonisation process, or a 15th-century *diario*, are not fumbling attempts with shifting levels of success to match up to modern standards of expression and exactitude. The blend of different temporal schemes was the

190 Gurevich, *Categories of Medieval Culture*, 34.

191 Chris Humphrey, "Introduction," in *Time in the Medieval World*, ed. Chris Humphrey and W.M. Ormrod (York: York Medieval Press, 2001), 1.

definitive standard performance. In this regard, Ad Putter's apt reminder that "the impression of 'medieval indifference to time' is only the reflex of our own indifference to medieval time"[192] is very pertinent.

It is natural that the inner workings of the medieval mind may sometimes seem foreign to us.[193] Although I do not suggest that we can hope to fully comprehend the variable ways people in 15th-century Rome perceived time, as they were possibly endless,[194] I would argue that the acknowledgement of the inherent readiness to accept a mixture of different temporal practices is one avenue through which we can understand the medieval mentality in a more nuanced fashion. The aporia we may feel before this characteristic shapelessness of medieval time perception should not bother us, as it did not bother the contemporaries.

Bibliography

Sources

Ingletto, Alba and Stefania Santi, eds, *Stefano Caffari. Memorie di una famiglia della Roma del Quattrocento*. Roma: Società Romana Storia Patria, 2009 (cited as *Stefano Caffari*).

Isoldi, Francesco, ed. *Il Diario Romano di Antonio di Pietro dello Schiavo. Dal 19 ottobre 1404 al 25 settembre 1417*. Rerum Italicarum Scriptores. Tomo XXIV – Parte V. Città di Castello, 1917 (cited as *Diario Romano*).

Lugano, Placido, ed. *I processi inediti per Francesca Bussa dei Ponziani (santa Francesca Romana) 1440–1453*. Studi e Testi 120. Città del Vaticano: Biblioteca Apostolica Vaticana, 1945 (cited as *Processi*).

Literature

Arnold, John H. and Caroline Goodson. "Resounding Community: The History and Meaning of Medieval Church Bells." *Viator* 43 (2012), 99–130.

192 Ad Putter, "In Search of Lost Time: Missing Days in *Sir Cleges* and *Sir Gawain and the Green Knight*," in *Time in the Medieval World*, ed. Chris Humphrey and W.M. Ormrod (York: York Medieval Press, 2001), 136.

193 Gurevich, *Categories of Medieval Culture*, 5.

194 Cf. Fraser, *Of Time, Passion, and Knowledge*, 5 who is of the opinion that "[s]*eeking a comprehensive view of time is a task comparable to putting together a jigsaw puzzle whose pieces are alive and moving*," and Humphrey, "Introduction," 2 who states that "*it is futile to try to define 'medieval time' in absolute terms: then, as now, time was a malleable and relative medium*."

Bartolomei Romagnoli, Alessandra. S. *Francesca Romana. Edizione critica dei trattati latini di Giovanni Mattiotti.* Città del Vaticano: Libreria Editrice Vaticana, 1994.

Berger, Peter and Thomas Luckmann. *The Social Construction of Reality. A Treatise in the Sociology of Knowledge.* London: Penguin Books, 1991 (1966).

Bertrand, Paul. *Les écritures ordinaires. Sociologie d'un temps de révolution documentaire (entre royaume de France et empire, 1250–1350).* Paris: Publications de la Sorbonne, 2015.

Bourgain, Pascale and Jean-Yves Tilliette, eds, *Le Sens du Temps. The Sense of Time. Actes du VIIe Congrès du Comité International de Latin Médiéval. Proceedings of the 7th Congress of the International Medieval Latin Committee (Lyon, 10–13.09.2014).* Genève: Librairie Droz, 2017.

Bourin, Monique. "Quel jour, en quelle année? A l'origine de la «révolution calendaire» dans le Midi de la France." In *Le Temps, sa mesure et sa perception au Moyen Âge. Actes du Colloque Orléans 12–13 avril 1991*, edited by Bernard Ribémont. Caen: Paradigme, 1992, 37–46.

Bradbury, Nancy Mason and Carolyn P. "Changing Times: The Mechanical Clock in Late Medieval Literature." *The Chaucer Review.* Vol. 43, No. 4 (2009), 351–75.

Brand, Paul. "Lawyers' Time in England in the later Middle Ages." In *Time in the Medieval World*, ed. by Chris Humphrey and W.M. Ormrod. York: York Medieval Press, 2001, 73–104.

Burke, Peter. "Reflections on the Cultural History of Time." *Viator* 35 (2004), 617–26.

Cherubini, Giovanni. "Stagioni, cicli, lavoro: il tempo tardomedievale." In *Spazi, tempi, misure e percorsi nell'Europa del Bassomedioevo. Atti del XXXII Convengo storico internazionale. Todi, 8–11 ottobre 1995.* Spoleto: Centro italiano di studi sull'Alto Medioevo, 1996, 43–61.

Cipolla, Carlo M. *Clocks and Culture 1300–1700.* New York and London: W.W. Norton & Co., 2003 (1967).

Clanchy, M.T. *From Memory to Written Record. England 1066–1307.* 3rd Edition. Sussex: Wiley-Blackwell, 2013.

Conway, Martin A. *Autobiographical Memory. An Introduction.* Philadephia: Open University Press, 1990.

Dohrn-van Rossum, Gerhard. *History of the Hour. Clocks and Modern Temporal Orders.* Translated by Thomas Dunlap. Chicago: The University of Chicago Press, 1996 (1992).

Dubuis, Pierre. "Le temps des paysans alpins au moyen âge: tour d'horizon et inventaire des problèmes." *Traverse: Zeitschrift für Geschichte = Revue d'histoire* 4 (1997), 63–72.

Epstein, Steven A. "Business Cycles and the Sense of Time in Medieval Genoa." *The Business History Review* 62, No. 2 (1988), 238–60.

Esch, Arnold. "Gli interrogatori di testi come fonte storica. Senso del tempo e vita sociale esplorati dall'interno." *Bullettino dell'Istituto Storico Italiano per il medio Evo* 105 (2003), 249–65.

Esch, Arnold. "I processi medioevali per la canonizzazione di Santa Francesca Romana (1440–1451)." In Alessandra Bartolomei Romagnoli and Giorgio Picasso, eds, *La canonizzazione di Santa Francesca Romana. Santità, cultura e istituzioni a Roma tra medioevo ed età moderna. Atti del Convegno internazionale Roma, 19–21 novembre 2009*. Firenze: Edizioni del Galluzzo, 2013, 39–51.

Esch, Arnold. "Santa Francesca Romana e la società romana del suo tempo." In *Francesca Romana. La santa, il monastero e la città alla fine del medioevo*, edited by Alessandra Bartolomei Romagnoli. Firenze: Edizioni del Galluzzo, 2009, 3–21.

Falvay, Dávid. "Memory and Hagiography: The Formation of the Memory of Three Thirteenth-Century Female Saints." In Lucie Dolezalová, ed., *The Making of Memory in the Middle Ages*. Leiden & Boston: Brill, 2010, 347–64.

Finucane, Ronald. *Miracles and Pilgrims. Popular Beliefs in Medieval England*. London: J.M. Dent & Sons Ltd., 1977.

Fraser, J.T. *Of Time, Passion, and Knowledge*. Princeton: Princeton University Press, 1990 (1975).

Gauvard, Claude. *"De Grace especial". Crime, État et société en France à la fin du Moyen Âge*. Paris: Publications de la Sorbonne, 1991.

Glennie, Paul and Nigel Thrift. *Shaping the Day. A History of Timekeeping in England and Wales 1300–1800*. Oxford: Oxford University Press, 2009.

Golinelli, Paolo. "Social aspects in some Italian canonization trials. The choice of witnesses." In *Procès de canonisation au Moyen Âge: aspects juridiques et religieux = Medieval canonization processes: legal and religious aspects*, edited by Gábor Klaniczay, 165–80. Rome: École française de Rome, 2004.

Goodich, Michael. "Reason or revelation? The criteria for the proof and credibility of miracles in canonization processes." In *Procès de canonisation au Moyen Âge: aspects juridiques et religieux = Medieval canonization processes: legal and religious aspects*, edited by Gábor Klaniczay, 181–97. Rome: École française de Rome, 2004.

Gurevich, A.J. *Categories of Medieval Culture*. Translated by G.L. Campbell. London: Routledge, 1985 (1972).

Hanska, Jussi, "The hanging of William Cragh: anatomy of a miracle." *Journal of Medieval History* 27 (2001): 121–38.

Humphrey, Chris. "Introduction." In *Time in the Medieval World*, edited by Chris Humphrey and W.M. Ormrod, 1–4. York: York Medieval Press, 2001.

Humphrey, Chris. "Time and Urban Culture in Late Medieval England." In *Time in the Medieval World*, edited by Chris Humphrey and W.M. Ormrod, 105–17. York: York Medieval Press, 2001.

Ingletto, Alba and Stefania Santi, eds, *Stefano Caffari. Memorie di una famiglia della Roma del Quattrocento*. Roma: Società Romana Storia Patria, 2009.

Katajala-Peltomaa, Sari. *Gender, Miracles, and Daily Life. The Evidence of Fourteenth-Century Canonization Processes*. Turnhout: Brepols, 2009.

Katajala-Peltomaa, Sari. "Recent Trends in the Study of Medieval Canonizations." *History Compass* 8/9 (2010): 1083–092.

Katajala-Peltomaa, Sari and Christian Krötzl. "Approaching Twelfth- to Fifteenth-Century Miracles: Miracle Registers, Collections, and Canonization Processes as Source Material." In *Miracles in Medieval Canonization Processes. Structures, Functions, and Methodologies*, edited by Christian Krötzl and Sari Katajala-Peltomaa, 1–39. Turnhout: Brepols, 2018.

Klaniczay, Gábor. "Proving sanctity in the canonization processes (Saint Elizabeth and Saint Margaret of Hungary)." In *Procès de canonisation au Moyen Âge: aspects juridiques et religieux = Medieval canonization processes: legal and religious aspects*, edited by Gábor Klaniczay, 117–48. Rome: École française de Rome, 2004.

Klaniczay, Gábor. "The Power of the Saints and the Authority of the Popes. The History of Sainthood and Late Medieval Canonization Processes." In *Church and Belief in the Middle Ages. Popes, Saints, and Crusaders*, edited by Kirsi Salonen and Sari Katajala-Peltomaa, 117–40. Amsterdam: Amsterdam University Press, 2016.

Kleinberg, Aviad. "Canonization without a canon." In *Procès de canonisation au Moyen Âge: aspects juridiques et religieux = Medieval canonization processes: legal and religious aspects*, edited by Gábor Klaniczay, 7–18. Rome: École française de Rome, 2004.

Krötzl, Christian. "*Sexaginta vel circa*. Zur Wahrnehmung von Alter in hagiographischen Quellen des Spätmittelalters." In *Alterskulturen des Mittelalters und der frühen Neuzeit. Internationaler Kongress, Krems an der Donau, 16. bis 18. Oktober 2006*, 109–15. Wien: Verlag der Österreichischen Akademie der Wissenschaften, 2008.

Landes, David S. *Revolution in Time. Clocks and the Making of the Modern World*. Cambridge, MA: Harvard University Press, 1983.

Larsen, Steen F., Charles P. Thompson, and Tia Hansen. "Time in autobiographical memory." In *Remembering our past. Studies in autobiographical memory*, edited by David C. Rubin, 129–56. Cambridge: Cambridge University Press, 1996.

Le Goff, Jacques. *Time, Work, & Culture in the Middle Ages*. Translated by Arthur Goldhammer. Chicago and London: The University of Chicago Press, 1980.

LePoidevin, Robin. "The Experience and Perception of Time." Accessed 21 September, 2019. https://plato.stanford.edu/entries/time-experience/.

Le Roy Ladurie, Emmanuel. *Montaillou. Cathars and Catholics in a French village 1294–1324*. Translated by Barbara Bray. London: Penguin Books, 1990 (1978).

Lett, Didier. "Les noms des hommes, des filles et des épouses dans les Marches d'après le procès de canonisation de Nicolas de Tolentino (1325)." *Mélanges de l'École française de Rome – Moyen Âge* 119/2 (2007), 401–13.

Lett, Didier. *Un procès de canonisation au Moyen Âge. Essai d'histoire sociale*. Paris: Presses universitaires de France, 2008.

Lugano, Placido. "Introduzione." In *I processi inediti per Francesca Bussa dei Ponziani (santa Francesca Romana) 1440–1453*, edited by Placido Lugano, v–xxxiv. Studi e Testi 120. Città del Vaticano: Biblioteca Apostolica Vaticana, 1945.

Mauskopf Deliyannis, Deborah. "Year-Dates in the Early Middle Ages." In *Time in the Medieval World*, edited by Chris Humphrey and W.M. Ormrod, 5–22. York: York Medieval Press, 2001.

Murray, Alexander. "Time and Money." In *The Work of Jacques Le Goff and the Challenges of Medieval History*, edited by Miri Rubin, 3–25. Woodbridge: The Boydell Press, 1997.

Paul, Jacques. "Expression et perception du temps d'après l'enquête sur les miracles de Louis d'Anjou." In *Actes des congrès de la Société des historiens médiévistes de l'enseignement supérieur public, 13e congrès, Aix-en-Provence, 1982, Temps, mémoire, tradition au Moyen-Âge*, 20–41. 1982.

Pellegrini, Letizia. "La sainteté au XVe siècle entre procès et droit canonique: avant et après Bernardin de Sienne." In *Procès de canonisation au Moyen Âge: aspects juridiques et religieux = Medieval Canonization Processes: Legal and Religious Aspects*, edited by Gábor Klaniczay, 309–26. Rome: École française de Rome, 2004.

Pellegrini, Letizia. "Testifying to Miracles: A Report on the Canonization Process of Bernardin of Siena." In *Miracles in Canonization Processes: Structures, Functions, and Methodologies*, edited by Christian Krötzl and Sari Katajala-Peltomaa, 105–27. Turnhout: Brepols, 2018.

Putter, Ad. "In Search of Lost Time: Missing Days in *Sir Cleges* and *Sir Gawain and the Green Knight*." In *Time in the Medieval World*, edited by Chris Humphrey and W.M. Ormrod, 119–36. York: York Medieval Press, 2001.

Robertini, Luca. "Tagebuch, I. Italien." In *Lexikon des Mittelalters VIII. Stadt (Byzantinisches Reich) bis Werl*, 423–25. Darmstadt: WBG.

Ryder, Alan. "The papal states and the kingdom of Naples." In *The New Cambridge Medieval History Volume VII c.1415–c.1500*, edited by Christopher Allmand, 571–87. Cambridge: Cambridge University Press, 1998.

Schmitt, Jean-Claude. *Les rythmes au Moyen Âge*. Paris: Gallimard, 2016.

Smoller, Laura A. "Miracle, Memory, and Meaning in the Canonization of Vincent Ferrer, 1453–1454." *Speculum* 73, No. 2 (1998): 429–54.

Smoller, Laura Ackerman. "Northern and southern sanctity in the canonization of Vincent Ferrer: The effects of procedural differences on the image of the saint." In *Procès de canonisation au Moyen Âge: aspects juridiques et religieux = Medieval Canonization Processes: Legal and Religious Aspects*, edited by Gábor Klaniczay, 289–308. Rome: École française de Rome, 2004.

Smyth, Karen Elaine. "Changing Times in the Cultural Discourse of Late Medieval England." *Viator* 35 (2004): 435–53.

Vauchez, André. *La sainteté en occident aux derniers siècles du Moyen Âge. D'après les procès de canonisation et les documents hagiographiques*. Rome: École française de Rome, 1988.

Vigueur, Jean-Claude Maire. *The Forgotten Story: Rome in the Communal Period.* Translated by David Fairservice. Roma: Viella, 2016 (2010).

Wetzstein, Thomas. *Heilige vor Gericht. Das Kanonisationsverfahren in europäischen Spätmittelalter*. Köln, Weimar, Wien: Böhlau Verlag, 2004.

Zinner, E. "Wurde die Räderuhr in Deutschland oder in Italien erfunden?" *Die Himmelswelt* 53 (1943): 17–22.

CHAPTER 3

The Jubilee of 1300 as an Instrument of Time Control and Papal Power

Jasmin Lukkari

1 Introduction

On 22 February 1300, Pope Boniface VIII (1294–1303) ordered, with the bull *Antiquorum habet*, that every 100th year from the birth of Christ should be a year of remission – soon widely called a "jubilee." Pilgrims could gain plenary indulgences by coming to Rome and visiting the basilicas of S. Pietro in Vaticano and S. Paolo Fuori le mura for 30 (Romans) or 15 (foreigners) days.

This is what we know for a fact, but the details and motivations of the event remain unclear. A vast amount of research has been done on this first Christian jubilee of 1300 because of its organiser, the controversial Pope Boniface VIII, and because of its undisputed importance on Church history.[1] It is not my intention to delve deep into these already much discussed political topics or theological aspects of the first jubilee. However, the first jubilee of 1300 contains many fascinating time-related features that have not been discussed elsewhere, to my knowledge.

In this analysis, mainly based on the account given by Cardinal Giacomo Gaetani Stefaneschi (c. 1270–1343), I will show how the pope tried to control various time-related features, and different practical and abstract time levels, with careful decisions interlaced in an interesting and complicated manner. In this piece I will answer the following questions: (1) how does Stefaneschi's confusing story about the background of the first Christian jubilee justify the foundation of this new religious tradition?; (2) on what basis were the extent of the

1 E.g., Angiola Maria Romanini, ed., *Roma anno 1300* (Roma: L'Erma di Bretschneider, 1983); Gloria Fossi, ed., *La Storia dei Giubilei. Volume primo 1300–1423* (Firenze: Giunti – BNL Edizioni, 1997); *Romei e Giubilei. Il pellegrinaggio medievale a San Pietro (350–1350)* (Milano: Electa, 1999); and *I giubilei nella storia della Chiesa* (Città del Vaticano: Libreria editrice Vaticana, 2001). Arsenio Frugoni, "Il Giubileo di Bonifacio VIII," *Bullettino dell'Istituto Storico Italiano per il Medio Evo e Archivio Muratoriano* 62 (1950), 1–121, Raffaello Morghen, *Bonifacio VIII e il giubileo del 1300 nella storiografia moderna* (Roma: Edizioni dell'Elefante, 1975), Alfonso Stickler, *Il Giubileo di Bonifacio VIII. Aspetti giuridico-pastorali* (Roma: Edizioni dell'Elefante, 1977), and Mario Fois, "Il Giubileo del 1300," *Archivium Historiae Pontificae* 38 (2000), 23–40.

 | DOI:10.1163/9789004436251_004

jubilee indulgence and the duration of the jubilee year decided?; (3) how were the practical time limits of the pilgrimage determined and controlled?; (4) was there any connection between the first Christian jubilee and the ancient Roman *ludi saeculares*?; and (5) how did Boniface VIII use the jubilee as an instrument for gaining more authority for the Church, the pope, and the city of Rome?

I have chosen to examine the time-related aspects of the first jubilee in order to reveal the very roots of this long Christian tradition. Indeed, a wider overview of all medieval jubilees is long overdue, since a study of their time-related features has not yet been done, even though the tradition established by Boniface VIII actively continued long after his death and still continues today.[2] Hopefully this article can serve as a starting point for a discussion of whether, and in what manner, the following medieval jubilees were also instruments of time control for the Papacy, and what other time-related features we may find in the jubilee tradition in the medieval context.

Our most comprehensive source for the jubilee of 1300 is the *De centesimo seu iubileo anno* of the aforementioned Cardinal Giacomo Gaetani Stefaneschi.[3] He was born in Rome around 1270 to an old noble family. Stefaneschi was made a cardinal by Pope Boniface VIII, and Stefaneschi dedicated two separate works to him: *De coronatione*, and the *De centesimo* analysed here.[4] The earliest

---

2 The next seven jubilees were organised in 1350, 1390, 1400, 1423, 1450, 1475, and 1500, after which they were, and are still being, regularly organised.

3 For the *De centesimo seu iubileo anno* of Jacopo Stefaneschi, I use the edition and translation of Paul Gerhardt Schmidt and Antonio Placanica, based on the codex Arch. Cap. S. Pietro. G.3, in *Iacopo Stefaneschi. De centesimo seu iubileo anno*, eds, Claudio Leonardi, Paul Gerhard Schmidt, trans. and comm. Antonio Placanica (Firenze: Sismel – Edizioni del Galluzzo, 2001). In my opinion, it is the most accurate edition and translation available. I will always refer to this edition, when citing "Stefaneschi, *De centesimo,*" and use its line numbers together with the chapter division and page markings identical to the original manuscript. There are two older editions: the first modern edition was compiled by D. Quattrocchi in 1900 (in *Bessarione* 4, 1899/1900, 7, 299–317). Only in 1997 did another edition, with an Italian translation, appear, by A. Ilari (in *La storia dei giubilei*, vol 1. Giunti – BNL Edizioni, 1997), but it still transmits the errors of Quattrocchi, and the translation is not as accurate as Antonio Placanica's.

4 His father Pietro Stefaneschi was a senator in Rome and a *podestà* in Florence. His mother was Perna Orsini. His exact year of birth is not known, and scholars varyingly set it at 1260 or 1270. See Hösl's convincing argument for the year 1270: Ignaz Hösl, *Kardinal Jacobus Gaietani Stefaneschi. Ein Beitrag zur Literatur- und Kirchengeschichte des beginnenden vierzehnten Jahrhunderts* (Berlin: Verlag von Emil Ebering, 1908), 11–13. He studied at the universities of Paris and Bologna. He was close to Cardinal Gaetano Orsini, who became Pope Nicholas III. Giacomo Stefaneschi got his additional name Gaetani from Nicholas III. He also had a good relationship with Celestine V and wrote the *Opus metricum* about his life. For more information and literature on Stefaneschi, see Claudio Leonardi, Paul Gerhard Schmidt, and Antonio

manuscript of *De centesimo* has been dated to the author's time, and was presumably written very soon after the jubilee and during Boniface VIII's reign.[5] The work has three parts: the account in prose of the events of the first jubilee; two poems; and three documents signed by the pope.[6] It is very clear that in the *De centesimo* Stefaneschi wanted to demonstrate the supreme authority of the pope and to support his theocratic endeavours that would culminate soon after the jubilee in the papal bull *Unam sanctam* of 1302. We need to keep this context in mind when analysing his text: it is a semi-official account of the events and probably what Boniface VIII also wanted to convey to posterity within the Church. The beginning of the text and the invention of the jubilee is a very confusing story with lots of incongruences, and we cannot be sure of any other facts except what has been summarised above. However, we must not let this bother us too much; I will concentrate on the temporal aspects within the jubilee campaign of Boniface VIII, and not on the veracity of the story.

In addition to Stefaneschi, there was another important commentator on the first jubilee at the papal court: Jean Lemoine (1250–1313).[7] He studied at the University of Paris and went to Rome in 1292. He was made a cardinal by Celestine V, and was one of the greatest canon lawyers of his time. Under Boniface VIII, he was a vice chancellor of the Roman curia, and according to a

Placanica, *Iacopo Stefaneschi. De centesimo seu iubileo anno. La storia del primo giubileo* (Firenze: Sismel – Edizioni del Galluzzo, 2001), vii–xix and 54–55 n. 46.

5 All known copies are Roman and have been singularly traced back to this original manuscript. The original was in Avignon for some time around 1368 and had some markings added to it, but was returned to the Vatican by the beginning of the next century. For a detailed analysis on the manuscript of *De centesimo* and its history, see: Maria Grazia Ciardi Dupré Dal Poggetto, *Il maestro del codice di San Giorgio e il cardinale Jacopo Stefaneschi* (Firenze: Edam, 1981), 47–49; Giovanna Ragionieri, "Ipotesi per uno 'scriptorium': codici miniati a Roma per il cardinale Jacopo Stefaneschi," in *Roma anno 1300,* ed. Angiola Maria Romanini (Roma: L'Erma di Bretschneider, 1983), 393–397; Emma Condello, "I codici di Stefaneschi: uno 'scriptorium' cardinalizio del Trecento tra Roma e Avignone?," *Archivio della Società romana di storia patria* 110 (1987), 21–61; Emma Condello, "I codici Stefaneschi: libri e committenza di un cardinale avignonese," *Archivio della Società romana di storia patria* 112 (1989), 195–218; Giovanna Ragionieri, "Jacopo Stefaneschi e il *De centesimo.* Un cardinale testimone del primo giubileo," in *La Storia dei Giubilei. Volume primo 1300–1423,* ed. Gloria Fossi (Firenze: Giunti – BNL Edizioni, 1997), 218–223. For a short analysis of the contents and context of *De centesimo,* see Leonardi, Schmidt, and Placanica, *Iacopo Stefaneschi. De centesimo seu iubileo anno. La storia del primo giubileo,* vii–xix.

6 The copies of the papal bulls *Antiquorum habet* and *Nuper per alias,* and the *Gratia non bullata* (=*Ad honorem dei*).

7 In scholarship, he is also named Iohannes Monachus, Giovanni Monaco, Jean Lemoyne, and Jean Le Moine.

contemporary, Johannes de Polliaco, Lemoine spoke with Boniface VIII daily for two years at least. He also took part in the consistory in which the publication of the bull *Antiquorum habet* was decided, and despite his apparent close relationship with the pope, he sometimes criticised his decisions.[8] Lemoine also wrote a commentary for Liber VI of the *Corpus Iuris Canonici*. Especially useful for the subject at hand, however, is Lemoine's scholastic commentary of the pope's jubilee bull *Antiquorum habet*.[9]

The great interest in the first jubilee is evident in other sources. There are several short descriptions by chroniclers and poets that are relevant in this context. Magister Bonaiuto da Casentino (d. 1312)[10] wrote a song for the jubilee, and a papal scribe Silvester[11] wrote a letter that was distributed outside Italy in order to advertise it. Also, the chroniclers Giovanni Villani (c.1280–1348)[12] and Guglielmo Ventura (1249/50–c.1322)[13] – writers that were not based in Rome – described their contemporary experience of the jubilee. Even the great poet of the era, and an enemy of the pope, Dante Alighieri (1265–1321),[14] mentioned the jubilee in his *Divina commedia*. The texts of Jean Lemoine, Bonaiuto da Casentino, and Silvester originate from the sphere of influence of Boniface VIII's papal court, and it is worth examining how these texts in particular relate to Stefaneschi's account.

2 Stefaneschi's Account of the Ending of the 13th Century and the Anticipatory Mood of the Christians

In this part of the text, I will examine what we can deduce from Stefaneschi's confusing story about an ambiguously defined multitude of people[15] who

8 Ludwig Hödl, "Der Jubiläumsablaß Papst Bonifaz' VIII. 1300–der Ernsfall der Buße," *Wissenschaft und Weisheit* 62/2 (1999), 295–296.

9 For Jean Lemoine's commentary on *Antiquorum habet*, I will use the edition by Ludwig Hödl, "Die Glosse des Kardinals Johannes Monachus zur Ablass-Epistel 'Antiquorum,'" in *Medioevo. Rivista di storia della filosofia medievale* 26 (2001), 161–202 (cited: Jean Lemoine, *Glossa*).

10 Magister Bonaiuto, cited in *Annales Caesenates* (RIS 14, col. 1119).

11 Silvester's letter is edited in Angelo Mercati, "La lettera dello scrittore pontificio Silvestro sul Giubileo del 1300," in *Cronistoria dell'anno santo 1925* (Roma, 1928), 1191–1198.

12 Giovanni Villani, *Nuova Cronica* 8,36 (RIS 13 col. 367–368).

13 Guglielmo Ventura, *Chronicon Astense*, cap. 26 (RIS 11, col. 191–192).

14 Lots of research exists on the uneasy attitude of Dante towards Boniface VIII. See e.g., Jo Ann H. Moran Cruz, "Dante, Purgatorio 2, and the Jubilee of Boniface VIII," *Dante Studies, with the Annual Report of the Dante Society* 122 (2004).

15 At the beginning of his story, Stefaneschi does not use the Latin word "*populus*" but writes in much vaguer terms, such as in the passive or by using the words "Romans" (cap. I, f. 4v, 35), "citizens and foreigners" (cap. I, f. 5r, 42), "some" and "many" (cap. I, f. 5r, 43 and 45),

demanded an indulgence, while the pope was not planning to solemnise the change of the century in any way. Stefaneschi's story begins with the account of how, at the end of the year 1299, a rumour circulated among the people that the pilgrims who visited the basilica of S. Pietro in Vaticano in Rome would obtain a plenary indulgence, that is, a complete remission of sins and, according to many, less time spent in Purgatory after death.[16] The pope and the curia were seemingly surprised by this, and tried to find any record of a similar event in the past in the papal registers. None was found, perhaps due to neglect, wars, or other disasters, according to Stefaneschi.[17] Nor have modern historians found any convincing evidence of a pre-existing jubilee tradition. Indeed, it seems certain that no Christian jubilees were organised before 1300.[18]

In Stefaneschi's story, the pope initially decided to do nothing. Then, on the evening of 1 January, 1300, a great number of Romans were gathering in the basilica of S. Pietro in Vaticano believing that it was the only day when they could obtain the indulgence:

> Surprisingly, almost for the whole day of the first of January the mystery of the new remission of sins remained occulted. However, when the sun set towards the evening, almost until the silence of the darkest midnight, for a moment it revealed itself to the Romans, who hurried in great numbers to the holy basilica of San Pietro in Vaticano, crowding and shoving each other closely around the altar, so that it was nearly impossible to

or "crowd" ("*turba*" cap. I, f. 5v, 47; cap. III, f. 7v, 91). I use here the general English word "people" for the lack of a better word in the midst of these constantly changing and vague forms of expression. Stefaneschi himself uses the word "*populus*" for the first time at the point in his story where the pope declares the jubilee (cap. III, f. 7v, 92), and only after this he starts to describe in a detailed manner who these people were (cap. IV–VI).

16 In the Roman Catholic tradition, an indulgence is granted by an ecclesiastical authority and is a remission of a defined amount of temporal punishment for sins of which the guilt has been forgiven. Usually in the Middle Ages, in order to get an indulgence, one needed to complete predefined tasks such as a pilgrimage to a certain church at a certain time. Indulgences were extremely common in the churches of Europe in the 13th century and during the jubilee. It is unclear how indulgences of this era were believed to affect the time in purgatory. For medieval indulgences in general, see: Nikolaus Paulus, *Geschichte des Ablasses im Mittelalter vom Ursprünge bis zur Mitte des 14. Jahrhunderts* (Paderborn: Druck und Verlag von Ferdinand Schöningh, 1922–1923); "Ablaß," in *Lexikon des Mittelalters* I (München: Artemis & Winkler Verlag, 1980), col. 43–46; "Indulgenze," in *Enciclopedia cattolica* V (Città del Vaticano, 1951), 1901–1907.

17 Stefaneschi, *De centesimo,* cap. I, f. 4r–4v, 26–31.

18 See also Frugoni, "Il Giubileo di Bonifacio VIII," 7 and 29. On the loss of documents in the Vatican archives during this period, see Leonardi, Schmidt, and Placanica, *Iacopo Stefaneschi. De centesimo seu iubileo anno. La storia del primo giubileo*, 47 n. 12.

> approach. As if they thought that the remission would expire or be much larger only on the day that was soon about to end.[19]

Therefore, according to Stefaneschi, there were many people – who are not precisely defined – who were certain that there would be a plenary indulgence for the turn of the century. Yet, the pope was completely taken aback by this idea and did not find any evidence for this idea, even in the historical archives.[20] So, how could the people have developed this idea of a great indulgence due to the new century completely independently from the pope and the leaders of the Church?

Stefaneschi cautiously hypothesises that the gathering of so many people at the basilica of S. Pietro in Vaticano could either be spontaneous, or due to the Holy Spirit, or perhaps because of something they had heard in the morning mass.[21] Still, after 1 January, the masses of Romans and foreigners continued to grow. In his opinion, this was partly due to the fact that 17 January was near, which was the day when the pope would parade an extremely popular relic, the Veil of Veronica, in the streets.[22] On the other hand, there were rumours spreading that the first day of January had not been the only day to obtain the indulgence, but pilgrims could still have it every day for the rest of the year 1300. According to some, one could get a plenary indulgence on the first day of January, and on other days of the year "only" a 100-year indulgence.[23] The pope

19 Stefaneschi, *De centesimo*, cap. I, f. 4v–5r, 32–38. "*Mira res, tota pene prima ianuarii dies novelle misterium remissionis occuluit, proclivo vero in vesperam sole, quoad medie fere noctis intempeste silentium parumper eo Romanis patefacto, frequentes ipsi ad sacram beati Petri basilicam properant, oppressi altare stipant alternatimque se, ut vix accedere fas esset, impediunt, tanquam ea brevi finienda die expirare gratiam vel maiorem autumarent.*" All the translations from Latin and Italian are my own, unless otherwise stated.

20 Also the chronicler Giovanni Villani mentions these rumours in his independent eyewitness story: *Nuova cronica* 8,36 (RIS 13 col. 367–368).

21 Stefaneschi, *De centesimo*, cap. I, f. 5r, 38–41.

22 Stefaneschi, *De centesimo*, cap. I, f. 5r–v, 45–48. The Veil of Veronica, also known as the "Sudarium" or simply "The Veronica," is a piece of cloth which has a face of Jesus on it, not made by human hand. This cloth was the one that St Veronica used to wipe the blood and sweat off Jesus' face, when he was walking with the cross to Calvary, and after this, the cloth had the imprint of his face. According to the story, she brought the relic to Rome and presented it to Emperor Tiberius. The first certain records of the veneration of the relic in Rome are from the 13th century, however. The date of the procession of the Veronica was not fixed and fell on 17 January in the year 1300. Cf. Leonardi, Schmidt, and Placanica, *Iacopo Stefaneschi. De centesimo seu iubileo anno. La storia del primo giubileo*, 48–49, n. 18. See also Holger Kaasik's article in this volume (p. 256–257).

23 Stefaneschi, *De centesimo*, cap. I, f. 5r, 43–45: "*Quibusdam prima centesimi die omnium culparum sordes deleri asserentibus, ceteris annorum centum indulgentiam fore.*" Also: Jean Lemoine, *Glossa* 21, 189. Other sources: Leonardi, Schmidt, and Placanica, *Iacopo Stefaneschi. De centesimo seu iubileo anno. La storia del primo giubileo*, 46–47, n. 10.

still did not speak to the people, but nor did he stop them from gathering or spreading rumours. According to Stefaneschi, this lack of denial for its part incited the belief that there really was a possibility for a great indulgence.[24]

In Stefaneschi's story, the first person to give a voice to this undefined multitude of people was a certain 107-year old man, who recounted that he was brought to Rome 100 years ago by his father and thus he had personally witnessed the last jubilee. According to him, the pilgrims could obtain an indulgence of 100 years for the whole year during the last jubilee. Also, his father had been confident that in the year 1300 there would be a similar indulgence in Rome and his son should not miss the occasion.[25] Also, in Beauvais in France,[26] two people of sufficiently old age were found claiming to have participated in the same jubilee a century ago. Furthermore, several old Italians emerged confirming this, some of them adding that to obtain the indulgence, it had been necessary to visit the basilica of S. Pietro in Vaticano not for one day but for three days.[27] It is remarkable that none of these old witnesses are identified as Roman. This only adds to the mystery of the matter.

Finally, Boniface VIII decided to act. Stefaneschi tells his audience that the pope consulted the college of cardinals and contemplated for a long time and with great care the best way to fulfil the demands of the people, whose spirit the pope allegedly did not want to extinguish.[28] To sum up, Stefaneschi wants us to believe that the pope and the curia had not expected this surprising demand for a great indulgence in 1300, and the fact that it was finally proclaimed almost two months retroactively seems to serve as sufficient proof of this for some scholars.[29] Seemingly for Stefaneschi, the people were offering the pope a great unexpected occasion to celebrate Christ's 1300th birthday in a magnificent and novel way and to bring thousands of pilgrims to Rome. Moreover, all

---

24 Stefaneschi, *De centesimo*, cap. II, f. 5v, 52–54.

25 Stefaneschi, *De centesimo*, cap. II, f. 5v–6r, 54–60.

26 Antonio Placanica notes that Stefaneschi used to be the canon of the church of Beauvais from 21 December 1291, and according to some hypotheses Stefaneschi might have brought this information to the curia himself: Leonardi, Schmidt, and Placanica, *Iacopo Stefaneschi. De centesimo seu iubileo anno. La storia del primo giubileo*, 49–50 n. 25.

27 Stefaneschi, *De centesimo*, cap. II, f. 6r, 63–67. Later in the story, when the jubilee has already been announced, Stefaneschi mentions another old man from Savoy who had also seen the last jubilee: *De centesimo*, cap. VII, f. 13r–v, 202–205. See also Guglielmo Ventura, *Chronicon Astense*, cap. 26 (RIS 11, col. 191): *"audivimus ab antiquis…"*

28 Stefaneschi, *De centesimo*, cap. III, f. 6v, 70–77.

29 Alfonso Stickler and Debra J. Birch deduce that the pope was clearly not prepared: Stickler, *Il Giubileo di Bonifacio VIII*, 14; Debra J. Birch, *Pilgrimage to Rome in the Middle Ages. Continuity and Change* (Woodbridge: The Boydell Press, 1998), 197. On other possible explanations, see my analysis below.

the ingredients for its celebration had already been suggested by the people, and the pope only needed to make the best possible combination of them to determine the magnitude, duration, and the requirements for the great indulgence. In his bull of 22 February, the pope confirmed – to the great joy of the people – that it was possible to gain a plenary indulgence during the whole year 1300, starting from the Christmas of 1299.[30]

This is what Stefaneschi wants to convey to his readers regarding the background of the first jubilee and the different ideas that the people and the pope had regarding the dawn of the new century. Yet the story seems unbelievable and confusing. Why would the pope want to give an impression that the people dictated to him when to give a plenary indulgence? In my opinion, this part of the story serves to justify the founding of this new religious tradition. Firstly, Stefaneschi surmises that the people may have been guided by the Holy Spirit; this was not a weak justification, because it was a common belief at this time that the Holy Spirit could sometimes steer the thoughts of the masses. Secondly, collective talk or rumours actually had great authority in medieval communities. Even medieval legal systems acknowledged the force of common opinion, and therefore collective talk equalled truth and could validate the hazy memories of eyewitnesses – just as is the case in Stefaneschi's story. Even the canon lawyer Jean Lemoine comments: "A rumour spoken by everyone is not at all false."[31] Indeed, the fact that the rumours described by Stefaneschi do not seem to have any precise source – and even the sources that are mentioned come from far away from the city of Rome – highlight the universality and power of this common opinion.

30 The bulls *Antiquorum habet* and *Nuper per alias*, and Stefaneschi, *De centesimo,* cap. III, f. 6v–7r, 77–86.

31 Jean Lemoine, *Glossa* 21, 189: *"sermo famatus ab omnibus non est omnino falsus."* Stefaneschi uses various words such as *"fama"* (cap. II, f. 6r, 67–68), *"vulgari"* (cap. II, f. 6r, 63), *"opinione"* (cap. I, f. 4v, 31; cap. I, f. 4r, 21), and *"rumor"* (cap. I, f. 4r, 23). For the significance of "common opinion" in the Middle Ages, see Thelma Fenster and Daniel Lord Smail, *Fama: The Politics of Talk & Reputation in Medieval Europe* (Ithaca: Cornell University Press, 2003), 3 and 8–10; Julien Théry: "Fama: l'opinion publique comme preuve judiciaire. Aperçu sur la révolution médiévale de l'inquisitoire (xiie–xive)," in *La Preuve en Justice de L'antiquité à Nos Jours*, ed. Bruno Lemesle (Rennes: Presses universitaires de Rennes, 2003) 119–147; Roberta Mucciarelli, "Bisogna essere molto prudenti con le voci perché fanno presto a trasformarsi in verità. Qualche considerazione su *fama* e *publica vox* nell'Italia comunale," in *Fama e pubblica vox nel medievo*, eds, Isa Lori Sanfilippo and Antonio Rigon (Roma: Istituto storico italiano per il medio evo, 2011), 25–46; Heather Kerr and Claire Walker, *Fama and Her Sisters. Gossip and Rumour in Early Modern Europe* (Turnhout: Brepols, 2015).

Following this line of thought, organising a jubilee and granting a plenary indulgence to everyone might have been such a dramatic gesture of power that Boniface VIII preferred to make it look like it was the people's collective wish – and not just by the few, for example the Romans, but collectively by countless Christians from various parts of Christendom. This aspect alone gave considerable authority to his new jubilee tradition.

Another obvious question arises: if the pope wanted to justify his jubilee indulgence by believing that there had been a similar indulgence also in 1200, why did he not simply forge some suitable documents to prove this more officially?[32] In this way, he could have made it look like he was continuing a tradition already invented by Innocent III (1198–1216) in 1200, or by another previous pope at an even earlier point in time. Regardless of their ecclesiastical power and dogmatic influence, the popes could not simply invent new religious traditions without a strong basis in some earlier tradition. Indeed, once successfully established by Boniface VIII, the future popes did not hesitate to harness the jubilee tradition at every possible occasion. Maybe it was not necessary to forge anything though, because the common opinion and the common will of the people were valid and legal enough to justify such a "new" tradition; as explained above, actual evidence in writing did not matter. Whatever the truth, Boniface VIII used the occasion of the end of the century, and the power conveyed by this powerful time-related concept, in a very intelligent manner.

In any case, if we consider the possibility that the people really "were" expecting a plenary indulgence for the end of the century, we must ask why they had this idea? One possible explanation can be found in an examination of the indulgence habits of this era. Indulgences had become increasingly numerous and popular before 1300. Already from the 12th century onward the popes started to grant indulgences of varying importance for those who made a pilgrimage to Rome – or to other churches and shrines, for that matter. This development continued during the 13th century, even to the extent that in the Fourth Lateran Council (1215) the rules for granting indulgences were made stricter, and their maximum extent was set to one year for attendance at the actual day of consecration of a church and to 40 days for attending the following anniversaries.[33] Nevertheless, the rules were not always obeyed, and already during

32 Forgery of historical documents was common: Horst Fuhrmann, "Einfluß und Verbreitung der pseudoisidorischen Fälschungen. Von ihrem Auftauchen bis in die neuere Zeit," in *Schriften der Monumenta Germaniae historica* 24, I–III (Stuttgart: Anton Hiersemann Verlag, 1972–74), LV, 236 + VIII, 237–624 + VI, 625–1127.

33 Fourth Lateran Council, *Constitution* 62. See Michele Maccarone, "L'indulgenza del giubileo del 1300 e la basilica di San Pietro," in *Roma anno 1300*, ed. Angiola Maria Romanini

the rest of the 13th century the popes granted generous indulgences of up to several years for those who made a pilgrimage to Rome, especially to those visiting the basilica of S. Pietro in Vaticano on certain days.[34] Through the whole century preceding the jubilee, pilgrims going to Rome expected great indulgences on a few exceptional occasions and regular smaller indulgences on a number of less important annual feasts.

At the same time, Rome was becoming an increasingly popular destination for pilgrimage partly because the Holy Land had become more and more difficult to access since the loss of the last Christian stronghold of Acre in 1291.[35] Also by the 13th century, several valuable relics, such as the Veil of Veronica, pieces and nails of the True Cross, the table of the Last Supper, and the Column of the Flagellation already attracted pilgrims to Rome. This could have made the people see the city of Rome expectantly as another Jerusalem, a new place where to gain salvation. Indeed, Boniface VIII presented the city in this way in 1300.[36]

In terms of existing traditions, there was, in fact, some pre-existing background for the people expecting an exceptional indulgence on 1 January. One of the various indulgences already in place in Rome in 1300 was an indulgence of three years and 40 days for those who visited the basilica of S. Pietro in Vaticano on the 1 January, which was the day of the circumcision of Jesus. This

(Roma: L'Erma di Bretschneider, 1983), 733; Brunero Gherardini, "Indulgenza: Teologia e Storia," in *I giubilei nella storia della Chiesa* (*Città del Vaticano: Libreria editrice Vaticana, 2001*), 87.

34 On the development of 13th century indulgences, especially in S. Pietro, see Frugoni, "Il Giubileo di Bonifacio VIII," 32–39; Maccarone, "L'indulgenza del giubileo del 1300 e la basilica di San Pietro"; Birch, *Pilgrimage to Rome in the Middle Ages*, 196. The most comprehensive basic scholarly work on the history of indulgences is *Geschichte des Ablasses* by Nikolaus Paulus. See also a concise explanation of the development of indulgences from the 10th to 13th centuries, based on Paulus, by Anna Holst Blennow, *The Latin Consecrative Inscriptions in Prose of Churches and Altars in Rome 1046–1263* (Roma: Società romana di storia patria alla Biblioteca Vallicelliana, 2011), 268–273.

35 New large-scale crusades were no longer organised by the pope to reconquer strongholds, but other types of crusades continued actively. See e.g. Lee Manion, "The Loss of the Holy Land and 'Sir Isumbras': Literary Contributions to Fourteenth-Century Crusade Discourse," *Speculum* 85/1 (January 2010), 65–90.

36 Frugoni, "Il Giubileo di Bonifacio VIII," 78–83; Maccarone, "L'indulgenza del giubileo del 1300 e la basilica di San Pietro," 731–732; Laura Onofri, "Roma come nuova Terrasanta," in *Roma Sancta. La città delle basiliche*, eds, Marcello Fagiolo and Maria Luisa Madonna (Roma: Gangemi Editore, 1985), 28–31; Franco Cardini, "L'eclisse di Gerusalemme. Fallimento della crociata in Terrasanta, nascita del giubileo," in *La Storia dei Giubilei. Volume primo 1300–1423*, ed. Gloria Fossi (Firenze: Giunti–BNL Edizioni, 1997), 56–69; Birch, *Pilgrimage to Rome in the Middle Ages*, 199.

indulgence was established 11 years before by Nicholas IV,[37] and it must have contributed to the unusually large number of people in the basilica, described by Stefaneschi, during the 1 January in 1300. Perhaps people expected that this time the usual indulgence would be much larger. As Stefaneschi suspects, the priest who held the morning mass might have said something inspiring about the new century that was about to begin, and the rumour among the people transformed into a common idea that the usual indulgence would this time be for 100 years, or that during the year 1300 a great indulgence could be gained not only on the 1 January but every day of the year.[38] We should not exclude the possibility that the pope and the curia took advantage of the large crowds already in Rome for the festivities of the changing year and invented this idea, or even ordered the rumours to be initiated. Boniface VIII might well have recognised that being a pope when a Christian century was ending was an opportunity too good to miss to strengthen the status of the Papacy.

However, Stefaneschi wanted to convey to his audience that the people were certain that there had been such a plenary indulgence already in 1200. If so, such ideas must have come from somewhere. Traditions were of importance in the Middle Ages, and it is easy enough to understand why Stefaneschi wanted to show the jubilee of 1300 as a continuation of an older tradition. Still, the only known source that suggests that a jubilee was held before 1300 is Alberic of Trois-Fontaines (d. c.1252), a Cistercian chronicler, who wrote in his *Chronica* that the year 1208 "was celebrated as the fiftieth year or the year of Jubilee and remission in the curia at Rome."[39]

Here though, in the context of Alberic's text, we should note that the word "jubilee" had various meanings before 1300. Among other things, the word was often used to mean a plenary indulgence usually granted for crusades. To quote Arsenio Frugoni, "the jubilee was, diffusely, an expression of reconciliation and liberation from sin."[40] Neither the jubilee bull *Antiquorum habet* nor the three surviving non-funerary inscriptions that mention the related plenary indulgence, contain the word "jubilee," but Stefaneschi and other chroniclers use it

37 A papal bull of Nicholas IV, 25 February 1289, in *Collectio bullarum* I, 214. Cf. also Fois, "Il Giubileo del 1300," 27.

38 Maccarone, "L'indulgenza del giubileo del 1300 e la basilica di San Pietro," 744–745.

39 MGH, Scriptores, XXIII, 889: *"Dicitur quod annus iste quinquagesimus sive Jubileus et remissionis in Curia Romana sit celebratus."* The translation is from Birch, *Pilgrimage to Rome in the Middle Ages*, 198. The "fiftieth year" is a reference to the Hebrew jubilee.

40 Frugoni, "Il Giubileo di Bonifacio VIII," 9: *"Il giubileo era dunque, diffusamente, figura di riconciliazione e di liberazione dal peccato."*

continuously. Indeed, later it also became commonly to mean the Holy Year itself.[41]

It is most probable, then, that the phrase by Alberic cited above refers to one of the two indulgences that Innocent III granted in the same year of 1208: one plenary for the Albigensian crusade, and another of one year for anyone participating in the parade for displaying the Veil of Veronica.[42] Neither of these two indulgences were anywhere near the greatness of the indulgence of 1300, which was plenary and could be gained for the whole year. Debra J. Birch suggests that later writers probably added the word "jubilee" to Alberic's text, but before 1300. If this was the case, it could still mean that the concept of a jubilee year connected with Rome did exist prior to 1300.[43]

The Veil of Veronica might well be connected to a communal memory hypothesised by Debra J. Birch and others. Parading the Veil of Veronica on the first Sunday after the Octave of Epiphany was a new and popular annual tradition established by Innocent III in 1207, a year before the aforementioned one-year indulgence, and it was still a big event at the beginning of the jubilee year 1300.[44] It could be that this new parade tradition with an indulgence, even though not plenary, was so impressive that Alberic – or the later modifiers of his text – who wrote decades later, wanted to record it in his *Chronica*. Perhaps also for contemporaries, who did not have access to literature but participated personally in the first parades of the Veil of Veronica, the indulgence and the event were impressive. The indulgence might then have been by chance associated with the newly started 13th century, and some vague memory of it

41 Frugoni, "Il Giubileo di Bonifacio VIII," 10 ff. For the three non-funerary inscriptions, see notes 139 and 140, and Kantola on p. 132–133. There is also a fourth funerary inscription (IChiR 12:569) from S. Lorenzo fuori le Mura that actually includes the word *iubileum* together with the year 1300. This inscription on the tomb of Guichard, a canon from Lyon who apparently died during his pilgrimage to Rome in January 1301, could also have been commissioned sometime – but judging from its style not a very long time – after his death, when the word "jubilee" was already commonly used. Cf. Jörg Garms, Roswitha Juffinger, and Bryan Ward-Perkins, eds, *Die mittelalterlichen Grabmäler in Rom und Latium vom 13. bis zum 15. Jahrhundert, 1. Band: Die Grabplatten und Tafeln* (Rom, Wien: Verlag der österreichischen Akademie der Wissenschaften, 1981), 97–98, nr. 21.2. I thank Urpo Kantola for all the remarks and readings concerning all the inscriptions mentioned in this article.

42 Frugoni, "Il Giubileo di Bonifacio VIII," 7; Maccarone, "L'indulgenza del giubileo del 1300 e la basilica di San Pietro," 733 and 731 n. 3.

43 Birch, *Pilgrimage to Rome in the Middle Ages*, 198, citing Herbert Thurston, *The Holy Year of Jubilee. An Account of the History and Ceremony of the Roman Jubilee* (London, 1900), 16.

44 Fois, "Il Giubileo del 1300," 27. Cf. also Maccarone, "L'indulgenza del giubileo del 1300 e la basilica di San Pietro," 732–733, 744; Frugoni, "Il Giubileo di Bonifacio VIII," 7. For more sources and literature, see Leonardi, Schmidt, and Placanica, *Iacopo Stefaneschi. De centesimo seu iubileo anno. La storia del primo giubileo*, 48 n. 18.

still lived in the communal consciousness when the year 1300 was approaching. Of course, the indulgence was granted a few years after the beginning of the 13th century, but it does not seem at all impossible that over the course of several generations such minor details were forgotten within the communal memory.

At the same time, with the increasing number of indulgences and pilgrims in Rome, there seems to have been an expectant general mood and growing eschatological tension among Christian scholars.[45] Starting from the 12th century, there was not only more awareness of changing centuries but also more attempts to determine the moment when the world would end. Negative developments in the Holy Land only incited this mood. During the 13th century, there were many who were inspired by the eschatological writings of Joachim of Fiore: among them the Franciscans, Thomas Aquinas, Peter John Olivi, and Bonaventure. Many interpreted certain passages from the Bible[46] to mean that the end would come soon or maybe even around 1300.[47]

Even those who were not worried about the potential end of the world were at least sensitive to the changes spanning over long time periods. They were, perhaps, expecting that there would be some kind of a revolution or a reform. According to Simona Cohen, during the 13th century's "Aristotelian revival [...] time was increasingly perceived in terms of the natural world." The concept of measuring and controlling time became more and more actual, at least in scholarly circles.[48]

In this context, Alexander von Roes (d. after 1288), who was a canon at S. Maria in Kapitol in Cologne, wrote a historical treatise, *Noticia Seculi*, in 1288,

---

45 Frugoni, "Il Giubileo di Bonifacio VIII," 17–18.

46 E.g. Apoc. 20,7 and Dan. 12,11.

47 Roberto Rusconi, "Milleniarismo e centenarismo: tra due fuochi," in *Roma Sancta. La città delle basiliche,* eds, Marcello Fagiolo and Maria Luisa Madonna (Roma: Gangemi Editore, 1985), 3–4. On eschatological thought in the Middle Ages, see e.g., Richard Kenneth Emmerson and Bernard McGinn, eds, *The Apocalypse in the Middle Ages* (Ithaca and London: Cornell University Press, 1992); Riccardo Quinto, "Giubileo e attesa escatologica negli autori monastici e nei maestri della *sacra pagina," Rivista di storia della filosofia medievale* 26 (2001), 25–82; Brett Edward Whalen, *Dominion of God* (Cambridge, MA: Harvard University Press, 2009).

48 Simona Cohen, *Transformations of Time and Temporality in Medieval and Renaissance Art* (Boston: Brill, 2014), 39–49, 93, 115. Convincing evidence of this is also apparent among Italian merchants: Steven A. Epstein, "Business Cycles and the Sense of Time in Medieval Genoa," *The Business History Review* 62.2 (summer 1988), 238–260. For infinitely different ways to understand and relate oneself to time in a Roman context, see Saku Pihko's contribution in this volume.

in which he was very aware of the changing eras and centuries, "*mutatio seculorum.*"[49] Alexander saw a repeating pattern in history, according to which considerable changes happen in the world every 100 years. He adopted this idea from *De semine scripturarum* (c.1205) of Arnaldus de Villa Nova (c.1240–1311/2), who had written that his century would end and the next would start in 1215. Alexander expected, like Arnaldus, that before the current century ended, in 1315,[50] Palestine would be recaptured, and the Church would be purged of heretics. Alexander already saw significant changes, in the way that the first half of his century had seen the rapid growth of papal power in both spiritual and temporal matters. He used the most recent years as a troubling example. There had been a disastrous flood of the North Sea in 1287 and again in 1288, and at the same time the Holy See was vacant for almost a year while the Mongols were causing trouble in the north and panic in Italy, and Christians were fighting with each other. Alexander perceived that he was living at the end of an era.[51] If so, there was need for a spiritual purification to wash away the sins of the troubled 13th century, and Rome was now the best place to look for this. It is likely that these ideas too were restricted to a small circle of scholars, to which Stefaneschi himself can to some extent be said to belong, but we should not reject the possibility that such concepts reflected at least partly the thoughts of the majority of the Christian population.

Therefore, the demands for a great indulgence in 1300, or at least Stefaneschi's idea to write about such demands, did not come out of nothing, even though a jubilee year had never been celebrated before. The tendency to grant indulgences had already been frequent in the 13th century, and the change of the century was an obvious place to expect another one.[52] Also, there was a growing eschatological mood in scholarly circles during the 13th century, and pilgrimage to Rome was rapidly increasing.

So far we can deduce that obviously the concept of time, or its passing, had become important by the early 14th century. The concept of time had numerous different levels and here we can see that, among other things, it was inseparably intertwined with religion: it was of course important to keep track of time as such, but it was also important for God's plan. We do not need to know

49 MGH, Scriptores, Staatsschriften, 1,1, 159 (*Noticia Seculi* 12,10).

50 This is an interesting reminder that there were different ways of seeing when eras changed. Alexander von Roes clearly did not think that long term changes in society automatically happen alongside the centuries counted from the nativity of Christ. On the development of the concepts of centuries and eras, see Ernst Lewalter, "Das Jahrhundert als Epoche (aetas, centuria, saeculum)," *Romanische Forschungen* 49 (1935), 343–351.

51 MGH, Scriptores, Staatsschriften, 1,1, 151–155 (*Noticia Seculi* 6–8).

52 Frugoni, "Il Giubileo di Bonifacio VIII," 13, 29.

how true Stefaneschi's story was to understand that the concept of the end of a Christian century was important to him. It was important to the pope, and to many other Christians, too, judging from the great number of pilgrims that arrived in Rome after the declaration of the jubilee.

Glimpses of this interest in changing eras and centuries can also be seen in other earlier sources. The justification for the jubilee indulgence harnessed the distant and obscure past, as shown above. Furthermore, the papal bull *Antiquorum habet* admits this quite plainly: the authority of the plenary jubilee indulgence was based on *antiquorum fida relatio*, the "trustworthy tradition of the ancients," and at this time that was deemed strong enough a justification to found what in practice was a new religious tradition.

3 Determining the Extent and the Chronological Limits of the Indulgence

When we examine the effectiveness of the jubilee indulgence itself, we encounter yet another even more abstract concept of time. How did the people understand the length of an indulgence in practice? For a long period, the concept of indulgences developed without a clear theological direction, until the first attempts were made to define them and to stop them from spreading without control in the 12th century. Essentially, an indulgence meant that anyone obtaining it had a defined amount of remission from temporal penance accumulated by sin. The plenary indulgences were mostly reserved for crusaders and were much rarer than limited indulgences. It was not actually meant that the indulgence cancelled the same amount of time in Purgatory after death, which itself was a place outside of worldly time.[53]

This is not the place to discuss the exact way in which indulgences functioned, but we can acknowledge that the rules for this were unclear even within the Church, and the people seemed to have their own ideas of what a plenary indulgence meant.[54] Many seemed to simply understand it to be directly connected to the time spent in Purgatory, and these ideas were not corrected

53 For the development and meaning of indulgences, see Frugoni, "Il Giubileo di Bonifacio VIII," 30 ff; Gherardini, "Indulgenza: Teologia e Storia," 84–88. For time in Purgatory, see Anca Bratu, "Le purgatoire entre le temps terrestre et le temps eschatologique: forms de representation du temps," in *Le Temps, sa mesure et sa perception au Moyen Age*, ed. Bernard Ribémont (Caen: Paradigme, 1992), 47–60; Jacques Le Goff, *La naissance du Purgatoire* (Paris: Gallimard, 1981).

54 See e.g., Karl Rahner, *Encyclopedia of Theology. A Concise Sacramentum Mundi* (New Delhi: Burns & Oates, 1975), 702 ff.

by the Church authorities.[55] This idea can be seen for example in a jubilee poem by magister Bonaiuto, talking about the indulgence of 1300 and addressed to Purgatory: "Cease this harsh suffering, at least as long as the jubilee year lasts."[56] In this way, many people living in that time may have felt that by obtaining the jubilee indulgence one could cancel the time they should spend in Purgatory after death. Moreover, this would take place with papal authorisation. Thus, granting plenary indulgences was an immensely powerful gesture of authority from the pope.

The bull *Antiquorum habet* itself does not define the extent of the jubilee indulgence in any way, other than saying it was "not only full and more extensive but, rather, completely plenary from all the sins committed,"[57] which should be understood as "from all the penance caused by the sins committed" as Stefaneschi and the canon lawyer Jean Lemoine explained. This phrase might have been made for maximum rhetorical effect though, and may not actually mean that this is a new and more extensive plenary indulgence, in terms of its effectiveness.[58] Finally, both Stefaneschi and Jean Lemoine simply state that the indulgence was effective as far as the authority of the pope extends.[59] The matter then becomes even more unclear, because *Antiquorum habet* confusingly instructed that the more devotedly and frequently a pilgrim carried out the pilgrimage, the more "efficient" the indulgence would become.[60] According to the Church, a plenary indulgence only cancelled the temporal penance, and the rest was for God to decide. Therefore, the ultimate

55 Gherardini, "Indulgenza: Teologia e Storia," 88; Etienne Doublier, "*Libra misericordiae* le indulgenze di Bonifacio VIII," *Rivista di storia della Chiesa in Italia* 64/2 (luglio–dicembre 2010), 355 n. 25.

56 *Annales Caesenates*, in RIS XIV, 1119b: *"suspende rigorem hunc, donec saltem jubileus transeat annus."* This seemed to be the common view also later: see Giovanni Rucellai (1403–1481), *Zibaldone*, col. 51rA.

57 Registres de Boniface VIII, II, 922–923, no. 3875: *"non solum plenam et largiorem immo plenissimam omnium suorum concedimus veniam peccatorum."*

58 Frugoni, "Il Giubileo di Bonifacio VIII," 49; Stickler, *Il Giubileo di Bonifacio VIII*, 22–23; Hödl, "Der Jubiläumsablaß Papst Bonifaz' VIII. 1300 – der Ernsfall der Buße"; Stefaneschi, *De centesimo*, cap. XIII, f. 22v–23r. According to Jean Lemoine, the phrase "*plenissima venia peccatorum*" actually means "*poenarum pro peccatis debitarum*": Jean Lemoine, *Glossa* 40, 199. See the comment in Leonardi, Schmidt, and Placanica, *Iacopo Stefaneschi. De centesimo seu iubileo anno. La storia del primo giubileo*, 51 n. 34, 80 n. 161–162, and 81 n. 167.

59 Stefaneschi, *De centesimo*, cap. XIII, f. 23r, 395–396; Jean Lemoine, *Glossa* 40, 199.

60 Stefaneschi, *De centesimo*, f. 30r, 30–31: *"Unusquisque tamen plus merebitur et indulgentiam efficatius consequetur qui basilicas ipsas amplius et devotius frequentabit."* Also: Jean Lemoine, *Glossa* 43, 201–202. Cf. comment in Leonardi, Schmidt, and Placanica, *Iacopo Stefaneschi. De centesimo seu iubileo anno. La storia del primo giubileo*, 74 n. 139, and Stickler, *Il Giubileo di Bonifacio VIII*, 31.

effect of the indulgence was primarily determined by the devotedness of the pilgrim.

Boniface VIII was relatively frugal with indulgences during his reign, which made the jubilee indulgence seem even more generous and rare. As mentioned above, indulgences had become very common during the 13th century, and the popes (including Boniface VIII) often exceeded the indulgence limitations set by the Fourth Lateran council. But during Boniface VIII's reign there were relatively few granted and were small compared to his predecessors: the indulgences for church visits rarely exceeded one year and 40 days.[61] Boniface VIII even cancelled the plenary indulgence granted by Celestine V for attendance at the church of Collemaggio for the anniversary of his coronation day on 29 August, in order to take control of plenary indulgences and to restore the authority of the Church that had suffered during Celestine V's ineffective reign.[62] Then, a little before the jubilee, between 1295–1297, Boniface VIII granted his first four plenary indulgences. Three of them were related to crusades: the first for the Holy Land, the second against Frederick III of Sicily, and the third against the Colonna family. The fourth was granted to Franciscan friars who went as missionaries among the Mongols.[63] In this light, the plenary indulgence of 1300 seemed even more impressive and generous, a real demonstration of the pope's *plenitudo potestatis*.[64] It also was a great moment for basilicas; never before had there been an indulgence of more than seven years connected with the basilica of S. Pietro in Vaticano.[65] Also, this indulgence was not about crusades or missionaries, but rather personal salvation for anyone, and the cancelling of the time of temporal penance.

Now, we will examine how the moment of the proclamation for *Antiquorum habet* was determined, why the repetitive frequency of 100 years was chosen for jubilees, and how it relates to the distant past and the future. Stefaneschi describes the proclamation day of *Antiquorum habet*:

---

61 For the significance of indulgences during the time of Boniface VIII, a complete list of his indulgences, and a summary of previous research, see Doublier, "*Libra misericordiae* le indulgenze di Bonifacio VIII." See also Frugoni, "Il Giubileo di Bonifacio VIII," 37–43; Maccarone, "L'indulgenza del giubileo del 1300 e la basilica di San Pietro," 742, and Axel Goria, "Le indulgenze a Roma nel 1300 (un elenco fin'ora sconosciuto)," *Studi medievali* 18 (1952).

62 Maccarone, "L'indulgenza del giubileo del 1300 e la basilica di San Pietro," 741; Registres de Boniface VIII, I, n. 815.

63 Maccarone, "L'indulgenza del giubileo del 1300 e la basilica di San Pietro," 747–748; Fois, "Il Giubileo del 1300," 32; Leonardi, Schmidt, and Placanica, *Iacopo Stefaneschi. De centesimo seu iubileo anno. La storia del primo giubileo*, 88 n. 207.

64 Frugoni, "Il Giubileo di Bonifacio VIII," 44.

65 Maccarone, "L'indulgenza del giubileo del 1300 e la basilica di San Pietro," 745.

> The day of the publication had arrived, set on the Feast of the Chair of S. Pietro. The gift to the Prince of the Apostles was published in his temple, which was more honourable than any other place. The pope and the cardinals rose on the pulpit, which was covered in silk and gold, and after a speech to the crowd they proclaimed the great gift of joy, a letter with the bull hung by a silken thread. And even though it bore the original text "given at the Lateran Basilica," the pope ordered to be written "at the basilica of S. Pietro." The gift was placed on the altar. A similar document was sent to the basilica of S. Paolo.[66]

The papal bull *Antiquorum habet* was declared around two months retroactively, and the reason that is usually offered is the pope's unpreparedness, but that is only what Stefaneschi wants to convey. In my opinion, declaring *Antiquorum habet* retroactively does not exclude the possibility that all was planned beforehand. It demonstrated in another way that the pope had control of every aspect of time. Also, if there had really been rumours about a jubilee indulgence already in January, the pope might have wanted to keep the people on edge for a while for maximum effect. This would make even more sense if the rumours were started on the orders of Boniface VIII and the curia, or even invented by Stefaneschi.

Be that as it may, the declaration day of the bull was carefully chosen to be 22 February, the feast of the Chair of St Peter, even though the text for *Antiquorum habet* had been composed already on 16 February.[67] This was not an uncommon procedure for bulls, but it seems obvious that the choice of 22 February underlined the importance of the indulgence and the supremacy of the basilica of S. Pietro in Vaticano, which had not been so obvious in the preceding centuries. Ordering the jubilee almost two months retroactively was ultimately a flamboyant demonstration of papal authority, since no one could have gained the indulgence before the 22 February without knowing the

---

66 Stefaneschi, *De centesimo*, cap. III, f. 7v, 88–94: *"Publicandi aderat tempus cathedre beati Petri celebritas. Offrendum principi in suo quam alibi honestius templo donum publicatur. Sericis aureisque amboni circumfultis, quo Romanus presul patresque conscenderant, orationeque ad turbam facta, grande leticie munus, littere serico appensis bullis disserentur. Que etsi Laterani prius datum contineret, illic tamen datum apud sanctum Petrum suis presul iussit litteris ascribi altarique impositum donum. Non impari forma doctoris gentium basilice missum."*

67 Federico Canaccini, *Al cuore del primo Giubileo. Bonifacio VIII e l'*Antiquorum habet (Città del Vaticano: Lateran University Press, 2016), 123. See also pages 121–138 for a previously unedited draft of the *Antiquorum habet* in letter form, dated to the 7 February 1300. This draft is also mentioned in Stefaneschi, *De centesimo*, cap. III, f. 6v, 73–77.

requirements. On the other hand, it is also possible that Boniface VIII was indeed late with the organisation of the jubilee, and thus invented a way to turn this disadvantage into an intelligent gesture of authority. In both cases, he demonstrated an ability to use the end of a century and time itself as an instrument of papal power.

As the motivation for celebrating a jubilee in 1300 relied on the imaginary jubilee of 1200, the papal decision actually created a tradition of celebrating jubilees over precisely 100-year intervals. It was known to Stefaneschi and Boniface VIII, and everyone familiar with the Bible, that, according to Leviticus 25 of the Old Testament, there had been a jubilee tradition in ancient Israel that had supposedly repeated itself at 50-year intervals. Stefaneschi explains that according to this tradition, every 50 years vendors recovered their property from buyers without having to pay for it. Also, slaves would be freed.[68] In the Old Testament, the Hebrew jubilee year is described as an exceptional sabbatical year that was marked by repose of the arable land and by social and economic features that were supposed to restore the social order back to its original state after 50 years.[69]

However, the Christian jubilee was not supposed to be an event of economic or social remission but, rather more importantly, an event of religious remission.[70] Indeed, the only directly common aspect between the Christian and Hebrew jubilee seems to be the name, "jubilee."[71] The origin of the word is *yôbel* in Hebrew. It used to mean "a ram" or a trumpet made from the horns of this animal; later it also came to mean the joyous sound made by a trumpet.[72] The word was translated into Latin, in which *iubilare* means "to call joyfully."[73] In the Middle Ages, Isidore of Seville's explanation was accepted: "*Jubilaeus*

68 Stefaneschi, *De centesimo*, cap. XII, f. 20v–21r, 350–353.

69 Roberto Zagnoli, "Il significato del Giubileo nella tradizione biblica," in *Romei e Giubilei. Il pellegrinaggio medievale a San Pietro (350–1350)* (Milano: Electa, 1999), 29–30. There is no evidence that this "law" was actually observed or that the jubilee years were actively followed. The mention in the Old Testament might be a prophetic reminder rather than an actual law. For more sources and literature, see Leonardi, Schmidt, and Placanica, *Iacopo Stefaneschi. De centesimo seu iubileo anno. La storia del primo giubileo*, 75–77 n. 145.

70 See Stefaneschi's comparison of the Hebrew and Christian jubilees in *De centesimo*, cap. XII, f. 20v–21r, 350–358, and Jean Lemoine, *Glossa* 19–23, 188–190.

71 Gustave Lambert, "Jubilé hebreu et Jubilé chrétien," *Nouvelle revue théologique* 82 (1950), 234–251. Cf. also Leonardi, Schmidt, and Placanica, *Iacopo Stefaneschi. De centesimo seu iubileo anno. La storia del primo giubileo*, 75 n. 145 for bibliography, sources, and explanations about the relationship between the Christian and Hebrew jubilees.

72 Zagnoli, "Il significato del Giubileo nella tradizione," 29.

73 Varro, *Ling.* 6.68. See all ancient uses of the Latin words *iubilare* and *iubeleus* in the *Thesaurus Linguae Latinae.* Cf. also Leonardi, Schmidt, and Placanica, *Iacopo Stefaneschi. De centesimo seu iubileo anno. La storia del primo giubileo*, 45, n. 1.

means a year of remission."[74] The word was therefore often connected with indulgences, as explained above. However, even though Stefaneschi and other sources also use the word jubilee with the event of 1300, the papal bull *Antiquorum habet* itself does not contain the word. The pope maybe did not want to use it in *Antiquorum habet* in order to underline the uniqueness and novelty of the Christian jubilee tradition in contrast to the Hebrew tradition.[75]

Still, for the pope it was important to justify the reason why he decided that the Christian jubilee would repeat itself every 100 years, and not, for example, every 50 years, as *Antiquorum habet* clearly states.[76] Stefaneschi affirmed that the reason is that the rarer the jubilee is, the more effect and prestige it would have as an event: "Rarely offered foods are consumed with more greed, since everyone strives to avoid that the generosity passes by in vain, for they do not hope to ever again touch such foods."[77] He continues that, according to the same logic, the jubilee indulgence cannot be granted more often because it would diminish the quantity of pilgrims to the Holy Land.[78] This is because, in his opinion, when great indulgences were too often available in Rome, the interest in other more laborious indulgences would decrease. In this manner, Boniface VIII controlled the jubilee's time-related parameters for a very practical reason, and, by making the jubilee so rare, the pope highlighted the special nature of the year 1300, and his presence and authority during it.

The canon lawyer Jean Lemoine tells what "he himself heard directly from the pope's mouth" (*"sicut ego ex ore ipsius audivi"*). In his opinion, too, the pope wanted to grant the people their wish, that is, a great indulgence in celebration of every 100th nativity of the Christ. Also, it seems that the jubilee indulgence was supposed to be a once in a lifetime opportunity. A 50-year interval would have been too frequent for this reason as well. Also according to Jean Lemoine, the plenary indulgences granted for crusades could have been obtained only once. For this reason, a 100-year interval was suitable because extremely few people would live longer than that.[79] The idea of the jubilee as an exceptional,

74 Giovanni Minucci, "La bolla di'indizione di Bonifacio VIII e il commento di Giovanni Monaco," in *I giubilei nella storia della Chiesa* (Città del Vaticano: Libreria editrice Vaticana, 2001), 226; Isid. *Etym.* 5.37.3: *"Jubilaeus interpretatur remissionis annus."*

75 Stickler, *Il Giubileo di Bonifacio VIII*, 16–18.

76 Registres de Boniface VIII, II, 922–923, no. 3875: *"et in quolibet anno centesimo secuturo."*

77 Stefaneschi, *De centesimo*, cap. XII, f. 21v, 360–362: *"Que raro exibentur alimenta, avidius sumuntur, dum quisque in vanum gratiam elabi, quam numquam sibi redituram sperat, refellat."*

78 Stefaneschi, *De centesimo*, cap. XII, f. 21v, 362–364.

79 Jean Lemoine, *Glossa* 21, 189. Cf. also analysis in Frugoni, "Il Giubileo di Bonifacio VIII," 53; Stickler, *Il Giubileo di Bonifacio VIII*, 17.

unique opportunity is reflected also in Stefaneschi's text. As noted above, he started his text by explaining that he feels it necessary to write, because so few people from his era would see the next jubilee.[80] Also, he reproaches those "unfortunate who have let the jubilee pass by, perhaps because of indolence or temerity, perhaps thinking they would live to see also the next jubilee."[81]

Stefaneschi and *Antiquorum habet* both express that the pope wanted this jubilee to be remembered forever, even though the next one would be after such a long time. Stefaneschi wrote already in the introduction that he is writing "to consign to the memory of future generations some knowledge of the hundredth-year celebration established as jubilee."[82] An event small or great would be equally insignificant if the memory of them was cancelled by oblivion.[83] He also wrote about the *Antiquorum habet* that to avoid the bull's contents being forgotten by future generations, he would include a copy of it in his own text.[84] Lastly, *Antiquorum habet* itself starts with the formulaic words "For the certainty of contemporaries and for the memory of posterity."[85] Therefore, regardless as to whether Boniface VIII had seen the jubilee coming in advance or not, he at least wanted to make sure that the jubilee tradition would not be forgotten "again." For me, it seems that Stefaneschi and the *Antiquorum habet* let it be known that the pope was taking the situation – or the concept of a century – into his control by establishing an eternal tradition and leaving a permanent mark on history. By his command, the jubilee would not be lost in time, and in this he indeed succeeded.

To sum up, the 100-year interval was chosen because it was long enough to give the jubilee an air of importance and rareness. Indeed, the once in a lifetime jubilee underlined the extraordinary role of Boniface VIII, whose reign fell at the turn of a century and who perceived the religious needs of the people. His reign seemed to be as auspicious and extraordinarily beneficial as the rare jubilee. Another important aspect of this long interval was its practicality – if the interval was shorter, jubilee indulgences might diminish the interest in other indulgences in other places. This would have been counterproductive for the Church. However, Stefaneschi does not insist that the 100-year interval is

80 Stefaneschi, *De centesimo*, f. 3v, 11–12.

81 Stefaneschi, *De centesimo*, cap. XV, f. 26v, 462–463: *"Infelices vero qui torpore vel temeritate dum alterius sibi forsan evum iubilei spondent, neglexerint."*

82 Stefaneschi, *De centesimo*, f. 3r, 6–7: *"Futurorum aliquid mandare memorie de centesimo, quem iubileum fore constat, anno…".*

83 Stefaneschi, *De centesimo*, f. 3r, 9, and f. 3v, 11–13.

84 Stefaneschi, *De centesimo*, cap. III, f. 7r, 83–86.

85 Registres de Boniface VIII, II, 922–923, no. 3875: *"Ad certitudinem presentium et memoriam futurorum."*

the only right way. He admits that in some other future era the opinion for a suitable jubilee interval might be different.[86] He was indeed correct, for Clement VI held the next jubilee in 1350, and immediately after this the interval was shortened first to 33 years and then to 25 years. The jubilee became a very popular tradition very quickly.

4 The Practical Time Limits of the Indulgence

The jubilee indulgence also had a more concrete and immediate time-related feature, for the bull *Antiquorum habet* instructed that the indulgence could be gained only by visiting both basilicas of S. Pietro in Vaticano and S. Paolo fuori le mura every day for 15 days, if one was a foreigner, or for 30 days if one was a Roman. The visits had to continue every day without interruption. If a pilgrim visited only one basilica instead of both, the pilgrimage would continue but the time limit must be prolonged. A pilgrim could visit the basilica of S. Pietro in Vaticano one day and the basilica of S. Paolo fuori le mura the next day, and this would all count together as one fulfilled day of the 15 or 30-day pilgrimage.[87] This was a sensible facilitation, because it was probably not an easy task for all pilgrims to walk the approximately 6 km distance between the two basilicas every single day as well as the distance to and from one's home or lodging.

Still, Alfonso Stickler has noted that "just" 15 or 30 days of pilgrimage inside Rome seems a very short and effortless time in exchange for a powerful plenary indulgence if we compare it to the earlier plenary indulgences that were usually granted for long and dangerous pilgrimages to the Holy Land. Stefaneschi justifies the difference of time during which the Romans and foreigners should frequent the basilicas by explaining that the Romans could stay at home during the jubilee, but the foreigners had to leave their work at home and travel for a long time even to arrive in Rome. In his opinion, it was a fair compensation that the Romans had to visit the churches for twice as long.[88] But why did the Romans only have to do the pilgrimage for twice as long as foreigners, who had come from every corner of the Christian world with great cost and toil?[89] The canon lawyer Jean Lemoine explains that the indulgence is not a question of

86 Stefaneschi, *De centesimo*, cap. XII, f. 21v, 364–365.

87 Registres de Boniface VIII, II, 922–923, no. 3875; Stefaneschi, *De centesimo*, f. 30r, 27–30, and cap. III, f. 7r, 80–81; Jean Lemoine, *Glossa* 40, 199–200.

88 Stefaneschi, *De centesimo*, cap. XII, f. 20r–20v, 337–346.

89 Stickler, *Il Giubileo di Bonifacio VIII*, 27–31; Jean Lemoine, *Glossa* 17, 186: *"de ultimis Galliae, Angliae, Scotiae, Hiberniae, Hispaniarum finibus cum magnis laboribus et expensis."*

personal merit but, rather, of divine pardon. Therefore, it is not the quantity of work that matters but the quality, i.e. devotion.[90] Therefore, this different time limit for Romans and non-Romans was set only to facilitate economic challenges that the pilgrims would need to suffer in order to come to Rome.[91] In their case, time was literally money. The time limit was not meant to be directly comparable to the greatness of the indulgence, for they both operated on completely different levels of time – one in practical worldly time, and another in some abstract divine and hardly measurable dimension. To link this to the discussion of the effectiveness of the indulgence above, even if it was clear that the indulgence would remove all temporal penance, and therefore would also operate on the worldly temporal level, it had at the same time an unknown and undefined effect after death in Purgatory.

Another practical time limit was set by the pope, when he ordered that the jubilee year would start – retroactively – on Christmas Day of 1299, "*millesimo trecentesimo a festo nativitatis Domini nostri Ihesu Christi.*"[92] In *Gratia non bullata* (otherwise known as *Ad honorem dei*),[93] Boniface VIII further proclaimed on Christmas Day of 1300 that the jubilee year had then ended, after the custom of the Roman Church, and it would not continue until 25 March, as "certain people," i.e. Pisans, would think.[94] Therefore, the practical time limits of the jubilee indulgence were very precisely defined by the pope. There was no doubt as to when the jubilee started and ended, and how many days each person needed to undertake the pilgrimage to obtain the indulgence.[95]

90 Jean Lemoine, *Glossa* 14–17, 185–187.

91 It was also not unheard of that an indulgence had different rules for different nationalities. Petrus Mallius wrote that, during the reign of Alexander III (1159–1181), the basilica of S. Pietro in Vaticano had only three indulgences in a year, and their greatness depended on how far a pilgrim had come from. Romans and those from nearby areas could have one year of remission, those from Lombardy, Tuscany, Apulia and other regions reachable by land could have two years, and those who arrived by ship could have three years. In this case, however, the pilgrimage distance did not affect the requirements of the indulgence, as in 1300, only its extent. These three indulgences were for Maundy Thursday, Ascension Day, and 18 November, which was the dedication day of the basilica. Cf. Petrus Mallius, *Descriptio Basilicae Vaticanae*, 385; Maccarone, "L'indulgenza del giubileo del 1300 e la basilica di San Pietro," 732–733.

92 Stefaneschi, *De centesimo*, f. 29v, 20–21 and f. 31v, 16–17.

93 A declaration – on the last day of the jubilee by the pope, but without a papal bull – which was valid only once on the occasion of that jubilee: Stefaneschi, *De centesimo* (*Gratia non bullata*), f. 33r–34r.

94 Stefaneschi, *De centesimo* (*Gratia non bullata*), f. 34r, 25–28. On different ways to start the year, see Leonardi, Schmidt, and Placanica, *Iacopo Stefaneschi. De centesimo seu iubileo anno. La storia del primo giubileo*, 90 n. 225.

95 Frugoni, "Il Giubileo di Bonifacio VIII," 58.

However, according to Stefaneschi, pilgrims were continuously asking for a reduction of the time limits. The pilgrims felt that the limits, judged adequate by the pope and his curia, were still too demanding. Stefaneschi explains that Boniface VIII used his papal authority "only" three times during 1300 to make exceptions to these precise time limits: on Maundy Thursday (7 April); on the basilicas' dedication day of 18 November; and on Christmas Day.[96] The first two concessions were granted by the pope himself at the Lateran Basilica. Any pilgrims present and listening to the pope immediately obtained the plenary indulgence, however the second concession excluded Romans and those from nearby areas. It is plausible to think that these concessions were also given for practical reasons: the city must have been bursting with too many pilgrims right before Easter and towards the end of the jubilee year.

The third concession, recorded in the aforementioned *Gratia non bullata*,[97] was granted by the pope at the basilica of S. Pietro in Vaticano on Christmas Day, which was the last day of the jubilee year. The pope granted the plenary indulgence to every pilgrim who had started the pilgrimage but for some legitimate reason had not finished it in time. These reasons included, among others, death or some serious impediment during the journey to Rome. It was unheard of that an indulgence could be applied to the dead in this manner, and it is indeed unclear and theologically problematic how this concession was supposed to work.[98] However, leaving the theological analysis aside for the moment, we shall examine the matter based only on Stefaneschi's literary depiction: "Boniface's exercise of papal power over the souls of those in the afterlife could not have been more clear." Indeed, Dante described in his *Divina commedia* the gates of Purgatory as rusty out of disuse.[99]

Stefaneschi further specifies that the last concession concerned only the jubilee of 1300 and not future jubilees, otherwise, in his opinion, everyone would just do the pilgrimage for one day before Christmas, out of laziness, and gain the indulgence too easily. Therefore, strict time limits were necessary. Also, the clergy[100] had a separate rule: for Romans 15 days would be enough, and for

96 Stefaneschi, *De centesimo*, cap. VII, f. 12r–12v, 181–192. Jean Lemoine confirms: *Glossa* 44, 202. Cf. also Goria, "Le indulgenze a Roma nel 1300 (un elenco fin'ora sconosciuto)," 151; Maccarone, "L'indulgenza del giubileo del 1300 e la basilica di San Pietro," 749; Fois, "Il Giubileo del 1300," 31–32.

97 Stefaneschi, *De centesimo* (*Gratia non bullata*), f. 33r–34r.

98 See e.g., Frugoni, "Il Giubileo di Bonifacio VIII," 54–58; Le Goff, *La naissance du Purgatoire*, 442–443; Jean Lemoine, *Glossa* 37 and 38, 197–198.

99 Cruz, "Dante, Purgatorio 2, and the Jubilee of Boniface VIII," 18; *Divina commedia*, Pur. 9,133–138.

100 With "*religiosi*" Stefaneschi probably means anyone ordained as a priest, or monks and nuns.

foreigners ten days. The remaining days would not be cancelled, but they would be converted into days to be used in masses and prayers for the dead.[101] Lastly, the pope declared that people had time until the following Easter to restore stolen or other goods belonging to others, otherwise the indulgence would not be gained.[102]

With these three concessions, the pope demonstrated that he fully controlled the worldly limits of time. More than that, he operated with ease on two different levels of time: the practical temporal time and the abstract time that existed in the afterlife and to which he had access through indulgences. He could give only one plenary indulgence, but with these adjustments of time limits he could respond to the requests of the people. In the end, it seems that it was because of the very requests of the people that the jubilee tradition had been initiated in the first place, and the same consistent theme of the pope responding to the needs of the faithful seemed to continue throughout the jubilee year. This situation is prominent in Stefaneschi's text: the people always have their own idea of how things should go. They knew that there had been a jubilee before, but the pope did not. They knew how and for how long the indulgence should work, and the pope did not. They felt that the time limits decided by the pope were too demanding. Stefaneschi intentionally presents Boniface VIII as a sensible and powerful pope who responded to all these requests in favour of the faithful.

The first jubilee year had thus ended, and it had been a huge success. Several sources attest how countless people arrived from distant lands as soon as the news of the jubilee indulgence had been received, until the last day of the jubilee. Stefaneschi himself described that pilgrims arrived from every part of western Europe in such great numbers that it looked like an army or a swarm. The estimates of the number of pilgrims in Rome during 1300 by other chroniclers varied between 200,000 and 2 million. A more trustworthy corpus of evidence are the records of the road tolls that clearly show a huge increase in travellers arriving on horseback into Italy from across the Alps via the Aosta Valley.[103] There have been some modern attempts to determine how many

101 Stefaneschi, *De centesimo,* cap. VII, f. 12v–13r, 192–197.

102 Stefaneschi, *De centesimo* (*Gratia non bullata*), f. 33v–34r, 20–28. Cf. also Frugoni, "Il Giubileo di Bonifacio VIII," 58–59; Leonardi, Schmidt, and Placanica, *Iacopo Stefaneschi. De centesimo seu iubileo anno. La storia del primo giubileo*, 90 n. 224.

103 Stefaneschi, *De centesimo,* cap. IV–VI, 8r–12r; Giovanni Villani, *Nuova Cronica* 8,36 (RIS 13 col. 367–68); Guglielmo Ventura, *Chronicon Astense* cap. 26 (RIS 11 col. 192–93); Dante, *Divina commedia,* Inf. 18,28–33. For more sources, see Birch, *Pilgrimage to Rome in the Middle Ages*, 199; Fois, "Il Giubileo del 1300," 34–36; Leonardi, Schmidt, and Placanica, *Iacopo Stefaneschi. De centesimo seu iubileo anno. La storia del primo giubileo,* 56 n. 50. For toll records, see Robert-Henri Bautier, "Le Jubilé romain de 1300 et l'alliance

pilgrims there were exactly, but the results are unsatisfactory; in the end, it is impossible to know the number of pilgrims. What we can determine is that the end of the century and the possibility it held for personal salvation through the Roman Church, was clearly deemed important not only in Rome but also more widely elsewhere in Italy and beyond. Whether these people were expecting a jubilee indulgence does not matter; we know for a fact that they enthusiastically embraced the opportunity for one.

5 The Ancient *saeculum* and *ludi saeculares* and Their Relation to the Jubilee of 1300

The question arises whether the model for this medieval celebration repeating itself every 100 years was taken from ancient Rome, where centennial *ludi saeculares* were celebrated, the last time more than 1000 years before Boniface VIII's papacy. Did the pope and the curia know about or were they thinking of the *ludi saeculares* when they decided to establish the jubilee tradition? Were they harnessing an ancient past as an instrument of power and to justify the first Christian jubilee? This possibility is not mentioned even once in the above cited sources, but some earlier scholars have nevertheless thought that Boniface VIII used the *ludi saeculares* as a model for his own concept of jubilees. However, these views are not based on solid evidence. For example, Gustave Contestin saw a connection between the Holy Door, which was later an important part of the Christian jubilees, and a door of a temple depicted in some undefined Augustan coins, where the emperor himself hits the door with a staff. He suggested that this image was used as a model for the Holy Door of the jubilees.[104] However, according to a letter of 1400, it seems that the opening of the Holy Door in the Lateran Basilica was introduced into the jubilee tradition only in 1350.[105] In any case, it is easy to agree with Arsenio Frugoni that a single

franco-pontificale au temps de Philippe le Bel et de Boniface VIII," *Le Moyen Age* 86 (1980), 190–192.

104 Gustave Contestin, "Origine historique des Jubilés," *Revue des Sciences Ecclesiastiques* 180 (1875), 280. It is very unclear to which coin he refers. See detailed analysis of Contestin's arguments in Frugoni, "Il Giubileo di Bonifacio VIII," 20–21. There are also other, earlier, scholars who saw Augustan *ludi saeculares* as a model for medieval jubilees: Lewalter, "Das Jahrhundert als Epoche (aetas, centuria, saeculum)," 22: *"Das Anno santo seit 1300, die Säkularfeier des Rienzo als Wiederneuerung augusteischer Säkularfeiern…".*

105 One of the merchants' letters kept in the Archivio Datini di Prato (filza 545, n. 505241), written in the jubilee year of 1400, mentions that the Holy Door was opened in that year and that this had happened also in the jubilee of 1350: see the picture and the transcription of the letter in Arnold Esch, "I giubilei del 1390 e del 1400," in *La Storia dei Giubilei.*

Augustan coin is not sufficient proof that the jubilee of 1300 – or even the later tradition of the Holy Door itself – was based on the *ludi saeculares*.

Yet, some scholars, including Pietro Fedele, have seen a passage in the *Noticia seculi* by Alexander von Roes as evidence that the memory of the *ludi saeculares* was still alive in the Middle Ages.[106] Alexander wrote:

> ...manifestum est, quod antiqui sapientes secundum unam divisionem etiam secula dividebant in *septem annorum hebdomades, id est in quadraginta et novem annos*, ut in anno *quinquagesimo* esset annus *iubileus*. In quo anno Romani plurima renovabant, considerantes, quod infra 50 annos tota res publica Romani imperii consuevit variari.[107]

In my view, Alexander seems to be referring to the Hebrew jubilee tradition of 50 years, consisting of seven sets of seven years; he is not talking about the *ludi saeculares* or an interval of 100 years. "*Antiqui*" refers to the Hebrews mentioned in the Old Testament and "*res publica Romani imperii*" might refer to his contemporary society, the recent changes of which he goes on to describe in the next phrase, or he is simply mixing up Roman and Hebrew traditions. Alexander is not talking about the ancient *ludi saeculares* at all.[108]

According to Arsenio Frugoni, the first clear mention that the jubilee tradition might have been based on the ancient *ludi saeculares* seems to be by Polydore Vergil (c.1470–1555), an Italian humanist scholar and priest, in 1499. Polydore wrote in his work *De rerum inventoribus:* "The pope ordered that every hundredth year should be celebrated as a jubilee year. He seemed to do this in order to lead the Roman people among the first from the meaningless *ludi saeculares* to the truly solemn and holy celebration."[109] After this he goes on to

---

Volume primo 1300–1423, ed. Gloria Fossi (Firenze: Giunti – BNL Edizioni, 1997), 288–289. Cf. Arnold Esch, "L'economia nei giubilei del quattrocento," in *I giubilei nella storia della Chiesa* (Città del Vaticano: Libreria editrice Vaticana, 2001), 343. There seems to be an erroneous, common view that the first Holy Door was opened in 1423 by Martin V, probably in reference to Giovanni Rucellai (1403–1481), *Zibaldone* col. 51rA–52vB, and Niccola della Tuccia (1400–1473/4), see Ignazio Ciampi, *Cronache e statuti della città di Viterbo* (Firenze: Cellini 1872), 117. Cf. Amedeo de Vincentiis, "Il giubileo di Martino V," in *La Storia dei Giubilei. Volume primo 1300–1423*, ed. Gloria Fossi (Firenze: Giunti – BNL Edizioni, 1997), 302.

106 Pietro Fedele, *Gli Anni Santi* (Torino: Società Editrice Internazionale, 1934), 11.

107 MGH, Scriptores, Staatsschriften, 1,1, 154.

108 Frugoni, "Il Giubileo di Bonifacio VIII," 22. Opposing suggestion in: MGH, Scriptores, Staatsschriften, 1,1, 154, n. 1.

109 Polydore Vergil, *De rerum inventoribus* 8,1: *"Pontifex Jubileum centesimo quoque anno servari mandavit, quod fecisse videtur ut populum cum primis romanum ab inani saecularium ludorum celebratione ad verum solemne sacrum reduceret."*

briefly describe the origins and the end of the *ludi saeculares* tradition. In the wake of Polydore, it became common to compare the *ludi saeculares* with the jubilee. In other words, the comparison between the two traditions appears in the extant literary sources only two centuries after the first jubilee. Arsenio Frugoni suggests that the scholarly atmospheres were quite different in 1300 and 1499. Certain texts from antiquity had already been read and admired before, but it was only by the middle of the 15th century that interest in the physical remains of antiquity increased significantly, and periodical jubilees made it possible for many to familiarise themselves with the physical city of Rome.[110] It was much easier or natural then, in the scholarly atmosphere of 1499, to link contemporary phenomena with similar phenomena in classical antiquity. Such discourses were not yet current in 1300.[111] Consequently, literary sources seem to give us no leads for finding any connection between the ancient Roman *ludi saeculares* and the jubilee of 1300.

But are there clear similarities in the contents of both traditions? The *ludi saeculares* were religious festivals that originally lasted for three days and had a placating or purificatory function.[112] According to legend, the first such celebration was held at the beginning of the republican era in the late 6th century BC, in order to get divine help for a plague in Rome.[113] The first verifiable *ludi saeculares* were held in 249 BC to placate the gods during the First Punic War, and then again in 149 or 146 BC[114]

After a long pause, the Emperor Augustus (r. 27 BC–14 AD) reinvented the *ludi saeculares* tradition in 17 BC, when he organised magnificent celebrations with new rituals emphasising the beginning of a new and fruitful golden age and a purification from the sins of the last century.[115] He established a new

110 Alessandro Rinaldi, "Cartografia urbana e anni santi," in *Roma Sancta. La città delle basiliche,* eds, Marcello Fagiolo and Maria Luisa Madonna (Roma: Gangemi Editore, 1985), 253–254.

111 Frugoni, "Il Giubileo di Bonifacio VIII," 21.

112 For general works on the ancient Roman *ludi saeculares*, see *Recherches sur les jeux séculaires* (Paris: Société d'édition "Les Belles lettres," 1934) of Jean Gagé and *De ludis saecularibus* (Amsterdam: Schippers, 1965) of Ioannes Baptista Pighi. After Pighi's work, the PhD thesis of Susan Bilynskyj Dunning, "Roman Ludi Saeculares from the Republic to Empire" (PhD diss., University of Toronto, 2016), is the most recent general research on the *ludi saeculares*. She is also preparing a monograph on the subject.

113 Zos. 2.1–3; Val. Max. 2.4.5; Plut. *Vit. Publ.* 21.1.

114 Gary Forsythe, *Time in Roman religion: One Thousand Years of Religious History* (New York, London: Routledge, 2012), 67–68.

115 For Augustus' *ludi saeculares* in particular, see Bärbel Schnegg-Köhler, "Die *augusteischen* Sakularspiele," *Archiv für Religionsgeschichte* 4 (2002), 156–170; John F. Hall, "The *Saeculum novum* of Augustus and its Etruscan Antecedents," in *Aufstieg und Niedergang der*

regular 110-year interval for future *ludi saeculares*.[116] The reinvention of the *ludi saeculares* tradition was an essential part of Augustus's golden age propaganda and his mission to reform the Roman calendar and thus take control of time itself. He understood how powerful a concept controlling time was, and how efficient a way it was to utilise the concept of the century to demonstrate this. After Augustus, the celebration of the ancient Roman *ludi saeculares* actively continued and the emperors did not miss a chance to celebrate a changing era. The Emperor Claudius (r. 41–54 AD) even started a competing tradition with a different 100-year frequency that ran uninterrupted side by side with the Augustan 110-year celebrations; this continued until the last verifiable centennial celebration by the Emperor Philip the Arab (r. 244–249 AD) in 247 AD.[117]

At first glance, we can notice some aspects in common between the ancient *ludi saeculares* and the medieval jubilee of 1300. The concept of a century and time was harnessed in both cases as a powerful tool of religious and political power by the current leader of Rome. Both events emphasised spiritual purification; also, both were centralised in the city of Rome. Lastly, both events were accompanied by panegyrical literature. For Augustus's *ludi saeculares*, the poet Horace (65–8 BC) composed the *Carmen saeculare*, a celebratory poem about the starting of a golden age, sung during the rituals.[118] Cardinal Stefaneschi, on the other hand, attached to the end of his *De centesimo* a poem called *Heroycum Carmen*.[119] This poem also praises the golden age that was starting with the jubilee of 1300. In his own words, he wrote the poem in heroic meter, that is hexameter, "according to the ancient style, which is less familiar to us than the modern style but more adapted to the truthfulness and to the solemnity of the event."[120] The poems starts: "A golden age rises with the hundredth circle

römischen Welt II.16/3, ed. Wolfgang Haase (Berlin, New York: Walter de Gruyter, 1986), 2564–2590.

116 Censorinus, *DN* 17.9.

117 The theory that Emperor Honorius organised *ludi saeculares* in 404 is convincingly argued to be extremely unlikely in Susan Dunning's thesis. It is also remarkable that, just as the intervals between the jubilees after 1300 were gradually shortened from 100 years to 25 years in little more than a century, the ancient *ludi saeculares* were also organised essentially twice per century immediately after Augustus' celebrations. The propaganda value of such time-related celebrations was indeed powerful.

118 Hor. *Carm. saec.* Also other writers of the Augustan era mentioned the beginning of a golden age: Verg. *Aen.* 6,780–800 and 8,324–325. For the *ludi saeculares* of Septimius Severus in 204, a similar *carmen saeculare* was composed: CIL VI, 32327–32328.

119 Other hexametric poems were also composed for the jubilee: Bonaiuto da Casentino in *Annales Caesenates* (col. 1119a–1120a) and Silvester de Adria's jubilee poem (see p. 72).

120 Stefaneschi, *De centesimo*, cap. XVI, f. 27r, 468–470: *"Quos vetustiori nec moderno, nec non minus nobis familiari, veritati tamen reique geste magis gravitati congruenti stilo…"*.

of the sun, and radiates celestial light."[121] Stefaneschi mentions that the solemnity of the jubilee required the use of the ancient style, as if to give it even more authority by creating connotations with the ancient past. Indeed, the poem even uses the pagan god Phoebus, that is Apollo, as a personification of the sun.[122] Stefaneschi may have admired famous ancient Roman poets such as Virgil and Lucan, who also used hexameter.[123] His admiration of ancient style is clear from his jubilee poem, but it does not necessarily mean that he was drawing parallels to the ancient *ludi saeculares*.[124] The most plausible reason for using the ancient style is that antiquity emanated authority and an impression of erudition.

There is an enormously long gap of more than a millennium between the first certain jubilee in 1300 and the last certain *ludi saeculares* – or actually the celebration of the city's 1000th founding day – of Philip the Arab in 247. These eras are separated by vast cultural, religious, and social differences. After the centennial celebrations of 247, the next possible year for organising the *ludi saeculares* would have been 314 during the reign of Constantine I "the Great" (r. 306–337), but neither Constantine nor his followers organized the celebrations, regardless of some complaints.[125] The *ludi saeculares* had been a very popular instrument of power for the preceding emperors, but Constantine did not need them anymore – he would soon focus on his new capital Constantinople, and he employed a much more current and efficient religious instrument

121 Stefaneschi, *De centesimo* (*Heroycum Carmen*), f. 27v, 1–2: "*Aurea centeno consurgunt secula Phebo / Et radiat celeste iubar...*".

122 Stefaneschi, *De centesimo* (*Heroycum Carmen*), f. 27v, 1. Again, in *Heroycum Succintum Carmen* f. 28v, 520, Phoebus is mentioned.

123 Ragionieri, "Jacopo Stefaneschi e il *De centesimo*," 218. Indeed, Antonio Placanica (Leonardi, Schmidt, and Placanica, *Iacopo Stefaneschi. De centesimo seu iubileo anno. La storia del primo giubileo*, 86 n. 197) notes that Virgil also used "*aurea saecula*" in *Aen*. 6.792–793 and 8.324–325.

124 There is no way of knowing whether the history of the *ludi saeculares* was known in 1300. It seems plausible, however, that Stefaneschi and other high-ranking members of the Church, who were educated in classical literature, knew something about it. At the very least, Augustus was familiar to everyone from the Bible. The fact that the ancient *ludi saeculares* are not mentioned in the medieval sources of the era could simply mean that it was not seen as an interesting or current theme at the time.

125 Zos. 2.4 and 7. There were also hopeful attempts to combine past *ludi saeculares* and Christianity: Oros. 7.20,2–3 (PL 31, 1115–16); Giuseppe Zecchini, "Feste e identità: dai ludi saeculares ai natali di Roma," in *Identità e valori. Fattori di aggregazione e fattori di crisi nell'esperienza politica antica*, eds, Alberto Barzanò, Cinzia Bearzot, Franca Landucci, Luisa Prandi, and Giuseppe Zecchini (Roma: "L'Erma" di Bretschneider, 2001), 208.

of power by officially allowing the Christian religion in the empire.[126] However, the *saeculum* rhetoric that had become so common during the 3rd century continued to be so even under Constantine's reign.[127] This seems to hint that the concept of a golden age and its power had separated itself from the *ludi saeculares* tradition and lived on successfully even during the rise of Christianity. Indeed, the golden age rhetoric is used very openly in Stefaneschi's jubilee poem. According to Arsenio Frugoni, it is only the fact that Christianity and ancient Roman religion are so incompatible that proves that the jubilee could not be based on the *ludi saeculares*.[128] This argument is not watertight in my opinion though, because so many non-Christian elements were readily recycled into Christian contexts, a glaring example of which is Stefaneschi's poem.

The general concept that there should be a religious purificatory celebration every 100 years, and a new leader who brings a golden age with him, seem to be the only obviously similar aspects in common between the *ludi saeculares* and the jubilee of 1300. Yet this concept is quite universal and repeats itself in many different contexts and time periods, also before the ancient Romans, such as in Etruscan, Greek, and Egyptian cultures.[129]

---

126 However, unlike the traditional Augustan *ludi saeculares*, the anniversary celebrations of Claudius, Antoninus Pius, and Philip the Arab were connected to the *Natalis Urbis Romae* celebration, which seemed to gradually lose its pagan religious elements and continued long into Late Antiquity. It seems that these centennial celebrations for the founding of the city could also have taken place, at least in theory, later into the Christian era, but we do not have any sources that confirm this. See the discussion in Michele Renee Salzman, *On Roman Time: the Codex-Calendar of 354 and the Rhythms of Urban Life in Late Antiquity* (Berkley, Los Angeles, Oxford: University of California Press, 1990), 155, 232–246. Interestingly, brought to my attention by Holger Kaasik, there is one Veronese calendar from the 9th century that includes a mention of a *Urbis rome natalis* on 21 April (Berlin, Staatsbibliothek zu Berlin – Preußischer Kulturbesitz, Ms. Phill. 1831, fol. 2v). On this, see also MGH, Antiquitates, Libri mem. 2,1: Der karolingische Reichskalender und seine Überlieferung bis ins 12. Jahrhundert. Teil 2, 800, 802 n. 21.

127 Porfyrius, *Panygyricus Constantini Carmina*.

128 Frugoni, "Il Giubileo di Bonifacio VIII," 22.

129 Ancient Romans, as we saw above, considered a *saeculum*, that is a century or an era, to be c.100 or 110 years long. According to Censorinus, for example, the length of the Roman *saeculum* was the length of the longest possible human life. Still, he wrote, no one knew the exact length of it: Censorinus, *DN* 17.1, 17.3–5 and 17.7. The accurate length of a *saeculum*, however, was not as important as its significance. For the Romans, *saeculum* was a modifier, with which they were able to understand their place in the cosmos. Changes and even disasters were simply an inevitable part of the flow of time: Jason P. Davies, *Rome's Religious History. Livy, Tacitus and Ammianus on their Gods* (Cambridge: Cambridge University Press, 2004), 213–221; Tac. *Hist.* 1.2, 2.37, and *Ann.* 3.55. If we go even further, the ancient Egyptians and Greeks also had similar ideas: see Pierre Brind'Amour, "L'Origine des Jeux Séculaires" in *Aufstieg und Niedergang der römischen Welt* II.16/2, eds,

The concept of a century or changing eras, something also connected with the jubilee of 1300, has very old roots in Mediterranean cultures. In the Middle Ages this concept also encompassed a similar idea that when a century or an era changed, human society or the world around it also changed or reformed. However, more than one idea of a century existed: as we discussed before, Alexander von Roes started his century from the year 1215, based on recent historical events.[130] Indeed, the changing of a calendrical century did not, in the minds of contemporary writers, automatically correspond with great cultural, social, and religious changes that would happen exactly at the same time – or *vice versa*. This is more of a modern illusion, as we are used to dividing history into calendrical centuries after and before Christ and trying to characterise them – sometimes artificially – with different qualities.[131] The concept of a century was therefore as flexible and as easily adaptable in the 13th century as it had been in antiquity. After a long hiatus in the sources, we can confirm that, at least from the year 1300 onwards, the universal concept of a century and time in general was readily utilised and transmuted by the Papacy, with the help of jubilees and without a break until the 21st century.

6 The Jubilee and Boniface VIII's Propaganda

For Boniface VIII, the jubilee was clearly an extraordinary opportunity to gain more authority.[132] This was especially necessary at the time because the year 1300 was preceded by a difficult period in papal history, and so it is helpful to examine the question of the jubilee as an instrument of power within its historical context.[133] The early 13th century, starting from the reign of Innocent III, had been a period when the political power of the popes was much stronger and more pronounced than towards the end of the century. In the last third of the 13th century, the average period of rule of the popes was very short.[134]

Hildegard Temporini and Wolfgang Haase (Berlin, New York: Walter de Gruyter, 1978), 1334–1417.

130 MGH, Scriptores, Staatsschriften, 1,1, 151–155 (*Noticia Seculi* 6–8).

131 Lewalter, "Das Jahrhundert als Epoche (aetas, centuria, saeculum)."

132 See e.g., Canaccini, *Al cuore del primo giubileo*, 89–93.

133 The following synopsis of papal history before Boniface VIII draws largely upon the chapter "Papacy" by J.A. Watt in *The New Cambridge Medieval History*, vol. V, ed. David Abulafia (Cambridge: Cambridge University Press, 1999), 107–162.

134 Between Urban IV (1261–1264) and Boniface VIII (1294–1303) only three popes reigned for more than four years: Gregory X (4 years and 131 days), Martin IV (4 years and 34 days), and Nicholas IV (4 years and 42 days). Four died or resigned during their first year: Innocent V (153 days), Adrian V (38 days), John XXI (254 days), and Celestine V (161 days).

The era was also marked by two periods of *interregnum* without a pope: the first, and longest in papal history, of almost three years, between 29 November 1268 and 1 September 1271; and the second for two years, between 4 April 1292 and 5 July 1294. These periods weakened the internal integrity of the Church and its authority, at the same time when it had powerful challengers from secular rulers, such as the king of France, Philip IV.

Boniface VIII was elected in 1294 when Celestine V resigned after a period of ineffective and weak rule. The Church was in dire need of a powerful leader after decades of difficulties and inconsistencies. Lots of expectations were directed at Boniface VIII, who had experience of decades in the curia and was a powerful personality. Boniface VIII was indeed an able and effective leader, but ultimately caused his own downfall through a series of diplomatic disasters. In this period, the popes were chosen from powerful Italian families, which sometimes led to difficult conflicts within the Church. A good example is the feud between the Gaetanis, the family of Boniface VIII, and the Colonnas. After having supressed the resistance of the Colonnas, Boniface VIII found a powerful enemy in Philip IV of France, a challenger to attempts to increase papal power outside of Italy.

The jubilee year of 1300 is situated in the middle of these political struggles. A little more than one year before the jubilee, on 15 October 1298, Boniface VIII's enemies, cardinals Giacomo and Pietro Colonna and others, had surrendered to the pope. But before he started his struggle with Philip IV, he managed to organise the jubilee. In another bull *Nuper per alias,* which was proclaimed on the same day as *Antiquorum habet*, the pope affirmed that certain "false and impious Christians" were not worthy of the plenary indulgence.[135] Those excluded were the Colonna family, anyone trading with the Saracens, his enemy King Frederick III of Sicily, and the Sicilians. Stefaneschi mentions the same *"inobedientes scismatici"* in his text and also attached a copy of *Nuper per alias* to the end of his *De centesimo* in order to ensure that it would not be forgotten.[136] To exclude certain people was theologically justifiable, since, for example, the Colonnas had already been excommunicated from the Church, but the reality of the political motivations should be taken into consideration here.[137] Again, we can see the practical aspect of the jubilee and how Boniface VIII could use it against his enemies – he alone decided who got to celebrate the new century with an indulgence.

135 Stefaneschi, *De centesimo,* f. 31v–32r, 23–25.
136 Stefaneschi, *De centesimo,* cap. III, f. 7r, 83–86.
137 Frugoni, "Il Giubileo di Bonifacio VIII," 46.

The jubilee was advertised and made public in various ways, and the news was meant to be efficiently disseminated throughout Western Christendom. A certain magister Silvester de Adria, one of many papal scribes, composed a letter which was apparently distributed throughout Italy and beyond the Alps. The letter was dated to the same day as *Antiquorum habet* and explained the newly established jubilee tradition, and it also included a copy of the text of *Antiquorum habet*. At the end of the letter, Silvester composed a short jubilee poem, as if to advertise the event with a memorable slogan:

> Every hundredth year is always a jubilee in Rome,
> the sins are forgiven for those who regret.
> This was declared and ratified by Boniface.[138]

There were also permanent and long-lasting ways to preserve the memory of the jubilee even after its conclusion in addition to the literature, such as Stefaneschi's text. The text of *Antiquorum habet* was also inscribed in marble onto a wall somewhere in the basilica of S. Pietro in Vaticano.[139] This inscription, and another official inscription[140] mentioning a structure that was erected in the "year 1300 when there were plenary indulgences in Rome," demonstrate

138 The letter of Silvester de Adria in Angelo Mercati, "La lettera dello scrittore pontificio Silvestro sul giubileo del 1300," in *Cronistoria dell'anno santo 1925* (Roma: Tipografia Poliglotta Vaticana, 1928), 1191–1198. The poem in p. 1198: *"Annus centenus Rome semper est iubilenus / Crimina laxantur cui penitet ista donantur /Hoc declaravit Bonifacius et roboravit."* This poem can be read today, for example, on the doorframe of the Siena Cathedral. On the letter of Silvester and its different versions, see also later studies: Edoardo e Elisabetta Mori, *Un documento cortonese sul giubileo del '300. La lettera di Silvestro* scriptor *pontificio (carta 6 del codice 101) con un saggio di Gerhard Wolf* (Cortona: Accademia etrusca, 2000), and Canaccini, *Al cuore del primo giubileo*, 118–120, 151–158.

139 IChiR 6:17. According to Forcella, the inscription is "fitted high up in the portico of the basilica near the door of the Holy Year." The exact original location remains unknown. For more sources and literature on this inscription, see Leonardi, Schmidt, and Placanica, *Iacopo Stefaneschi. De centesimo seu iubileo anno. La storia del primo giubileo*, 87 n. 206.

140 The inscription IChiR 1:5, (see Appendix 1, p. 348, n. a), reads: *"mandato s(ancti) d(omi)ni n(ost)ri pontificis Bonifatii p(a)p(e) VIII d(omi)n(u)s Riccard(us) de Anibale (et) Gentil(is) de filiis Ursi alme urb(is) | senatores ill(u)stres hoc opus marmoreu(m) addideru(n)t ann(o) D(omi)ni MCCC, quo Rome fuit indulge(n)tia om(n)iu(m) peccato(rum)."* There is also a verse inscription (IChiR 1:4) from the year 1300 by the same senator Riccardo Annibaldi as in IChiR 1:5. However, IChiR 1:4 does not refer directly to the jubilee, but mentions *ludi* and *pena remissa* together. Could the choice of the word *ludi*, "games," which is hardly an accurate word to describe the jubilee celebrations, be a poetic way to connote ancient centennial celebrations, i.e. *ludi saeculares*, into the context of the first jubilee? It is worth mentioning that, according to the survey by Urpo Kantola, direct references to the jubilees in extant inscriptions are nonexistent after 1300, until the jubilee of 1450. On

that the pope wanted to insure that the jubilee tradition established by him would leave a permanent mark in both time and space, and in this way it would not be forgotten in the near or distant future.[141] Boniface VIII may also have commissioned relevant art soon after the jubilee. Probably the most famous piece of art usually connected to the jubilee is a fresco, commonly attributed to Giotto (1267–1337), depicting Boniface VIII proclaiming a bull in the Lateran Basilica. Only a fragment remains of this painting inside the Lateran, but it had commonly been thought to represent the proclamation of the jubilee of 1300 on 22 February,[142] although this interpretation has recently been proved unlikely. However, the painting could represent the proclamation made by the pope in the Lateran Basilica on Maundy Thursday during the jubilee year.[143] A combination of various ways of communicating papal messages is a logical thought, but the interpretation of the fragment remains too uncertain to safely argue that the painting was part of Boniface VIII's jubilee propaganda, together with the inscription and the advertisement letter.

---

time-related features in medieval Roman inscriptions, see Kantola's contribution in this volume.

141 Among other ways to spread the word and memory of the jubilee, there were also the donations made by Boniface VIII. For example, there is a bronze bell (inv. nr. 32) in the Museo Arcivescovile di Ravenna, that bears the following text from right to left: *"an(n)o D(omi)ni MCCC t(empor)e d(omi)ni Bonifacii p(a)p(ae) o(c)t(avi)."* Boniface VIII apparently donated this bell in 1300 to the church of Primaro and Hospice located north of Ravenna, which was an important stop-over for pilgrims going to Rome from northern and eastern Europe.

142 E.g. Frugoni, "Il Giubileo di Bonifacio VIII," 111; Charles Mitchell, "The Lateran Fresco of Boniface VIII," *Journal of the Warburg and Courtauld Institutes* 14/1 (1951), 1–6.

143 Maccarone, "L'indulgenza del giubileo del 1300 e la basilica di San Pietro," 750; Francesco Gandolfo, "Bonifacio VIII, il Giubileo del 1300 e la Loggia delle Benedizioni al Laterano," in *Romei e Giubilei. Il pellegrinaggio medievale a San Pietro (350–1350)*. (Milano: Electa, 1999), 220–228; Ragionieri, "Jacopo Stefaneschi e il *De centesimo*," 218. The loggia of the Lateran Basilica is depicted in the painting, but Boniface VIII proclaimed *Antiquorum habet* in the basilica of S. Pietro in Vaticano. Even the theory that the painting depicts the Maundy Thursday of 1300 becomes less certain if we consider Francesco Gandolfo's analysis, according to which the loggia part of the basilica was probably not constructed until soon after the jubilee year, so it did not exist in 1300. Also, the suggestion that the painting depicts the proclamation of *Antiquorum habet* becomes even more doubtful when we consider that the reconstruction of the whole painting relies on a copy made by Jacopo Grimaldi in the 17th century in a manuscript (Ms ambrosiana 1622, f. inf 227). For a discussion on how the painting might not refer to the jubilee at all, see Silvia Maddalo, "Alcune considerazioni sulla topografia del complesso lateranense allo scadere del secolo XIII: il patriarchio nell'anno del Giubileo," *Roma anno 1300,* ed. Angiola Maria Romanini, (Roma: L'Erma di Bretschneider, 1983), 621–632; Agostino Paravicini Bagliani, "Bonifacio VIII, la Loggia di giustizia la Laterano e i processi di scomunica," *Mélanges de l'École française de Rome – Moyen Âge* 112 (2005), 377–428.

Using the jubilee, and thus time, as an instrument of power must have been a very tempting opportunity for such a politically oriented pope as Boniface VIII. Indeed, with *Nuper per alias* he once again demonstrated his control over time – in Stefaneschi's opinion the jubilee happened almost automatically, but Boniface VIII could still freely manipulate its various rules and temporal aspects. The jubilee of 1300 can also be seen as a coherent part of the development of the pope's use of power, which started with the schism with the Colonnas, continued with the jubilee, and ended with the struggle against the French and the bull *Unam sanctam* on 18 November 1302.[144] If we look at the jubilee in this light, it seems even more probable that the pope did not only organise the jubilee because the people requested it, but also for his own ends. He was a person of great political experience, with an interest in expanding his authority. It seems plausible that he saw the great potential in an event like the turning of the century, and consciously seized it. As a pope, he was certainly interested in the peoples' spiritual well-being, but the pope could not afford to concentrate solely on spiritual matters. This had been proven by Celestine V, who had been an extremely spiritual pope but whose reign had been a detriment to the Church because of his inability to wield sufficient authority and to handle practical matters.

Whatever Boniface VIII's exact intentions were, the jubilee of 1300 made a huge impression on his contemporaries, but it did not help the pope against his political adversaries. The impact of the jubilee was pivotal and long-lasting in other ways in Church history. It certainly demonstrated that controlling time in various ways in the context of a jubilee could be a very powerful instrument of authority for the Papacy and an immensely popular way of striving towards salvation for the people.[145] Indeed, the first jubilee became a model for future popes, who did not miss chances to organise them frequently. Furthermore, for the first time a plenary indulgence had been available to almost

144 Morghen, *Bonifacio VIII e il giubileo del 1300 nella storiografia moderna*, 38–40. Cf. Tomaž Mastnak, "The Fall of the Papal Monarchy and the Rise of Territorial Power," in *Crusading Peace. Christendom, the Muslim World, and Western Political Order*, ed. Tomaž Mastnak (Berkley, Los Angeles, London: University of California Press, 2002) for detailed analysis of *Unam sanctam* and how its significance should not be exaggerated as any real attempt by the pope to take away the king's jurisdiction. For more about the historical context of *Unam sanctam*, see Walter Ullmann, "Boniface VIII and his Contemporary Scholarship," *The Journal of Theological Studies, New Series* 27/1 (April 1976), 58–87.

145 As for the economic aspects of the first jubilee, the pilgrims certainly brought lots of money to the City, but it is impossible to tell if this was one of the motivations of Boniface VIII. Cf. Fois, "Il Giubileo del 1300," 36–38.

every Christian individual. Before, plenary indulgences had always served some specific practical end for the Church, such as the crusades.[146] Boniface VIII succeeded in finding a way to bring thousands of individual Christians to Rome; indeed, during the jubilee, the whole of Western Christendom was looking at Rome and the basilicas of S. Pietro in Vaticano and S. Paolo fuori le mura. The special status of Rome and these two basilicas was recognised, and even in the future, the city of Rome remained the only place where jubilees were organised. No exception was made even during the Avignon Papacy – the next jubilee was organised in Rome in 1350 by the order of the Avignon pope Clement VI, but without his presence.

7 Conclusions

In this article, I have demonstrated how Cardinal Stefaneschi, in his *De centesimo* dedicated to Boniface VIII, constructed an elaborate semi-official story around the first Christian jubilee of 1300. I have detected at least three overlapping time levels in the way the jubilee is presented in *De centesimo* and a few other sources: the abstract level of semi-mythical time of the distant past that could be used – albeit obscurely defined – as a justification for the new jubilee tradition. In addition, the first jubilee had a practical level of worldly time, and lastly, through indulgences, it had another abstract level of unmeasurable future time, after death. Pope Boniface VIII was masterfully able to demonstrate his power throughout all these levels.

The abstract time of the distant, unclear, past is evident in Stefaneschi's ambiguous story of the establishment of the jubilee tradition by Boniface VIII. Stefaneschi alleged that the pope and the curia had not even thought about solemnising the ending of the century before the people started to demand a centenary indulgence. But this is probably just a part of a strategy to justify a new tradition that had certainly not existed before – invoking the common opinion of the people and the Holy Spirit and resorting to the obscurity of the distant past was enough to sufficiently justify a new tradition. Despite his power and authority, it was difficult even for the pope to establish new religious traditions from scratch.

From all this, we can certainly deduce that by 1300, the concept of time, its passing, and the changing of the centuries had become important to a great

146 Stickler, *Il Giubileo di Bonifacio VIII*, 36–37.

number of Christians. The jubilee indulgence of 1300 was anticipated by increasingly numerous indulgences and pilgrims in the city of Rome throughout the preceding century. Indeed, after the very successful first jubilee of 1300, the tradition was continued by other popes at even more frequent intervals.

I have also briefly examined whether the jubilee of 1300 was based on much more ancient traditions than suggested by Stefaneschi – that of the ancient Roman centennial celebrations called the *ludi saeculares*. Some earlier scholars have seen a connection, but concrete evidence remains non-existent. The only common elements we know for certain between the jubilee of 1300 and the ancient *ludi saeculares* was the general idea that a changing of centuries or eras should be celebrated with religious purificatory rituals in the city of Rome, organised by the current religious leader(s) of the city.

The more practical time-related features of the 1300 jubilee can be discerned by the various time limits recorded in the papal bull *Antiquorum habet* and commented upon in our other sources. The pope decided that the jubilee should repeat on every 100th nativity of Christ. This highlighted the rarity and prestige of the event and at the same time of his own papacy. He also chose to announce this bull on 22 February, which was the feast of the Chair of St Peter, an important feast for the pope as Peter's successor. This underlined the importance of the jubilee but also of the basilica of S. Pietro in Vaticano, something which had not been obvious in the previous centuries.

Boniface VIII defined the practical time limits for the indulgence with extreme precision. These time limits were defined differently depending on whether the pilgrim came from far away or was a Roman. Boniface VIII also manipulated these time limits by conceding exceptions three times even to those who had not yet fulfilled the required duration of their pilgrimage. He was literally giving away time for free, for an indulgence cancelled all the time of the temporal penance before death.

In the last of these concessions, recorded by Stefaneschi as *Gratia non bullata*, on the last day of the jubilee year, that was Christmas Day, the pope remarkably also granted the indulgence to anyone who had died during their pilgrimage. This was something that had not been done before, for indulgences could not be applied to souls already in Purgatory. While the theological complexities of this arrangement have not been my focus in this article, the gesture itself was a great demonstration of power from the pope. At least symbolically, he was extending his control over even the abstract and unmeasurable time beyond death.

Lastly, Boniface VIII was able to use the jubilee, and thus time, as a tool against his enemies, at least for a short period. The jubilee was a part of his long struggle against the Colonna family, the king of France, Philip IV, and other enemies. With the bull *Nuper per alias* he excluded his enemies from the jubilee celebration – he alone could decide who would get the indulgence and time off from temporal penance. The ending of a century was also a great opportunity for this politically able pope to gain more authority for the city of Rome and to establish a lasting legacy. Indeed, Boniface VIII efficiently advertised the jubilee outside Rome and Italy as well, with literary works originating from the papal court, with inscriptions, and possibly even with works of art depicting the event. He clearly wanted the first jubilee never to be forgotten and lost in time, and in this he succeeded.

Bibliography

Sources

Manuscripts

Berlin, Staatsbibliothek zu Berlin – Preußischer Kulturbesitz
Ms. Phill. 1831
Milano, Biblioteca Ambrosiana
Ms Ambrosiana 1622, f. inf 227
Prato, Archivio di Stato
Fondo Datini, filza 545, n. 505241
Città del Vaticano, Biblioteca Apostolica Vaticana
Archivio del Capitolo di S. Pietro, capsa 1, fasc. 1, nn. 8–9
Archivio del Capitolo di S. Pietro, G.3

Printed Sources

Alberic de Trois Fontaines. *Chronica Alberici Monachi Trium Fontium a Monacho Novi Monasterii Hoiensis interpolata*. In *Monumenta Germaniae Historica*. Scriptores, Staatsschriften, 23, edidit Georgius Heinricus Pertz, 631–950. Hannoverae: Impensis Bibliopolii Aulici Hahniani, 1874.

Alexander von Roes. *Noticia Seculi*. In *Monumenta Germaniae Historica*. Scriptores, Staatsschriften 1,1: Die Schriften des Alexander von Roes und des Engelbert von Admont. Teil 1: Alexander von Roes, hergestellt von Herbert Grundmann und Hermann Heimpel, 149–171. Stuttgart: Anton Hiersemann, 1958.

Bonaiuto da Casentino. Cited in *Annales Caesenates*, in *Rerum Italicarum Scriptores* 14, col. 1119–1120. Italian translation "Carme giubilare" by Modestino Cerra in Arsenio Frugoni. *Pellegrini a Roma nel 1300. Cronache del primo Giubileo*, edited by Felice Accrocca, 153–156. Casale Monferrato: Piemme, 1999.

Censorinus. *De die natali. Censorini de die natali ad Q. Caerellium: accedit anonymi cuiusdam epitoma disciplinarum (fragmentum Censorini)*, edited by Klaus Sallmannn. Leipzig: BSB B.G. Teubner Verlagsgesellschaft, 1983. (Cited as Censorinus, *DN*.).

Corpus Inscriptionum Latinarum VI: *Inscriptiones urbis Romae Latinae.* Collegerunt G. Henzen, I.B. De Rossi, E. Bormann, Chr. Huelsen, et M. Bang. (Cited as CIL.).

Collectio Bullarum Sacrosanctae Basilicae Vaticanae Notis auctus, & illustratus – 1: A Sancto Leone Magno Ad Innocentium VI. Productus. Romae: Jo. Maria Salvioni Typographus Pontificius Vaticanus in Archigymnasio Sapientiæ, 1747.

Dante Alighieri. *La Commedia secondo l'antica vulgata I–III.* A cura di Giorgio Petrocchi. Milano: Mondadori, 1966–67.

Der karolingische Reichskalender und seine Überlieferung bis ins 12. Jahrhundert. Teil 2. In *Monumenta Germaniae Historica.* Antiquitates, Libri memoriales 2,2. Hergestellt von Arno Borst. Hannover: Hahnsche Buchhandlung, 2001.

Forcella, Vincenzo. *Iscrizioni delle chiese e d'altri edificii di Roma dal secolo XI fino ai nostri giorni,* volumi 1–14. Roma: Tipografia delle scienze matematiche e fisiche, 1869. (Cited as IChiR.).

Horace, *Carmen saeculare.* In *Odes and Epodes.* Edited and translated by Niall Rudd, 262–268. Loeb Classical Library 33. Cambridge, MA: Harvard University Press, 2004. (Cited as Hor. *Carm. saec.*).

Isidore of Seville, *Etymologiae* (also known as *Origines*). In *Isidori Hispalensis episcopi Etymologiarum sive Originum libri XX.* Edited by Wallace Martin Lindsay. Oxford University Press, 1911. English translation in *The Etymologies of Isidore of Seville.* Translated by Stephen A. Barney, W.J. Lewis, J.A. Beach, and Oliver Berghof. Cambridge: Cambridge University Press, 2006. (Cited as Isid. *Etym.*).

Jean Lemoine, *Glossa.* In Ludwig Hödl. "Die Glosse des Kardinals Johannes Monachus zur Ablass–Epistel 'Antiquorum.'" *Medioevo. Rivista di storia della filosofia medievale* 26 (2001): 161–202.

Les registres de Boniface VIII; recueil des bulles de ce pape. 4 vols. Ed. G. Digard, M. Faucon, A. Thomas, and R. Fawtier. Paris: E. de Boccard, 1884–1939.

Niccola della Tuccia. *Cronache di Viterbo e di altre città.* In *Cronache e statuti della città di Viterbo.* A cura di Ignazio Ciampi. Firenze: Cellini, 1872.

Orosius. *Historiae Adversus Paganos libri septem.* In *Patrologia Latina* 31. Ed. Jacques-Paul Migne, col. 664–1174. Paris, 1846. English translation in A.T. Fear. *Orosius. Seven Books of History against the Pagans.* Translated Texts for Historians 54. Liverpool: Liverpool University Press, 2010. (Cited as Oros.).

Petrus Mallius. *Descriptio basilicae Vaticanae aucta atque emendata a Romano presbitero.* In *Codice topografico della città di Roma.* III, ed R. Valentini and G. Zucchetti. Fonti per la Storia d'Italia 90. Roma, 1946.

Plutarch. *Vitae Parallelae, Publicola.* In *Plutarch. Lives, Volume I: Theseus and Romulus. Lycurgus and Numa. Solon and Publicola.* Translated by Bernadotte Perrin. Loeb Classical Library 46. Cambridge, MA: Harvard University Press, 1914. (Cited as Plut. *Vit. Publ.*).

Polydore Vergil, *De Rerum inventoribus*. In *Beginnings and Discoveries: Polydore Vergil's De inventoribus rerum. An Unabridged Translation and Edition with Introduction, Notes and Glossary*. Edited by Beno Weiss and Louis C. Pérez. Biblioteca Humanistica & Reformatica 56. Leiden and New York: Brill, Hes & De Graaf, 1997.

Porfyrius. *Panygyricus Constantini Carmina*. In *Publilii Optatiani Porfyrii Carmina*. Edited by Giovanni Polara. Torino: G.B. Paravia, 1973.

Rucellai, Giovanni di Pagolo. *Zibaldone*. A cura di Gabriella Battista. Memoria Scripturarium 7. Testi in volgare 3. Firenze: Sismel – Edizioni del Galluzzo, 2013.

Stefaneschi, Iacopo. *De centesimo seu iubileo anno. La storia del primo giubileo (1300)*. A cura di Claudio Leonardi testo critico di Paul Gerhard Schmidt, traduzione e note di Antonio Placanica. Firenze: Sismel – Edizioni del Galluzzo, 2001.

Tacitus. *Histories: Books 1–3*. Translated by Clifford H. Moore. Loeb Classical Library 111. Cambridge, MA: Harvard University Press, 1925. (Cited as Tac. *Hist.*).

Tacitus. *Histories: Books 4–5. Annals: Books 1–3*. Translated by Clifford H. Moore and John Jackson. Loeb Classical Library 249. Cambridge, MA: Harvard University Press, 1931. (Cited as Tac. *Hist. or Ann.*).

Valerius Maximus. *Facta et dicta memorabilia. Memorable doings and sayings, Volume I*. Edited and translated by D.R. Shackleton Bailey. Loeb Classical Library 492. Cambridge, MA and London: Harvard University Press, 2000. (Cited as Val. Max.).

Varro. *On the Latin Language, Volume I: Books 5–7*. Translated by Roland G. Kent. Loeb Classical Library 333. Cambridge, MA: Harvard University Press, 1938. (Cited as Varro, *Ling.*).

Ventura, Guglielmo. *Chronicon Astense*. In Rerum Italicarum Scriptores 11. A cura di Lodovico Antonio Muratori, col. 139–168. 1727.

Virgil. *Eclogues. Georgics. Aeneid: Books 1–6*. Translated by H. Rushton Fairclough, revised by G.P. Goold. Loeb Classical Library 63. Cambridge, MA: Harvard University Press, 1916. (*Aeneid*. Cited as Verg. *Aen.*).

Virgil. *Aeneid: Books 7–12. Appendix Vergiliana*. Translated by H. Rushton Fairclough, revised by G.P. Goold. Loeb Classical Library 64. Cambridge, MA: Harvard University Press, 1918. (Cited as Verg. *Aen.*).

Villani, Giovanni. *Nuova Cronica*. In Rerum Italicarum Scriptores 13. A cura di Lodovico Antonio Muratori. 1728.

Zosimus. *Zosime, Histoire nouvelle*. Edited and translated by François Paschoud. Paris: Les Belles Lettres, 1971–1989. 2nd edition 2000. (Cited as Zos.).

The Cited Papal Bulls of Boniface VIII

Antiquorum habet (22 February 1300), attached at the end of Stefaneschi's *De centesimo*, f. 29r–30v. Also in *Les registres de Boniface* VIII, II, 922–923, no. 3875. The original manuscript digitized in Biblioteca Apostolica Vaticana, Archivio del Capitolo di S. Pietro, capsa 1.

Nuper per alias (22 February 1300), attached at the end of Stefaneschi's *De centesimo*, f. 31r–32v.

Ad honorem dei (25 December 1300), attached at the end of Stefaneschi's *De centesimo* with the title "*gratia non bullata,*" f. 33r–34r.

Unam sanctam (18 November 1302), in *Les registres de Boniface* VIII, III, col. 888–890.

Literature

Bautier, Robert-Henri. "Le Jubilé romain de 1300 et l'alliance franco-pontificale au temps de Philippe le Bel et de Boniface VIII." *Le Moyen Age* 86 (1980), 189–216.

Birch, Debra J. *Pilgrimage to Rome in the Middle Ages: Continuity and Change*. Woodbridge: The Boydell Press, 1998.

Blennow, Anna Holst. *The latin consecrative inscriptions in prose of churches and altars in Rome 1046–1263*. Roma: Società romana di storia patria alla Biblioteca Vallicelliana, 2011.

Bratu, Anca. "Le purgatoire entre le temps terrestre et le temps eschatologique: forms de representation du temps." In *Le Temps, sa mesure et sa perception au Moyen Age. Actes du colloque Orléans 12–13 avril 1991*, edited by Bernard Ribémont, 47–60. Caen: Paradigme, 1992.

Brind'Amour, Pierre. "L'Origine des Jeux Séculaires." In *Aufstieg und Niedergang der römischen Welt* II.16/2, edited by Hildegard Temporini and Wolfgang Haase, 1334–1417. Berlin and New York: Walter de Gruyter, 1978.

Canaccini, Federico. *Al cuore del primo Giubileo. Bonifacio VIII e l'Antiquorum habet*. Città del Vaticano: Lateran University Press, 2016.

Cardini, Franco. "L'eclisse di Gerusalemme. Fallimento della crociata in Terrasanta, nascita del giubileo." In *La Storia dei Giubilei. Volume primo 1300–1423*. Edited by Gloria Fossi, 56–69. Firenze: Giunti – BNL Edizioni, 1997.

Ciardi Dupré Dal Poggetto, Maria Grazia. *Il maestro del codice di San Giorgio e il cardinale Jacopo Stefaneschi*. Firenze: Edam, 1981.

Cohen, Simona. *Transformations of time and temporality in Medieval and Renaissance art*. Boston: Brill, 2014.

Condello, Emma. "I codici di Stefaneschi: uno 'scriptorium' cardinalizio del Trecento tra Roma e Avignone?" *Archivio della Società romana di storia patria* 110 (1987), 21–61.

Condello, Emma. "I codici Stefaneschi: libri e committenza di un cardinale avignonese." *Archivio della Società romana di storia patria* 112 (1989), 195–218.

Contestin, Gustave. "Origine historique des Jubilés." *Revue des Sciences Ecclesiastiques* 180 (1875), 280–295.

Cruz, Jo Ann H. Moran. "Dante, Purgatorio 2, and the Jubilee of Boniface VIII." *Dante Studies, with the Annual Report of the Dante Society* 122 (2004), 1–26.

Davies, Jason P. *Rome's Religious history. Livy, Tacitus and Ammianus on their Gods.* Cambridge: Cambridge University Press, 2004.

de Vincentiis, Amedeo. "Il giubileo di Martino V." In *La Storia dei Giubilei. Volume primo 1300–1423.* Edited by Gloria Fossi, 294–311. Firenze: Giunti – BNL Edizioni, 1997.

Doublier, Etienne. "*Libra misericordiae* le indulgenze di Bonifacio VIII." *Rivista di storia della Chiesa in Italia* 64/2 (2010), 347–380.

Dunning, Susan Bilynskyj. "Roman *Ludi Saeculares* from the Republic to Empire." PhD diss., University of Toronto, 2016.

Emmerson, Richard Kenneth and Bernard McGinn, eds, *The Apocalypse in the Middle Ages.* Ithaca and London: Cornell University Press, 1992.

Enciclopedia cattolica, V. Città del Vaticano: Ente per l'Enciclopedia cattolica e per il Libro cattolico, 1951.

Epstein, Steven A. "Business Cycles and the Sense of Time in Medieval Genoa." *The Business History Review* 62, no. 2 (1988), 238–260.

Esch, Arnold. "I giubilei del 1390 e del 1400." In *La Storia dei Giubilei. Volume primo 1300–1423.* Edited by Gloria Fossi, 279–293. Firenze: Giunti – BNL Edizioni, 1997.

Esch, Arnold. "L'economia nei giubilei del quattrocento." In *I giubilei nella storia della Chiesa,* 341–358. Città del Vaticano: Libreria editrice Vaticana, 2001.

Fedele, Pietro. *Gli Anni Santi.* Torino: Società Editrice Internazionale, 1934.

Fenster, Thelma and Daniel Lord Smail, eds, *Fama: The Politics of Talk & Reputation in Medieval Europe.* Ithaca: Cornell University Press, 2003.

Fois, Mario. "Il Giubileo del 1300." *Archivium Historiae Pontificae* 38 (2000), 23–40.

Forsythe, Gary. *Time in Roman religion: one thousand years of religious history.* New York and London: Routledge, 2012.

Fossi, Gloria, ed. *La Storia dei Giubilei. Volume primo 1300–1423.* Firenze: Giunti – BNL Edizioni, 1997.

Frugoni, Arsenio. "Il Giubileo di Bonifacio VIII." *Bullettino dell'Istituto Storico Italiano per il Medio Evo e Archivio Muratoriano* 62 (1950), 1–121. (Also in: *Istituto di storia e di arte del Lazio meridionale. Centro di Anagni. Documenti e studi storici anagnini* 13 (1996).)

Fuhrmann, Horst. "Einfluß und Verbreitung der pseudoisidorischen Fälschungen. Von ihrem Auftauchen bis in die neuere Zeit." In *Schriften der Monumenta Germaniae historica* 24. I–III, LV, 236; VIII, 237–624; VI, 625–1127. Stuttgart: Anton Hiersemann Verlag, 1972–1974.

Gagé, Jean. *Recherches sur les jeux séculaires.* Paris: Société d'édition "Les Belles lettres," 1934.

Gandolfo, Francesco. "Bonifacio VIII, il Giubileo del 1300 e la Loggia delle Benedizioni al Laterano." In *Romei e Giubilei. Il pellegrinaggio medievale a San Pietro (350–1350). Roma, Palazzo Venezia 29 ottobre 1999 – 26 febbraio 2000,* 219–228. Milano: Electa, 1999.

Garms, Jörg, Roswitha Juffinger, and Bryan Ward-Perkins, eds, *Die mittelalterlichen Grabmäler in Rom und Latium vom 13. bis zum 15. Jahrhundert, 1. Band: Die Grabplatten und Tafeln*. Rom, Wien: Verlag der österreichischen Akademie der Wissenschaften, 1981.

Gherardini, Brunero. "Indulgenza: Teologia e Storia." In *I giubilei nella storia della Chiesa atti del Congresso internazionale in collaborazione con l'École française de Rome sotto il patrocinio del Comitato centrale per il giubileo del 2000: Roma, Istituto Patristico Augustinianum, 23–26 giugno 1999*, 81–89. Città del Vaticano: Libreria editrice Vaticana, 2001.

Goria, Axel. "Le indulgenze a Roma nel 1300 (un elenco fin'ora sconosciuto)." *Studi medievali* 18 (1952), 144–151.

Hall, John F. "The *Saeculum novum* of Augustus and its Etruscan Antecedents." In *Aufstieg und Niedergang der römischen Welt* II.16/3. Edited by Wolfgang Haase, 2564–2590. Berlin, New York: Walter de Gruyter, 1986.

Hödl, Ludwig. "Der Jubiläumsablaß Papst Bonifaz' VIII. 1300 – der Ernstfall der Buße." *Wissenschaft und Weisheit* 62 no. 2 (1999), 292–308.

Hödl, Ludwig. "Die Glosse des Kardinals Johannes Monachus zur Ablass-Epistel 'Antiquorum.'" *Medioevo. Rivista di storia della filosofia medievale* 26 (2001), 161–202.

Hösl, Ignaz. *Kardinal Jacobus Gaietani Stefaneschi. Ein Beitrag zur Literatur- und Kirchengeschichte des beginnenden vierzehnten Jahrhunderts*. Berlin: Verlag von Emil Ebering, 1908.

Kerr, Heather and Claire Walker, eds, *Fama and Her Sisters. Gossip and Rumour in Early Modern Europe*. Turnhout: Brepols, 2015.

Lambert, Gustave. "Jubilé hebreu et Jubilé chrétien." *Nouvelle revue théologique* 82 (1950), 234–251.

Le Goff, Jacques. *La naissance du Purgatoire*. Paris: Gallimard, 1981. English translation: *The Birth of Purgatory*. Chicago: University of Chicago Press, 1986.

Leonardi, Claudio, Paul Gerhard Schmidt, and Antonio Placanica. *Iacopo Stefaneschi. De centesimo seu iubileo anno. La storia del primo giubileo (1300)*. Firenze: Sismel – Edizioni del Galluzzo, 2001.

Lewalter, Ernst. "Das Jahrhundert als Epoche (aetas, centuria, saeculum)." *Romanische Forschungen* 49 (1935), 343–351.

Maccarone, Michele. "L'indulgenza del giubileo del 1300 e la basilica di San Pietro." In *Roma anno 1300. Atti della IV Settimana di Studi di Storia dell'Arte Medievale dell'Università di Roma "La Sapienza" (19–24 maggio 1980)*. Edited by Angiola Maria Romanini, 731–752. Roma: L'Erma di Bretschneider, 1983.

Maddalo, Silvia. "Alcune considerazioni sulla topografia del complesso lateranense allo scadere del secolo XIII: il patriarchio nell'anno del Giubileo." In *Roma anno 1300. Atti della IV Settimana di Studi di Storia dell'Arte Medievale dell'Università di Roma "La Sapienza" (19–24 maggio 1980)*. Edited by Angiola Maria Romanini, 621–632. Roma: L'Erma di Bretschneider, 1983.

Manion, Lee. "The Loss of the Holy Land and "Sir Isumbras": Literary Contributions to Fourteenth-Century Crusade Discourse." *Speculum* 85 no. 1 (2010), 65–90.

Mastnak, Tomaž. "The Fall of the Papal Monarchy and the Rise of Territorial Power." In *Crusading Peace. Christendom, the Muslim World, and Western Political Order.* Edited by Tomaž Mastnak, 229–278. Berkeley, Los Angeles and London: University of California Press, 2002.

Mercati, Angelo. "La lettera dello scrittore pontificio Silvestro sul giubileo del 1300." In *Cronistoria dell'anno santo 1925,* 1191–1198. Roma: Tipografia Poliglotta Vaticana, 1928.

Minucci, Giovanni. "La bolla di'indizione di Bonifacio VIII e il commento di Giovanni Monaco." In *I giubilei nella storia della Chiesa atti del Congresso internazionale in collaborazione con l'École française de Rome sotto il patrocinio del Comitato centrale per il giubileo del 2000: Roma, Istituto Patristico Augustinianum, 23–26 giugno 1999,* 224–235. Città del Vaticano: Libreria editrice Vaticana, 2001.

Mitchell, Charles. "The Lateran Fresco of Boniface VIII." *Journal of the Warburg and Courtauld Institutes* 14 no. 1 (1951), 1–6.

Morghen, Raffaello. *Bonifacio VIII e il giubileo del 1300 nella storiografia moderna.* Roma: Edizioni dell'Elefante, 1975.

Mori, Edoardo and Elisabetta Mori. *Un documento cortonese sul giubileo del '300. La lettera di Silvestro* scriptor *pontificio (carta 6 del codice 101) con un saggio di Gerhard Wolf.* Cortona: Accademia etrusca, 2000.

Mucciarelli, Roberta. "Bisogna essere molto prudenti con le voci perché fanno presto a trasformarsi in verità. Qualche considerazione su *fama* e *publica vox* nell'Italia comunale." In *Fama e pubblica vox nel medievo. Atti del convegno. Ascoli Piceno, Palazzo dei Capitani, 3–5 dicembre 2009.* Edited by Isa Lori Sanfilippo and Antonio Rigon, 25–46. Roma: Istituto storico italiano per il medio evo, 2011.

Muntoni, Francesco. *Le monete dei papi e degli stati pontifici I.* Roma: P. & P. Santamaria, 1972–1973.

Onofri, Laura. "Roma come nuova Terrasanta." In *Roma Sancta. La città delle basiliche.* Edited by Marcello Fagiolo and Maria Luisa Madonna, 28–31. Roma: Gangemi Editore, 1985.

Paravicini Bagliani, Agostino. "Bonifacio VIII, la Loggia di giustizia la Laterano e i processi di scomunica." *Mélanges de l'École française de Rome – Moyen Âge* 112 (2005), 377–428.

Paulus, Nikolaus. *Geschichte des Ablasses im Mittelalter vom Ursprünge bis zur Mitte des 14. Jahrhunderts.* Paderborn: Druck und Verlag von Ferdinand Schöningh, 1922–1923.

Pighi, Giovanni Battista. *De ludis saecularibus populi romani quiritium, libri sex.* Amsterdam: Schippers, 1965.

Quinto, Riccardo. "Giubileo e attesa escatologica negli autori monastici e nei maestri della *sacra pagina.*" *Rivista di storia della filosofia medievale* 26 (2001), 25–82.

Ragionieri, Giovanna. "Ipotesi per uno 'scriptorium': codici miniati a Roma per il cardinale Jacopo Stefaneschi." In *Roma anno 1300. Atti della IV Settimana di Studi di Storia dell'Arte Medievale dell'Università di Roma "La Sapienza" (19–24 maggio 1980)*. Edited by Angiola Maria Romanini, 393–402. Roma: L'Erma di Bretschneider, 1983.

Ragionieri, Giovanna. "Jacopo Stefaneschi e il *De centesimo.* Un cardinale testimone del primo giubileo." In *La Storia dei Giubilei. Volume primo 1300–1423*. Edited by Gloria Fossi, 217–223. Firenze: Giunti – BNL Edizioni, 1997.

Rahner, Karl, ed. *Encyclopedia of Theology. A Concise Sacramentum Mundi.* New Delhi: Burns & Oates, 1975.

Rinaldi, Alessandro. "Cartografia urbana e anni santi." In *Roma Sancta. La città delle basiliche.* Edited by Marcello Fagiolo and Maria Luisa Madonna, 253–264. Roma: Gangemi Editore, 1985.

Romanini, Angiola Maria, ed. *Roma anno 1300. Atti della IV Settimana di Studi di Storia dell'Arte Medievale dell'Università di Roma "La Sapienza" (19–24 maggio 1980)*. Roma: L'Erma di Bretschneider, 1983.

Rusconi, Roberto. "Milleniarismo e centenarismo: tra due fuochi." In *Roma Sancta. La città delle basiliche.* Edited by Marcello Fagiolo and Maria Luisa Madonna, 3–9. Roma: Gangemi Editore, 1985.

Salzman, Michele Renee. *On Roman Time: the Codex-Calendar Of 354 and the Rhythms Of Urban Life in Late Antiquity.* Berkeley, Los Angeles and Oxford: University of California Press, 1990.

Schnegg-Köhler, Bärbel. "Die *augusteischen* Säkularspiele." *Archiv für Religionsgeschichte* 4 (2002), 156–170.

Stickler, Alfonso. *Il Giubileo di Bonifacio VIII. Aspetti giuridico-pastorali.* Roma: Edizioni dell'Elefante, 1977.

Théry, Julien. "Fama: l'opinion publique comme preuve judiciaire. Aperçu sur la révolution médiévale de l'inquisitoire (xiie–xive)." In *La Preuve en Justice de L'antiquité à Nos Jours.* Edited by Bruno Lemesle, 119–147. Rennes: Presses universitaires de Rennes, 2003.

Thurston, Herbert. *The Holy Year of Jubilee. An account of the History and Ceremonial of the Roman Jubilee.* London: Ams Pr, 1900.

Ullmann, Walter. "Boniface VIII and his Contemporary Scholarship." *The Journal of Theological Studies, New Series* 27 no. 1 (1976), 58–87.

Watt, J.A. "The Papacy." In *The New Cambridge Medieval History V.* Edited by David Abulafia, 107–162. Cambridge: Cambridge University Press, 1999.

Whalen, Brett Edward. *Dominion of God.* Cambridge, MA: Harvard University Press, 2009.

Zagnoli, Roberto. "Il significato del Giubileo nella tradizione biblica." In *Romei e Giubilei. Il pellegrinaggio medievale a San Pietro (350–1350). Roma, Palazzo Venezia 29 ottobre 1999–26 febbraio 2000*, 29–33. Milano: Electa, 1999.

Zecchini, Giuseppe. "Feste e identità: dai ludi saeculares ai natali di Roma." In *Identità e valori. Fattori di aggregazione e fattori di crisi nell'esperienza politica antica. Bergamo, 16–18 dicembre 1998*. Edited by Alberto Barzanò, Cinzia Bearzot, Franca Landucci, Luisa Prandi, and Giuseppe Zecchini, 195–208. Roma: "L'Erma" di Bretschneider, 2001.

CHAPTER 4

Time Set in Stone: Temporal References in the Non-funerary Epigraphy of Rome (1000–1527 AD)

Urpo Kantola

1 Introduction: Theme and Previous Studies

The non-funerary epigraphy[1] of medieval and early Renaissance Rome has not attracted much scholarly attention in the past, so in the study at hand, I will examine various aspects of time presented in this material. The main point to be discussed here is how such inscriptions were furnished with temporal points of reference to their present time and, when indicated, to their past or future. I attempt to answer these questions by providing an overview of these references in the source material and discussing how these practices varied between inscription genres and contexts. As the epigraphic presentation of temporal ideas is largely limited simply to a date, the questions posed here are: why was the decision made to set (or not to set) a certain kind of date, literally, in stone?; who made the decision?; and how do the choices made vary within different time periods? This study comprises discussion on the following themes: firstly, the numeral formats used; secondly, the years and longer periods indicated with precise and imprecise expressions; thirdly, day/month level indications; and fourthly, the duration and distance in time used; with brief inquiries into eternity, historical narration, and familial past. In terms of these themes, my systematic approach to formulaic variation, which, it seems, has

1 With epigraphy, I refer here to inscriptions on stone, or (rarely) on metal, mosaic, or writing on painted images, all of which are usually intended for broadly public display. This somewhat restricted definition derives from the primary source editions used here, mainly Forcella (see below); for a broader discussion, see Robert Favreau, *Épigraphie médiévale* (Turnhout: Brepols, 1997), 5–27. In the transcriptions, I follow the markings of *L'Année Épigraphique*, the relevant features being as follows: (a) = interpreted abbreviation; [a] = conjecture in lacuna; [. .] = lacuna of two unknown letters; [- - -] = lacuna of unknown length; {a} = letter in text omitted by editor; ‹a› = editor's addition; ⌜ a ⌝ = editor's correction; + = fragmentary but unknown letter; ạ = editor's interpretation of a fragmentary letter; | = line division (with line number in lower case every fifth line). In addition, I use / for verse division in metric inscriptions, when it is not identical with line division.

 | DOI:10.1163/9789004436251_005

not been carried out on this material before, ought to justify itself through its usefulness for evaluating undated or suspect source material.[2]

As to the temporal range of the present study itself, the main discussion begins at 1000 AD. This choice is principally dictated by Vincenzo Forcella's 14 volume corpus *Iscrizioni delle chiese e d'altri edificii di Roma dal secolo XI fino ai giorni nostri*, published between 1869–1884. Although preceding years are no less interesting, the lack of a comprehensive edition for these earlier examples makes the material too laborious to be fully included in this discussion. Nevertheless, at times 10th century material is also examined for the sake of comparison. The period to be discussed covers the mid-15th century revolution from medieval to Renaissance or "humanist" epigraphy, which had a profound effect on both the visuals and the contents of inscriptions,[3] and which also introduced or revived certain inscription genres, such as honorific monuments for living individuals set up in public spaces. For these reasons, discussed below in more detail,[4] I will – in the context of this article and the material from Rome – use the term "medieval epigraphy" to refer to the texts related to the period 1000–1450 and "Renaissance epigraphy" or "humanist epigraphy" to those from the period 1450–1527. Here, 1527 AD is taken as the end point for two reasons: first, it has been chosen in other studies of early Renaissance epigraphy from Rome for practical reasons,[5] whereas 1500, for instance, is arbitrary; and secondly, for this reason, the present study communicates better with these previous works.[6] In addition, even though many of the inscriptions, dated vaguely to the 16th century by Forcella, probably belong to a date later than

2 This applies especially to the lost inscriptions that either survive as later copies or are known from transcriptions found in manuscripts; see below.

3 The visual appearance contains, of course, various aspects: the inscribed monument (*Inschriftenträger*) itself, layout and paleography of the text, etc. For changes in contents, see, e.g., Iiro Kajanto, *Classical and Christian: Studies in the Latin Epitaphs of Medieval and Renaissance Rome* (Helsinki: Suomalainen tiedeakatemia, 1980), *passim*, and below, in the second part of this study.

4 See below p. 112–113.

5 Iiro Kajanto, "Dating in the Latin Inscriptions of Medieval and Renaissance Rome," *Arctos: Acta philologica Fennica* 11 (1977), 42; and Iiro Kajanto, *Classical and Christian*, 8 regarding his funerary material, refers to the 1527 Sack of Rome as the end of the "High Renaissance" in Rome, which briefly affected epigraphic production, and "as the end of the period in which Latin epigraphy passed from the Middle Ages to the Renaissance." Furthermore, he thinks epigraphy was thereafter more affected by the cultural resonances of the Counter-Reformation.

6 See the previous note. Elsewhere, Anna Cavallaro, Anna Esposito, and Andreas Rehberg, "Donne di pietra. Immagini, vicende, protagoniste delle sepolture romane del Rinascimento," *Mélanges de l'École française de Rome, Moyen age* 127, no. 1 (2015), 5–6 set the temporal limit to 1527 without specifying the reasons.

1527, they have been included for comparison, since there are likewise vaguely dated cases in the previous centuries.

As stated above, the core material derives from Forcella's vast corpus.[7] However, as others have noticed before, such a massive work, conducted relatively quickly in the 19th century, does not quite meet the standards of publication of present-day scholarship.[8] Inaccuracies of various kinds are all too frequent, and, what affects the present study perhaps the most, the given dates for the inscriptions are often unnecessarily vague or overly precise, or refer to a date given in the text, even though the inscription had been put up much later.[9] Given the volume of the work required, there has not been any full-scale attempt to ameliorate this situation for the Roman epigraphy of the period – after Angelo Silvagni's plans, begun in the interwar period, regrettably "proved abortive," as Iiro Kajanto puts it.[10] Nevertheless, studies of some particular groups of material have been published, with an emphasis on epitaphs.[11] Among these, Kajanto's inquiries into temporal features in Roman epitaphs of the discussed period provide a valuable point of reference. Elsewhere, I have found Anna Holst Blennow's and Peter Cornelius Claussen's works[12] to be of considerable value, on consecrative inscriptions and Roman marble sculptors,

7 Below, I will refer to this corpus plainly with "[volume number]:[inscription number]," e,g. 12:444 refers to vol. 12, inscription no. 444. Where a later publication has replaced Forcella's reading of a single inscription, I refer to the newer one; a *comparatio numerorum* of these and Forcella's lemmata is given in Appendix 2.

8 Angelo Silvagni, "Intorno alla pubblicazione delle iscrizioni cristiane, antiche e medioevali, di Roma e dell'Italia," *Rivista di archeologia cristiana* 5 (1928), 138; Kajanto, *Classical and Christian*, 7; see also Anna Holst Blennow, *The Latin Consecrative Inscriptions in Prose of Churches and Altars in Rome 1046–1263: Edition with Translation and a Commentary on Language and Palaeography* (Rome: Società romana di storia patria, 2011), 12.

9 E.g. 12:444, an 18th century inscription on the epistyle of SS. Vincenzo ed Anastasio alle Tre Fontane: for the date, see Appendix 1.

10 Kajanto, *Classical and Christian*, 7. The concrete result of this, regarding Rome, is Silvagni's *Monumenta Epigraphica Christiana* vol. 1 which contains photos of several inscriptions up to the 12th century, with references to some earlier publications and contemporary locations, and a brief description of the contents. This presents six 11th–12th century non-funerary inscriptions not published by Forcella or others (see Appendix 2), and forms, with Nicolette Gray, "The Paleography of Latin Inscriptions in the Eighth, Ninth and Tenth Centuries in Italy," *Papers of the British School at Rome* 16 (1948), 38–162, the base for the 10th century material referred to in this study.

11 Especially Jörg Garms, Tassilo Blittersdorff, and Hanns Jäger-Sunstenau, *Die mittelalterlichen Grabmäler in Rom und Latium vom 13. bis zum 15. Jahrhundert: Bd. 1. Die Grabplatten und Tafeln* (Wien: Österrreichische Akademie der Wissenschaften, 1980). This only goes up to 1450 AD, and regrettably there seems to be no *comparatio numerorum* with Forcella.

12 Blennow, *Latin Consecrative Inscriptions*; Peter Cornelius Claussen, *Magistri doctissimi romani: Die römischen Marmorkünstler des Mittelalters* (Stuttgart: Steiner-Verlag-

respectively. In addition, there are a multitude of works focusing on single churches and other ecclesiastic buildings, and a discussion of local epigraphs is sometimes included.[13]

Since around 40 % of the material is only known through transcriptions by 15th–18th century scholars, or as later lapidary copies or versions, I estimate the source value of the inscriptions on two levels, using a very coarse-grained critical approach. Firstly, countless examples show that the transcriptions (and later lapidary copies) frequently disregard the original forms, e.g. abbreviations and numerals, which are presented in a way the scholar found suitable himself.[14] This is clearly visible even as late as the work of Luigi Gaetano Marini (1742–1815), published with comments by Angelo Mai in 1831.[15] Forcella's corpus is, even if far from perfect, a significant improvement in this regard. Therefore,

---

Wiesbaden, 1987). For the latter's *Die Kirchen der Stadt Rom im Mittelalter*, with contributions of Darko Senekovic and Daniela Mondini, the four volumes published thus far (2002–2012) offer a great deal of useful information, but I have only been able to use the three first volumes for the present study – and only in haste, thus perhaps having missed some points. (To my satisfaction, I notice that the approach to inscriptions has ameliorated with every volume.) In addition, improvements and additions to the Forcellan corpus have already been provided by Mariano Armellini, *Le chiese di Roma dal secolo 4 al 19*, republished with additions by Carlo Cecchelli (Roma: Ruffolo 1942 [original 2nd edition by Armellini, 1891]), and in the volumes of Richard Krautheimer et al., *Corpus basilicarum christianarum Romae: the Early Christian Basilicas of Rome (IV–IX cent.)* (Città del Vaticano: Pontificia Istituto di Archeologia Cristiana 1937–1977); unfortunately, I have been unable to consult most of these works here.

13 For numerous examples, see the bibliography in Blennow, *Latin Consecrative Inscriptions*, 320–331. In addition, catalogues by Ivan Di Stefano Manzella, *Index inscriptionum Musei Vaticani. 1, Ambulacrum Iulianum sive "Galleria lapidaria"* (Città del Vaticano: Ex Officina Libraria Pontificia, 1995), and Giorgio Filippi, *Indice della raccolta epigrafica di San Paolo fuori le mura* (Città del Vaticano: Monumenti, Musei e Gallerie Pontificie, 1998) contain useful information about published and unpublished inscriptions, but as they are not editions themselves, they only serve as a starting point for a proper edition.

14 This can be assessed through innumerable cases where earlier transcriptions of surviving inscriptions are available. Forcella often comments on the discrepancies, but not systematically, and sometimes an inscription was lost to, or not seen by him, but has been found again later. For instance, Blennow records an inscription on an altar (*The Latin Consecrative Inscriptions*, 48–53 no. 8), which was lost in 1970, but found again and acquired by the Musei Capitolini; however, Forcella publishes it erroneously twice (4:1311[bis], 11:704), basing himself only on various earlier publications. Another inscription (4:146) was likewise unseen by Forcella who relied on an 18th century account. Claussen, *Magistri doctissimi Romani*, 17, mentions a better reading by the early 17th century scholar Francesco Gualdi (Cod. Vat. lat. 8254, fol. 351) and, more importantly, rediscovered the inscription and re-edits it.

15 Luigi Gaetano Marini and Angelo Mai, "[Inscriptiones Christianae]," in *Scriptorum veterum nova collectio e Vaticanis codicibus edita vol. 5*, ed. Angelo Mai (Roma: Typis Vaticanis 1831), 37–510.

in this study, I consider the recorded forms of the lost inscriptions, i.e. the material not seen by Forcella or a later scholar, as being not fully trustworthy, and I present the data deriving from this group with caution.[16] To those that have survived at least until Forcella's time, I refer to as the surviving inscriptions.

Secondly, even if the presented forms cannot be counted upon, the contents are often recorded reliably enough to be used. Yet, we also have a surviving group of inscriptions that are thought to be later copies of earlier, usually medieval inscriptions. In some cases, an old, for example 15th–16th century, manuscript transcription of the original, may prove or disprove the exactness of the copy, but several copies or adaptations survive without further evidence, and in some individual cases it may not necessarily be possible to distinguish which of the two is accurate.

In all probability, a thoroughly conducted scrutiny of all the available source material and post-Forcellan studies, would help us with a more critical evaluation of the inscriptions already published by Forcella and, to some extent, allow us to include others missed by or unknown to him.[17] In the preparation of the present study, such a laborious task has not been possible. For this reason, I prefer to attempt to simply present an overview of the dating practices in the non-funerary epigraphs of the given period, firstly to produce a general idea of what epigraphy yields to answer the questions posed in the present study, and secondly to contribute to the scholarship of this material, hopefully easing future efforts towards a more systematic critical approach of the corpus. In

16 Forcella's inspection is rarely indicated *expressis verbis*, but presumable when he uses the present tense in the description or mentions a location without basing it on an earlier scholar. However, in several cases his readings are suspect or erroneous, and sometimes he seems to have missed inscriptions that should have been visible to him, such as one presented by Claussen, *KSRM* 1, 139: it is on a column brought in the 18th century from S. Bartolomeo all'Isola to SS. Bonifazio ed Alessio and placed next to 13:1210. I wonder whether Forcella was that negligent, or if in some cases he relied on some other source, such as (perhaps inadequate) assistants, without checking the epigraph himself.

17 Two examples ought to illustrate the current situation well enough: the main material used by Blennow, *Latin Consecrative Inscriptions*, consists of 43 inscriptions, four of which Forcella has missed or excluded erroneously, and five have emerged only after Forcella. In addition, Blennow examines and refutes from her material eight inscriptions published less critically by Forcella. In a similar manner, but with more drastic figures, Claussen, *Magistri*, presents 25 Roman inscriptions found in the Forcellan corpus against 30 others (one is included in Blennow's material as well, and another is in Angelo Silvagni's *Monumenta epigraphica christiana saeculo XIII antiquiora quae in Italiae finibus adhuc exstant iussu Pii XII pontificis maximi*, vol. 1 (Civitas Vaticana: Pontificium institutum archaeologiae christianae, 1943) (= *MEC* 1); many others are, of course, published elsewhere too.) Perhaps the difference in proportions derives from the fact that Claussen's inscriptions are often short and less conspicuous, and therefore easier to miss.

respect of the latter point, the material used here is collected in three appendices: one a list of non-funerary inscriptions found in Forcella's volumes; another for inscriptions from later publications; and the third appendix provides a topographical summary of the material in the previous two.

1.1 *Material*

Moving to the actual discussion, the usable material[18] may be divided roughly into four categories of types of temporal reference: (a) 377 inscriptions that bear a purposeful set date with the precision of at least a year; (b) 17 inscriptions that show purposeful temporal references, but lack an exact year; (c) 193 inscriptions that show vague or secondary temporal references, mainly through individuals mentioned by name; (d) 48 inscriptions without any relevant temporal references. This last figure is certainly too low, since Forcella recorded various "minor" inscriptions, such as artisans' signatures or picture labels, rather sporadically.[19] However, this last group is of little significance here, as it only shows where time was not relevant enough for it to be recorded. This group consists of: 15 epigrams for various, mostly embellishing purposes; 15 notices or labels; seven relic catalogues; five dedications; a decree;[20] a donation record; a measuring table; an unfinished papal epistle; and two inscriptions of unknown nature. The existence and nature of temporal references in an inscription are often inherent to its genre, although some genres show more internal variation, and production practices may have undergone changes over time.

To take a look at the material in general, one quarter comes from three major places: the Vatican (57 inscriptions[21]), the Capitoline hill (44[22]+7 from S. Maria in Aracoeli), and the Lateran with its hospital of SS. Salvatore (34+16); a fourth place is S. Maria in Cosmedin (14). Any relevant comparison between

---

18 Eight inscriptions in Forcella, and numerous inscriptions published or catalogued elsewhere, are too fragmentary to be of any use here.

19 Cf. above note 17 for signatures; for image labels and other minor inscriptions with an embellishing function, the contributions by Claussen et al. in the *KSRM* volumes demonstrate alone, with clarity, that Forcella has omitted a significant amount of this sort of material.

20 This questionably datable 16th century inscription containing two communal decrees (1:53) in fact shows no chronological indicators, which is interesting in terms of the genre, but is beyond the scope of this study.

21 The inscriptions known to have been imported from other locations from Rome (often removed in restoration works) are not included among these.

22 These are edited by Forcella under "Piazza e palazzi Capitolini," and even though some of these inscriptions may have been brought there from elsewhere, the inscription genres (honorific monuments, decrees) mostly suggest that they indeed originate from the Capitoline hill.

the material from these sites is likely to be unfruitful because, firstly, the genre and the period seem to be the primary defining factor in what was put in an inscription not its location, and secondly, the inscription genres in these locations differ from each other.

As we have said, temporal differentiation is more relevant. Table 4.1 presents how the material of this study is distributed over time. The parallel figures of funerary inscriptions are given on the right;[23] the figures in square brackets show the relative frequency of inscriptions for the shorter periods – these figures will be utilised in the following tables as well;[24] the time categories comprising of only vague or uncertain dates are marked with an asterisk;[25] and the lost inscriptions contain all cases where the original is lost, that is also those surviving as a copy on stone.

These figures give us an idea of epigraphic tendencies, showing three longer periods which correlate with historical events. The first three centuries have relatively few epitaphs surviving, indeed inscriptions of other genres surpass their numbers, the latter often part of the active rebuilding and refurnishing of churches in the 12th and 13th centuries. In the period 1300–1449, however, non-funerary epigraphy decreases. These times saw the Avignon Papacy, the Western Schism, and other unrest in the first third of the 15th century, and resources were directed elsewhere, away from ecclesiastic or civic building projects in Rome. But this did not seem to affect the number of epitaphs, since those known from this period are clearly more abundant than those from the previous three centuries.[26] Also, for the present discussion, we can see clear changes in the epigraphic habit, with a differentiation between the medieval and Renaissance periods, it changing in the mid-15th century. Looking only at the figures, the production of epitaphs rocketed during the years 1450–1527,[27] and

23 These approximate figures derive from the data gathered by me in preparation for this study.

24 The factors are 100/50 = 2 for 1400–1449 and 1450–1499, and 100/28 = 3,57 for 1500–1527.

25 "15th c." and "16th c." contain inscriptions dated only vaguely to the respective centuries, except for two that are dated to the mid-15th century, 6:67[bis p. 40] and 13:1029, and 8:1 from 1517 or the following two or three decades.

26 The reason for this is unknown to me, but in any case, the difference in figures seems too large to be explained as a mere accident of survival. The matter remains to be examined elsewhere.

27 In 1528–1600 the growth stabilised, and the figures remain more or less at the level reached in the first three decades. Regarding the reliability of these kinds of figures in general, Kajanto, *Classical and Christian*, 8 notes that "[d]uring the Renaissance and even later" renovation works in churches in Rome removed any elements considered to be of "Gothic style" without remorse. In his opinion, in comparison to the Renaissance period, much fewer medieval inscriptions have survived as in the records. This may be true but perhaps

TABLE 4.1 Temporal distribution of medieval and Renaissance inscriptions.

Period	Non-funerary	Non-funerary	Non-funerary	Non-funerary	Epitaphs	Epitaphs
		Compensated value	Per year	Lost		Per year
11th c.	31		0.31	6	34	0.34
12th c.	91		0.91	29	12	0.12
13th c.	124		1.24	48	58	0.58
14th c.	59		0.59	24	324	3.24
1400–1449	19	38	0.38	11	119	2.38
1450–1499	153	306	3.06	59	588	11.76
15th c.*	5		0.05	4	154	1.54
1500–1527	96	343	3.43	45	617	22.04
16th c.*	39		0.39	20	274	2.74
uncertain*	17			9	13	
Total	634		1.20	255	2193	4.15

non-funerary epigraphy was invigorated as well. New genres came along or were revived, such as that of the public honorific monument, and the style of inscriptions in general changed radically, principally due to the imitation of classical epigraphy, which is obvious from the lettering, the layout, and the language used. Regarding the language, this change is apparent in the following discussion, since in several discussed features there is a, sometimes sharp, distinction between medieval and the Renaissance usage.

2 Temporal References

2.1 *Primary Temporal References*

2.1.1 Official Dating Formulas: Documentary Sources vs Epigraphy

The formal dating nomenclature used in documentary material provides a relevant comparison with the epigraphic material: the *"große Datierung"* and

not to the extent Kajanto thinks. In my view, the early Renaissance material was not bound well enough to survive, since no such tendency can be observed from the changes in proportions of lost inscriptions in table 4.1.

"kleine Datierung" used in papal documents and notarial dating.[28] The relevance of these naturally derives from the fact that most persons involved in epigraphy were generally acquainted with such documentary dating formulae, and thus some overlap is to be expected *a priori*. Interestingly, though, it seems rare that such documentary principles were fully implemented epigraphically according to the contemporary fashion: for example, out of 146 inscriptions with year and day mentioned, only four follow a papal dating formula,[29] and 27 show only an approximate notarial formula. Using slightly laxer criteria, 25 other inscriptions use something that resembles one of these. Even in inscriptions originating from papal activity, the dating usually follows neither the *"große"* nor the *"kleine"* formula exactly, and occasionally only rather loosely. This can be explained at least partially through the monumental nature of many epigraphs. A document typically recorded a transaction of administrative or legal significance to be archived and consulted later, and plain and formulaic language suited this purpose. In contrast, stylistic considerations carried more weight when drawing up a monumental inscription, and in many cases there was no problem in varying or embellishing the expression at the expense of preciseness or formulaic uniformity. This is particularly obvious in the case of metric inscriptions. Furthermore, in 230 cases, a date denoting only the year was thought to be sufficient.

Yet, these formal documentary paradigms did have some impact on epigraphic dating practice: the same elements were the most common ones used, and they were combined in a single inscription according to need and habit. Some dating elements, which are absent from the documentary material, remained less common in epigraphy, or were heavily context-related, such as

28 As Thomas Frenz, *Papsturkunden des Mittelalters und der Neuzeit* (Stuttgart: Franz Steiner, 2000²), 23 and 25 presents them with their respective evolutions: *"Große"*: "*dat(um/a)* + the name of the *datarius* + Roman date (i.e. day indicated by its relationship to *kalendae*, *nonae* or *idus*) + indiction + incarnation year [Roman numbers] + papal year [pope's ordinal number written out in full, year with Roman numbers]"; *"Kleine"*: "*dat(um/a)* + place in the locative case + *apud* [church] (or locative *Laterani*) + Roman date [Roman numbers before 1431; thereafter written out in full] + (only from 1188 onwards) *pontificatus nostri* [number written out]." Frenz seems to imply that the papal chancery's practices were fairly uniform and the changes in them were distinct. The notarial dating formulae seem to comprise of "incarnation year + indiction + reigning pope + current date," where the three latter elements may be in another order, and number formats vary between written out and Roman numbers; sometimes the papal year is included as well.

29 *"Große"* in an altar dedication by Callistus III: Blennow, *Latin Consecrative Inscriptions*, 62–64 no. 12 – however, as this is known only from a 15th century manuscript, the number formats may have been "normalised" in the transcription. *"Kleine"* in 4:821, 8:780, and 11:44, all originating from a papal document.

using a liturgical day as a point of reference (referred to hence as "liturgical dating"). Nevertheless, these can be found in inscriptions of papal origin, too. It is noteworthy that expressions in inscriptions did not need to be systematic. One could even split the date into three separate sentences, as shown in a consecrative inscription from 1256 in SS. Martina e Luca, with "AD" and the pope, the consecration day, and the archpriest of the church at that time appearing separately.[30]

2.1.2 Formats of Numeric Expression

Many temporal references on stone are numeric: that is dates and lengths of time. The stylistic aspects of the numeric expression in them change in time and in context, which is interesting in itself. But more importantly, the formats on stone offer a point where the difference between diplomatic and notarial practice and epigraphic expression may be assessed. Whereas documentary practices were somewhat fixed, epigraphic realisations show a more varied use of numerals, some written out in full, some using Roman, and some Arabic numbers.

Graph 4.1 presents the occurrences of these three types of number formats. More than one may occur in the same inscription, the total number of which is indicated below the time period. Graph 4.2 shows the division of inscriptions with uniform numbering (e.g. all Roman), with varying numbering in different expressions (e.g. *"anno d(omi)nice incarnationis* MCLXXVIIII, *indictione duodecima … in d(omi)nica* IIII *post octavam pasche"*[31]), and with mixed numeric expressions (e.g. *"mill(esim)o* CCCXXXVIII*"*[32]). Here, as well as in all the following tables, the relative quantities for the periods shorter than 100 years[33] are shown in square brackets for the total numbers.

Three relevant features emerge from the statistics. Firstly, as graph 4.1 shows, Roman numerals are unsurprisingly present in most cases from the 12th century onwards, but the more telling feature is the consistent reduction in the proportion of written-out numeration after the 13th century. This accords with the data of graph 4.2 that the most variation in numeration type occurs in the 12th and 13th century but greatly diminishes in the following centuries. Second,

---

30 Blennow, *Latin Consecrative Inscriptions*, 194–198 no. 42, L. 1–2 & 5: *"anno Do(mini)* MCCLVI, *d(omi)nus Ale(xande)r p(a)p(a) |* IIII *(…) |$_{5}$ co(n)sec(ra)v(it) ecc(lesi)a(m) ista(m)*, and LL. 12–17 *consecrat(i)o aut(em) | hec ec(clesi)a fuit i(n) media* XLa, *q(ua)n(do) | est stat(i)o ad s(an)c(t)um Cosma|$_{15}$tum in Silice. in t(em)p(or)e ar|chip(re)sbitero A(n)drea(!) fuit hec | ecc(lesi)a co(n)secrata."*

31 Blennow, *Latin Consecrative Inscriptions*, 112–116 no. 24.

32 9:244.

33 As presented above in note 24.

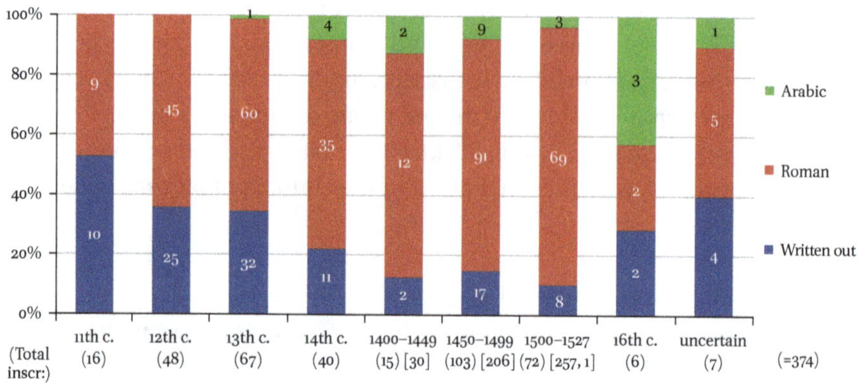

GRAPH 4.1 Occurrence of different number formats (one or more per inscription).

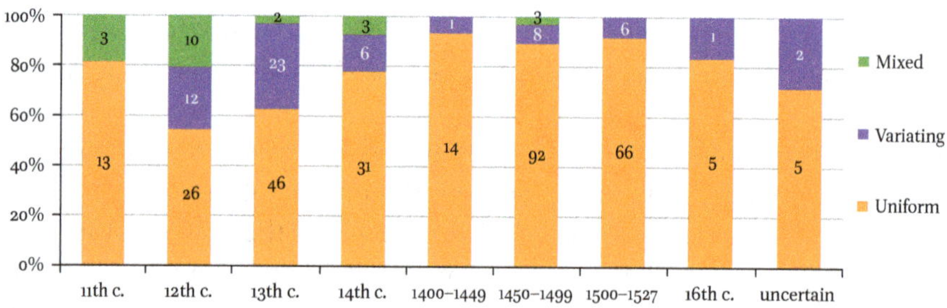

GRAPH 4.2 Uniformity of numeral expressions in single inscriptions.

among the inscriptions with multiform numbers, mixed expressions are found throughout all the time periods, but a closer look at the data reveals a clear peak between 1111–1127 with eight cases.[34] Cases of varying but unmixed numerals appear more evenly from 1100 to 1299. Thirdly, Arabic numbers appear in the 14th century but remain rare – the transcription of a lost 12th century altar dedication[35] records Arabic numerals, but this is of course highly unreliable. Most

34 8:717(2); 9:149; Blennow, *Latin Consecrative Inscriptions*, 46 note 95; 43–48 no. 7; 55–58 no. 10; 72–76 no. 15; 84–88 no. 17; Claussen, *KSRM* 1, 135–136. This mixed usage seems to center around altar dedications in the 11th and 12th centuries: the last four of the previous ones are such dedications, as are the three 11th century cases from between 1073–1090 (Blennow 19–26 no. 2; 33–35 no. 4; 35–39 no. 5), and the two later 12th century cases, both from 1196 (Blennow 127–134 no. 28; 134–139 no. 29). However, such mixed usage does not occur again until almost a century later in 1291 (Claussen, *KSRM* 2, 341), and neither this nor any of the cases thereafter are altar dedications.

35 Blennow, *Latin Consecrative Inscriptions*, 139–141 no. 30.

of the 13th–15th century cases are in building inscriptions and were mostly set up by popes, and thus do not distinguish themselves from other comparable material in any other visible manner. However, all the 16th century occurrences here are in vernacular donation inscriptions pertaining to S. Maria della Consolazione.

One further point of interest is numerals embedded in epigraphic poems, where they are written out in full in 14 out of 16 cases, most of which date back to before 1450.[36] The metre naturally dictates the composition, resulting in unusual or unique expressions, such as in 5:1353 (S. Maria in Monticelli), lines 4–7: "*cu(m) foret annus | ternus millenus deciesq(ue) quaterq(ue) decenus | et quartus decimus patris huius pontificatus | et sextu(m) sole(m) Maius revocaret in orbem.*"[37] This and ten other cases are straightforward;[38] elsewhere, in two versions of a building inscription with a list of relics, the Forty Martyrs are referred to with "XL,"[39] and a poem in a flood inscription from Palazzo Caetani has both a poetic dating "*bis denos menses decimo peragente Leone | idibus huc Tyberis unda Novembris adest,*" and a separate prosaic counterpart "*MDXIIII, die XIII Novembris.*"[40] This last example is an interesting piece of evidence also because of the double dating: the epigram gives a papal year (by months) and a Roman month date, whereas the prosaic part records the same with an AD year, and a running month date. The former is not only more poetic by expression but also, at this date, obviously more prestigious and stylish, with the pope and the Ides. Nevertheless, the author thought it was good to have at hand a prosaic, more readily informative counterpart, too. Lastly, two 14th century poems exhibit a deviation from written-out numerals: one has Roman-only numbering but with the suffixes written out, and the other mixes Roman numbers with verbal expression.[41]

Since 16th century vernacular inscriptions and metric inscriptions are the only subgroups that present any relevant difference attributable to geography,

---

36 11th century: four; 12th century: two; 13th century: four; 14th century: two; early 15th century: one; early 16th century: one. For the references, see below. Only the cases where the numbers appear in metric composition are counted here. In 16 other inscriptions with a metric part the date is expressed separately in a prose section.

37 I.e. 6 May 1143, the pope being Innocent II.

38 1:4; 1:21; 4:1245; 5:345; 5:1353; 6:26; 10:364; 10:540 with André Guillou, *Recueil des inscriptions grecques médiévales d'Italie* (Rome: École française de Rome, 1996) no. 117; 11:262; 11:321; 12:p. 508 = Mondini, *KSRM* 3, 348–349. One 10th century case is similar: Giovanni Battista Giovenale, *La basilica di S. Maria in Cosmedin* (Roma: P. Sansaini, 1927), 72 no. 69.

39 8:4; 9:818.

40 13:436, now lost.

41 12:177; 9:623.

inscription type, or author, we may conclude the following: the 11th–13th centuries show a more general use of both written-out numerals and internally variating formats, whereas later on Roman numbering and internally uniform usage gained ground. Arabic numbers are found from the 14th century onwards, but only in a small minority of cases. Unsurprisingly, numbering practices do not differ greatly from the documentary material in general, but the variation shows that there was no system behind the epigraphic representation of numbers.

2.1.3 Anno Domini Years

Inscribing a text on durable material in a (more or less) public space included the idea of leaving the contents on display for the future. In the majority of inscriptions (377) in our material, group A, those with a year date, or another method referring to a certain period, was used purposefully for fixing the inscription – or a recorded event – in time.[42] The Christian era, which I will refer to as AD years in general,[43] is by far the most frequently used method. But, importantly, when it occurred alongside other dating methods, it was systematically placed first, and was thus considered the primary year dating element, whereas the others had only a complementary role. These others were either years of papal reign, or imprecise references to the reigning pope, or references to other persons – mostly imprecise and referring to lower-ranking ecclesiastics;[44] these will be discussed in detail later. The following graph 4.3 presents the usage of AD years both alone and with complementary elements, and the other methods only as they occur in single inscriptions.

The most interesting features shown by the statistics are, first, the rise of "Only AD" dating from the 12th to the 13th century, and second, the outline of AD years used with complementary methods: from dominance in the 12th and 13th centuries to a drastic fall in the 14th century, and ultimately to a comeback

42 Not included in this: fragmentary 8:354 probably had an AD year date but nothing is left of it, and two 16th century inscriptions: 2:1080 which had an AD date but year numbers, have perished, and 7:624, which Forcella dates vaguely to this century, but the inscription mentions two historical dates, "*sub anno MCCXVIII*" and "*sub anno MCCXVII*" (cf. below note 185).

43 The variation in this (*anno Domini, anno a Christo nato, anno ab incarnatione Domini* etc.) will be discussed below.

44 I leave the indiction aside here and discuss it separately below, because it has substantial informative value only when appearing with a person dating, but without an AD year.

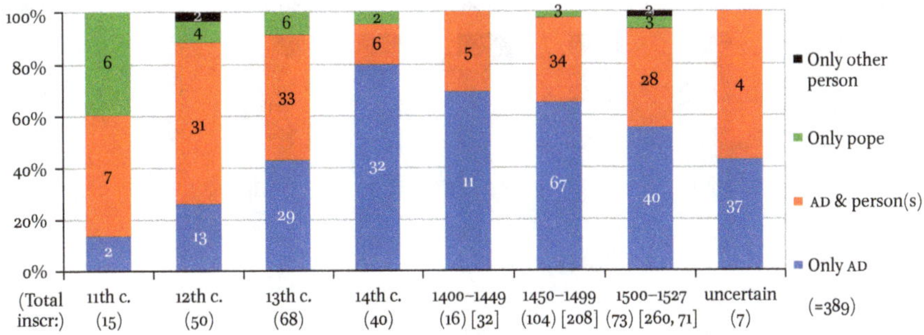

GRAPH 4.3 Year indications per inscription.

in the Renaissance period. The 12th–13th century development seems to be simply a general change in habit, since it happens regardless of any relevant connection to a particular genre or author. The 14th century drop is probably connected to the Avignon Papacy (1309–1376/77), and among the six cases of combined dating methods only one, concerned with the founding of a hospital with testament funds, is from the Avignon period and has the papal reign mentioned.[45] The pope was a more distant figure, and perhaps more importantly, papal building activity was low. However, even the few examples of papal authorship of inscribed objects lack a formal temporal reference to the pope's reign, except for one bull,[46] and papal dates also occur rarely in the first and last decades of the 14th century. If there was any general tendency towards using "Only AD" in the 14th century, the combined methods appear again in one-third of the cases in both halves of the 15th century, and in the early 16th century the proportion grows further. Again, no context-related distribution seems to be discernible.[47]

The AD years were expressed in various manners throughout our period, and the preferred forms did not remain the same. The following graph 4.4 presents the distribution of the typical variants; some minor variation within the

45 9:244 (Ospedale di S. Giacomo, 1338). The only other one from this period, 8:338 (Ospedale del SS. Salvatore al Laterano, 1348) dates itself not with a pope, but with guardians and priors. The pre- and post-Avignon cases are 1:4, 8:342, 11:544, and Blennow, *Latin Consecrative Inscriptions*, 210–215(2).

46 6:22 (1341, Vatican), and 8:30(1) & (2) (1369, Lateran, reliquaries for Sts Peter and Paul); the bull 8:31 (1372, Lateran).

47 From the end of the 15th century onwards, one might speculate on some connection between the frequency of papal dates and changes in the epigraphic representation of popes; cf. p. 152.

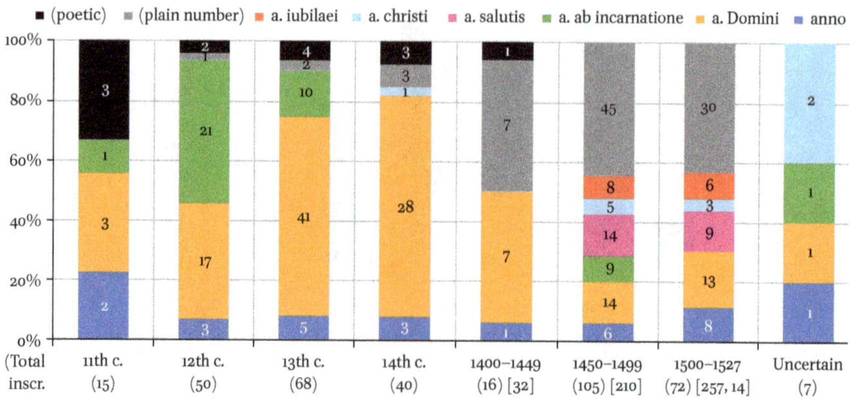

GRAPH 4.4 Expressions of AD years per inscription.

categories has been ignored,[48] and three partially lost cases have been excluded.[49]

"*Anno Domini*" is obviously the most frequent medieval formula. It endures in the Renaissance period as well, but decreases in favour of a plain year number, which had been rare previously. The incarnation year found its most widespread use in the 12th century, predominantly in inscriptions by high ecclesiastic authors, and then declined down to no exemplars in the 14th century. Its brief re-emergence in the late 15th century is limited to papal statutes,[50] except

48 Compare to Kajanto, "Dating," 45–46. Examples: under *"anno salutis":* 1:24 "*anno salutis humanae*"; 13:1142 "*anno salutis Christianae*"; and others, with even 7:860 "*humanae reparationis anno,*" which in any case carries approximately the same idea; under *"anno ab incarnatione"* (usually with *Domini* or *eiusdem*, referring to the preceding invocation): 1:1629 "*an(no) incarnationis d(omi)nice*"; Blennow, *Latin Consecrative Inscriptions*, 127–134 no. 28 "*anno inca(r)nationi(s) ei(us)de(m)*"; also two 13th century cases, 1:1098 and Blennow, *Latin Consecrative Inscriptions*, pp. 226–228, with "*anno nativitatis D(omi)ni,*" and a late 15th century 6:1173 "*anno natalis Chr(ist)i*"; under "*anno Domini*": Kajanto, *Papal Epigraphy*, 60–61 "*anno Dom(ini) Iesu Christi*"; Blennow, *ibid.* p. 190–194 no. 41(1) *anno d(omi) nic(e)*(?). For the odd last one, with its erroneous congruence, there are the following possibilities: (1) "*anno d(omi)nic(o),*" which would be untypical; (2) "*incarnationis*" has been omitted; (3) the last letter is written in error: "*anno D(omi)ni{c}.*" Note that inscription (2) on the same stone begins with "*a(nno) D(omi)ni.*" Certain philological interest lies in "*sub anno Domini*" (under *"a. Domini"*), all medieval (1290–1420), in 2:191, 4:1144, 5:436, 6:22, 8:338, 8:342, 11:544, 12:13, 13:586; and "*sub anno*" (under *"anno"*), apparently later only, in Iiro Kajanto, *Papal Epigraphy in Renaissance Rome* (Helsinki: Suomalainen tiedeakatemia, 1982), 82 (1473), 7:624 (16th century, AD years in historical references), and Blennow, *ibid.* p. 41 (date unknown).

49 2:1080; Claussen, *KSRM* 1, 226; unpublished (Filippi, *San Paolo fuori le Mura*, 41 nos. 51434+51440: my reading from *ibid.* fig. 94).

50 1:1196–1197, 1:1629, 3:507, 4:821, 8:42, 8:780, 11:44. Note that the subsequent four cases from 1494–1521 have either "*anno salutis*" or a plain number: 4:89, 4:280, 8:352, 8:785.

for one which has the unique, but thematically related, expression "*anno natalis Chr(ist)i* MCCCCLXXI."[51] One earlier papal documentary inscription has the year indicated as "*anno (Christ)i*," but since this is in a 1518 copy of a 1394 inscription, the formula does not necessarily correspond with the original one.[52]

Regarding other, less frequent formulae, the late 15th century shows the most variation, and sees the introduction of "*anno salutis*" as well as the Christian jubilee into the dating formulae.[53] The most varied poetic expressions faded away as medieval style epigraphy reached its end, the last one being from 1427;[54] metric inscriptions by no means disappeared, but in the Renaissance period the AD years were expressed in a separate prose section. Generally speaking, no other genre-related or other distinctive usages seem to emerge, and the few exceptions, where an inscription presents more than one expression for AD years, do not contribute much to the discussion.[55]

2.1.4 Popes and Papal Reign

Another means for denoting a particular year (or a longer period) is through persons, restricted mostly to popes. The following table shows the distribution of this use between, firstly, imprecise references to a papal reign (e.g. "*tempore Paschalis II papae*"), precise references to a year of a papal reign (e.g. "*anno domini Honorii secundi papae V*"), and to inscriptions with a reference only to another person (e.g. "*Antonius Militius, Stephanus Teulus, Sanus Corona co(n)s(ule)s*"); the first two categories may include other persons as well. I have counted three references with short "*sedes vacans*" periods as imprecise papal

51 6:1173 (S. Spirito in Sassia): record of restoration work by a monastic official.

52 Blennow, *Latin Consecrative Inscriptions*, 210–215(2). As one may observe in graph 4.4, the expression is otherwise unknown before the Renaissance period – the earliest certain case is in 6:57 from the Vatican, 1451. Note that all the other medieval papal documentary epigraphs date the year through papal reign only, also in 3:721 (S. Salvatore in Pensilis, 1285), where the AD year is mentioned in an epigram placed below a papal brief; but the 1394 part of Blennow's inscription seems indeed to have had the AD year (cf. Blennow, *ibid.*).

53 For the jubilee, cf. below and, more generally, Jasmin Lukkari's article in this volume.

54 5:345.

55 8:30 parts (1) and (2) have parallel dates referring to the respective actions of the pope and the emperor, in part 1 respectively with "*anno*" and "*anno Domini*," and in part 2 with a plain number and "*anno Domini.*" 10:867 has "*anno Domini*" in a historical context, and a plain number in the dating formula of the inscription. 9:622 has "*anno*" and "*anno Domini*" twice in a historical context, the latter serving as a "*terminus post quem*" for the inscription (1391). In 1:28 a plain number refers to the year the honorand was in office, and "*anno*" to the one the monument was erected in.

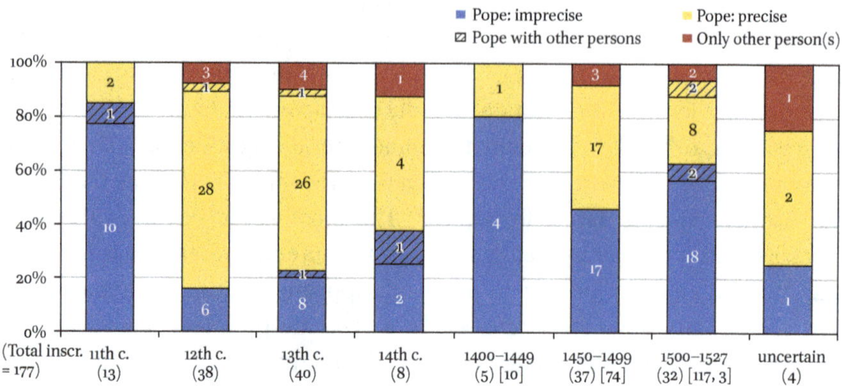

GRAPH 4.5 Dating through persons in general.

references,[56] and omitted one faulty expression.[57] It is good to keep in mind that most cases include an AD year as well.

It seems striking at first that imprecise papal dating types grow in number in the Renaissance period, but they often occur in specific genres. The first is early Renaissance honorific inscriptions, that mostly show either an AD year only (9), or an AD year and an imprecise reference to papal reign (14), whereas three inscriptions make use of other options: only one of them has both the AD year and the papal year.[58] Another genre with often an unspecified papal reference are private building or donation inscriptions, whereas inscriptions deriving from the papal chancery have, unsurprisingly, a precise papal year.[59] In the earlier material, patterns are seldomly distinguishable. The numerous 12th and 13th century altar dedications normally include a papal year in the dating formula, but, as Blennow has pointed out, a few (mostly 11th century) dedications also have an imprecise reference, or lack one at all.[60] Among metric and

56 4:1265, 13:423–424. It is not entirely certain whether the expressions would have been more specific had the vacancies been prolonged, but the indications are imprecise in the first place. The vague reference to the period of 1268–1271 in 1:1098 is surely an imprecise one, since the given AD date falls into its third year.

57 7:237 (1400, S. Venanziano, a combined building inscription and epitaph): "*pa(n)tif(icatu)*(!) | *B(onifatii) pape noni, ind(ictione) no(na)*," with rather odd abbreviations.

58 Nine inscriptions with AD only: ten (1:11, 1:13, 1:19–20, 1:23, 1:30, 1:33, 1:41, 1:48(1)); 14 with AD and an imprecise papal reference (1:12, 1:14–15, 1:17, 1:21, 1:25, 1:28, 1:31–32, 1:34–37, 1:42); one with AD and precise (1:24); one with only imprecise papal reference (1:29); one without a primary dating element (1:40, cf. below).

59 Cf. above note 50.

60 Blennow, *Latin Consecrative Inscriptions*, 248–251.

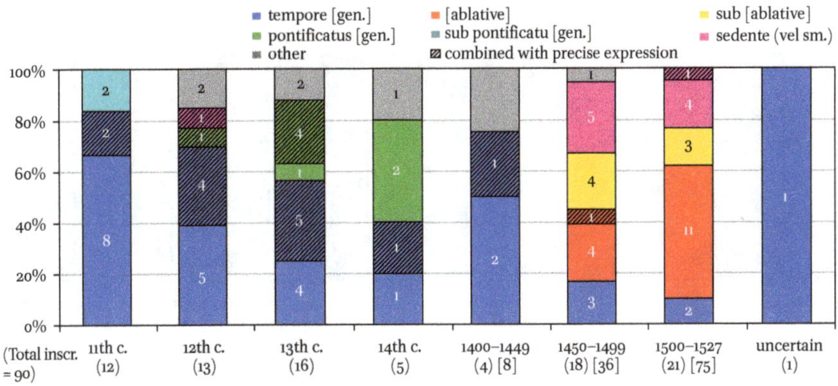

GRAPH 4.6 Structure of imprecise papal references.

non-metric flood inscriptions (ten; seven of them posterior to 1450), some trends may be seen: three out of four poetic examples have an imprecise papal reference, versus one out of five prosaic ones; one with a reference to "*sedes vacans*" is indecisive.[61] This may again relate to the ease of adding a precise element into a prose composition. The one precise poetic reference is atypical, since the papal reign is presented in months: the already mentioned 13:436 "*bis denos menses decimo peragente Leone.*"[62]

The graphs 4.6 and 4.7 provide a closer examination of the nature of both imprecise and precise papal references, and graph 4.8 collates the papal titulature used in both of these. The areas marked with black oblique lines point out the overlap between them, that is, the cases where an imprecise indication is given first, but followed (not necessarily immediately) by the papal year.[63] One further inscription has both, but these denote different events.[64] The "other" cases mostly consist of rare poetic expressions, and for the rest, of

61 Metric 6:1093, 13:427, 13:434; prosaic 13:430; "*ecclesia vacante*" 13:424.

62 Above p. 117 "*bis denos menses*" with note 40.

63 This is to keep the statistics simple enough to be legible. The structures of the ambivalent cases are: "[ablative]+*anno* [gen.]" one (late 15th century) / "*pontificatus* [gen.]+*anno* [gen.]" five (one × 12th; four × 13th century) / "*sedente* (vel sim.)+(*anno*) *pontificatus*" two (12th; beginning 16th century) / "*tempore* [gen.]+(*anno*) *pontificatus*" 2 (11th, 14th century) / "*tempore* [gen.]+*anno* [gen.]" nine (three × 12th; five × 13th; one × early 15th) / "*tempore* [gen.]+other" two (11th).

64 Mondini, *KSRM* 3, 308 no. 111 S. Lorenzo in Lucina, 1112 or slightly later. The precise temporal reference is in the dating formula, LL. 2–3 "*anno vero XII do(m)pni | Paschalis II p(a)p(e),*" and the imprecise one refers to a past event, L. 25 "*te(m)pore Gelasii s(e)c(un)di p(a)p(e).*" This has been counted once in tables 4.6 and 4.9, but the two references are examined separately in tables 4.7 and 4.8.

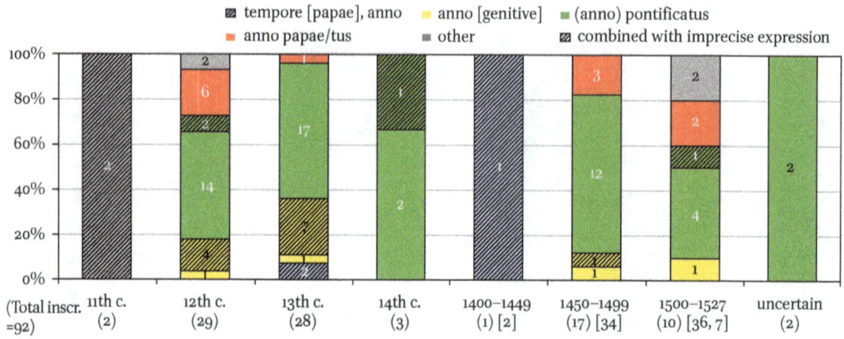

GRAPH 4.7 Structure of precise papal references.

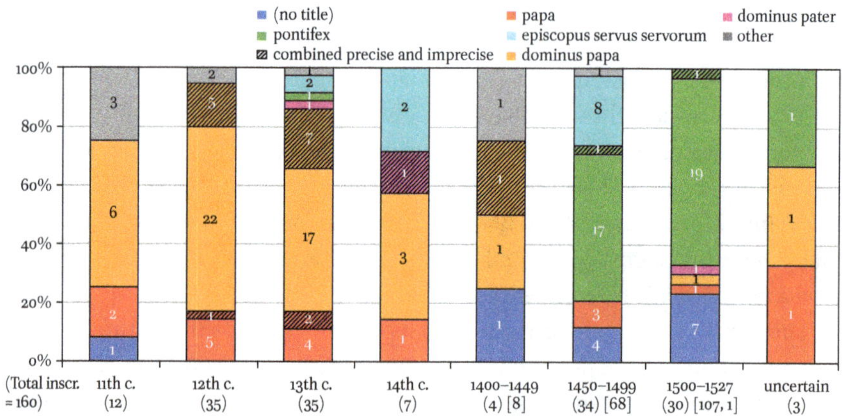

GRAPH 4.8 Titles of popes in papal temporal references.

other singular occurrences. Again, minor variation has been disregarded in the following tables.[65]

In general, imprecise temporal references and the titles of popes change considerably from the 1450s onwards, whereas the distribution of different precise expressions of time remains relatively stable. Reference to the pontificate is the predominant precise formulation, but not among the imprecise expressions, since five out of eight of the latter cases are followed by a more definite expression. Among the imprecise references, "*tempore*" is used the most in the medieval period, both alone and in connection with a precise reference, but in the Renaissance period it is found more often in stand-alone use. Not attested before the Renaissance period is the plain ablative absolute, "*sub* + [ablative]," and "*sedente*"; for the latter, though, there is a solitary 12th century variant

65 E.g. *tempore* and *temporibus* are interchangeable in this context.

"*p(rae)sidente in ap(osto)lica sede.*"[66] The structure of precise papal references shows little variation over the centuries, but "*anno*" and "*anno* + [pope's name in the genitive]" mostly coexist with a preceding imprecise element. A unique case, a building inscription in S. Lorenzo fuori le Mura from 1198–1216, presents the only example of a period described using two successive pontiffs. The building project was finished, as was the inscription, in the time of Innocent III, but apparently begun under his predecessor: LL. 1–4 "*tempọrib(us) | Celestini* III *et Innoce|nt[i* I]ỊI *Rom[a]ṇor(um) p̣[on]tifi|cụ(m).*"[67]

Regarding papal titles, the most relevant observation is that the most frequent medieval papal title "*dominus* [name] *papa/pater*" survives into the Renaissance epigraphy but becomes mostly replaced by either "*pontifex*" – nearly always as the clearly classicising "*pontifex maximus,*" first attested epigraphically in 1447–1455 or 1451[68] – only by name without a title. In our medieval material, the sole instance of "*pontifex*" in a temporal reference is the above-mentioned double papal reference "*Rom[a]ṇor(um) p̣[on]tifi|cụ(m).*"[69] Beside all these, "*papa*" exists as a minor case throughout the centuries. Otherwise, titulature does not present much of interest, with "*episcopus servus servorum*" naturally restricted only to inscriptions that derive from or imitate a papal epistle.

2.1.5 Dating a Period through Other Persons

In 15 inscriptions,[70] the year dating formula is supplemented by a reference to persons other than a pope. In addition, there is a case in which the emperor

66 Blennow, *Latin Consecrative Inscriptions*, 76–84 no. 16.

67 Mondini, *KSRM* 3, 339. For the reading, see Appendix 2.

68 6:53 and 6:57. For the title in general, see Kajanto, *Papal Epigraphy*, 62–63. Cf. also *ibid.* for 6:46 from Eugene IV's time (1431–1447), where Forcella read "*pontif(ex) max(imus),*" but according to Kajanto the correct reading is "*p(a)p(a).*" The temporal references without "*maximus*" are "*pontifex*" 8:785, "*dominus ... pontifex*" 1:15, "*pientissimus pontifex*" 8:504(2), and "*pont(ifex) d(ominus)*" 13:431 (which could be interpreted also as "*pont(ificatu) d(omini) Alexandri VI,*" but this is less likely because of the immediately following "*an(no) e(ius) IIII*"); other such cases are "*pontifex*" 2:662, "*Romanus pontifex*" 1:1197, 8:42, and "*summus pontifex*" 6:75, 6:124. For the medieval attestations regarding the last one, see the following note.

69 See also a 10th century case in hexametre with "*te(m)pore pontificis no|ni sum(m)ique Ioh(ann)is,*" Giovenale, *S. Maria in Cosmedin*, 72 no. 69 LL. 16–17; "*venerabilissimo pontifice*" in Blennow, *Latin Consecrative Inscriptions*, 220–221 of a dubious date, perhaps latter half of the 15th century; and *ibid.* 204–210, a falsification of perhaps a late medieval date (see Appendix 2). Otherwise, "*pontifex*" appears in at least 16 other 12th–14th century inscriptions, but not in primary temporal references. Compared to the previous note, the title is interestingly "*summus pontifex*" in five inscriptions (2:645, 6:12 and Blennow 105–110 no. 22, 139–141 no. 30 and 161–164 no. 35).

70 Also, in one 994 inscription from S. Adriano al Foro Romano, Marini and Mai, "[Inscriptiones Christianae]," 17 no. 2: LL. 14–15 "*te(m)p(o)re b(eati) Ioh(ann)is VX p(a)p(ae),*

takes the place of the pope in the formula, and in another a papal chancellor acts as the only temporal reference.[71] The supplementary dating elements follow the structure "*tempore / temporibus* + [genitive]," except for two out of three Renaissance cases, where the ablative absolute, or "*sub* + [ablative]" are used. The person(s) may also appear alone (7 cases), or a pope may appear in the dating (6) or be otherwise mentioned (3).[72] Regarding the temporal distribution, these inscriptions cover our whole period, but the 13th century stands out with six cases of this type.

One other feature connects practically all these cases: almost every person appearing in the dating formulas is an ecclesiastic, mostly belonging to the church or institution the inscription in question is from. Where a papal reign is referred to, the other persons are all the more supplementary in nature, except for the following case: 1:1098, from S. Maria ad Martyres: "*in n(omine) D(omi)ni am(en). anno na|tivitatis eiusdem* MC|CLXX, *indictione* XIII, | *me(n)se Iunii, die s(e)c(un)da, apo*|$_5$*stolica sede vacant|e, t(em)p(o)r(e) d(omi)ni Pandulphi d|e Sebura archip(res)b(yte)ri ec|clesie s(an)c(t)e Marie Rot|unde, et p(res)b(yte)ri Petri, p(res)b(yte)*|$_{10}$*ri Deodati, Petri Bars|ellone Romani Iaco|bi Romani Petri Corr|adi Pauli Ioh(ann)is Petri | et Tebaldi de Alp(er)inis,* |$_{15}$ *eiusdem eccl(es)ie cl(er)icis(!) | facte fuerunt nole | et nolarium.*" The reference to a papal reign is negative, as the Holy See is vacant, and this may have contributed to the choice to include other priests in the date. However, another reason is present, too: *eiusdem ecclesie clerici* on L. 15 refers, with all probability, to the previously listed personnel, and perhaps the names of the lower-ranking men are included in the dating formula only because their participation was intended to be recorded in any case.[73]

Primary temporal references through office-holders of the lay administration are absent in the material. An expression in 12:442, a record of the restoration of S. Stefano Rotondo for the Sixtine jubilee of 1475, LL. 6–7 "*operanti*

m(ense) | I(a)n(uario?), die XXVII, *ind(ictione)* VII, *te(m)porib(us) dom(ini) Leoni‹s› archipresbyteri (...).*"

71 Cf. also Claussen, *KSRM* 2, 341 (below p. 148 with note 163) with an interconnected historical reference to a miracle and a church consecration.

72 In these cases, the pope obviously provides the inscription with a temporal landmark; see below for secondary temporal references.

73 The exact number of individuals is uncertain: the ones referred to as *Barsellone, Iacobi, Iohannis, Tebaldi de Alperinis*, both *Romani*, and three *Petri* have been omitted from the name index of Forcella's volume 1, and since all names are in the genitive, it is uncertain whether some of the names are patronyms, e.g. "*Petri Barsellone, Romani Iacobi, Romani Petri, Corradi Pauli, Ioh(ann)is Petri*" – or whether each one belongs to a different cleric. The ones Forcella has included in the name index have been presented as single names without patronyms.

Gregorio de Ruvere | sacrarum aedium Urbis curatore," comes close to this, in that at least no clerical position is recorded for Gregorius. Non-papal personal references show only sporadic use, and recurring cases are found only for the Ospedale del SS. Salvatore al Laterano, where 8:338 (from 1348) and 342 (1397) are dated through "*tempore guardianatus*" of two men in both. Nevertheless, their status seems to vary, since in the former inscriptions the guardians are "*priores,*" but in the latter "*nobiles viri.*"

It is not surprising that nearly all references are to men. The only exception is an abbess in 10:867, which narrates briefly (in a slightly confused manner) on St Gregory of Nazianzus and his remains in S. Maria in Campo Marzo in 1505: "*situm est sub hoc di|vino altari corp(us) divi | Gregorii ep(iscop)i Naz‹i›anzeni et | patriarche in Co(n)stantinopoli: sub* $|_{5}$ *Tehodosio(!) seniore, an(n)o D(omi)ni ‹C›CCXC | obiit; tra(n)slatum inde corp(us) Constantino|polim, deinde in urbe Roma et in hac ecc(lesi)a condi|tum est et nup(er) sub Iulio s(ecu)ndo, MDV, Iunii | VO(?) XXV, postremo eodem an‹n›o, m(en)se Iulii XXIIII,* $|_{10}$ *sub abbatissa Martia de Palosiis.*" What the successive acts of replacement in 1505 precisely refer to is unclear. Nevertheless, the *sub* references to Julius II and the abbess seem to be not only temporal,[74] but to attribute the authorship of these actions to them as well.

In the same inscription, what is also noteworthy is the rare historical reference to Emperor Theodosius I (379–395), supplemented with an (erroneous) AD year. In a comparable manner, another inscription in S. Croce in Gerusalemme from 1521–1523, erected by one Bernardinus Lupi Carvaial, fixes a historical event in time through King Roderic of the Visigoths (c. 710–712): "*post cladem H[ispa]niae a Mahumetanis Aphricanis sub Roderico rege illatam,*" and an event of the recent past is framed with references to the Hispanic monarchs Ferdinand II and Isabella I, and Pope Innocent VIII.[75]

An additional case of special interest, not included above, presents a comparable usage of *sub* + ablative in an even rarer future reference: a building inscription from the same church and by the same Bernardinus, from 1520.[76]

---

74 Whether the references to persons are to be distinguished temporally from each other is also uncertain, since it is probable that both held their respective posts during the entire period. Note also that on LL. 4–5 the expression "*sub | Tehodosio senior*" is temporal only.

75 8:504(3) L. 1 for Roderic, and LL. 2 and 4 for the others, for whom see below p. 147–148.

76 8:505 "*ad gloriam dominicae crucis et spem recuperationis | sacri gloriosi sepulchri Hierosolymae sub s(ancto) in Christo | patre et d(omino) n(ostro) Leone X, Hetrusca gente, familia Medices, | clarissimo ac serenissimo, d(omino) Carolo V, Hispaniarum et* $|_{5}$ *utriusque Siciliae ac Hierusalem universaeque insulae | Atlanticae seu terrae Sanctae Crucis et Hispaniensis | ac aliarum insularum Africae et Indiae regi | catholico, archiduce Austriae (et) c(etera), reverend(us) in Chri|sto pater d(ominus) Bernardinus Lupi Carvaial episc(opus)* $|_{10}$

The invocation includes the "hope of recapturing the Holy Sepulchre in Jerusalem" under Leo X and Emperor Charles V, also King of Spain, which refers to both when and by whom the action should be effectuated. The author of the inscription was both the titular patriarch of Jerusalem and a Spaniard, which probably explains the hope being vested in Charles V. Yet another case, a pseudo-historical narration, again from S. Croce in Gerusalemme, and of uncertain but perhaps Renaissance date,[77] makes use – in addition to the year 1003 AD – of Emperor Otto III (996–1002) as a temporal reference with the use of *tempore*. Popes Silvester II (999–1003) and Sergius III (904–911) are also mentioned, and the ancient past is structured around Romulus and Tullus Hostilius.

The two exceptional cases without a pope mentioned remain to be addressed. The one with an imperial reign given, 8:5 from the Lateran, is dated with "*anno hoctavo do(mi)n(i) Henrici imp(erato)ris, indic(tione) septima, m(ense) Iuniu(s)(!), d(ie) V.*" The inscription has been left unfinished, and its exact nature and date are unclear, as well as whether this is Henry IV or V; it would be the latter and the date 1114 if we accept Angelo Silvagni's suggestion.[78] Regardless of which one of the emperors is referred to, this decision to use only an imperial reign obviously relates to a pro-imperial position in the investiture controversy. But this single case in an unfinished inscription does not affect

---

Sabinensis, cardi{di}nalis s(anctae) Crucis in Hierusalem, | patriarcha Hierosolymitanus, ante capellam | hanc ulteriori sacrae capellae s(anctae) Crucis in Hierusa|lem adiungendam curavit, anno salutis Christ(ianae) MDXX."

77 8:498, lost. Forcella, referring to Lorenz Schrader (*Monumentorum Italiae, quae hoc nostro sæculo & à Christianis posita sunt* (Helmaestadii: Typis Iacobi Lucij Transyluani, 1592), vol. 2 p. 128), tells us the inscription was formerly "*in un marmo incastrato nella parete destra presso l'ingresso della cappella sotterranea di S. Elena,*" but "[s]*crive il* [*Raimondo*] *Besozzi* (*La storia della Basilica di Santa Croce in Gerusalemme* [*1750*] *ecc. p. 61*) *che fu tolta via perchè riconosciuta falsa la storia in questa narrata.*" Because of Schrader's testimony, the inscription must predate 1592, but, in the Forcellan transcription, the style seems again to be of a Renaissance date (e.g. the lack of medieval abbreviations, the spelling "*foeminarum*"). Thus, the date perhaps lies somewhere between the 1430s and 1592, if the transcription is accurate enough; otherwise the inscription could date back to an even earlier date (cf. Kajanto, *Classical and Christian*, 20 note 93, who regards Schrader as "an author of notorious unreliability"). I wonder if the same Bernardinus Lupi Carvaial, clearly fond of historical narrations and long inscriptions, could have been behind this one as well.

78 According to Forcella, the inscription is "*principio di decreto a Enrico IV*" and dates from 1099, whereas Silvagni describes it as "*initium inscriptionis commemorativae a. 1114 non amplius incisae.*" Both of these years are the 7th indiction, but they do not concord with the 8th year of either Henry IV (1084–1106) nor Henry V (1111–1125). However, 1114 could be possible if we assume that Henry V's reign is counted from his father's death (7 Aug 1106), and the indiction is the Roman one (with the new indiction year beginning on 25 December; see below).

the papal see's dominant position as one of the cornerstones of the worldly temporal framework in Rome.

Lastly, an ecclesiastic official appears once as the primary dating element: the lack of other date indicators derives, once again, from the metric form. Papal chancellor Cencius, the future Honorius III, funded a building project at S. Lorenzo fuori le Mura around 1194–1198, which was recorded with a two-verse hexameter on an architectonic upper framing of the *confessio*: "*Cencius hoc fieri, cum cancellarius esset, | d[eque suo f]ecit, Lau[r]enti, Stephane, vobis.*"[79] Luckily the subordinate clause expresses the temporal reference more clearly than some prosaic construction would do, e.g. with apposition: *Cencius cancellarius fecit fieri,* etc. However, even if the time aspect is clearer for us here, this by no means necessitates any conclusion that prosaic constructions with only a name and title were thought of as less of a temporal reference.

2.1.6 The Indiction

The indiction cycle of 15 years sometimes figures in epigraphic sources, always in a supplementary role to another year date. Iiro Kajanto commented on the use of indictions in Roman epitaphs as "of considerable frequency,"[80] and Robert Favreau refers to this in stating that the indiction was used relatively frequently in Rome compared to elsewhere.[81] As we shall see below, in our non-funerary material, the indiction had fallen out of use by the 15th century, and already by the 14th century exemplars remain scarce (although the total amount of material from this period is relatively limited in any case).[82]

79 Mondini, *KSRM* 3, 337–339 and 442. I find the suggestion for supplying the lacuna "*d*[- - -]*ecit*" quite possible, and it is metrically more plausible than some other suggestions, for which see *ibid.* p. 338 note 83. For the identity of Cencius (and his dissimilarity with another, almost contemporary chancellor, Cencius), see *ibid.* and Claussen, *KSRM* 1, 139–140 notes 30–31. Moreover, Mondini (*ibid.* p. 338 note 86) suggests that the simple choice of title, only *cancellarius*, may have a tone of modesty about it, which is possible, but this could be attributed to the requirements of metric expression as well. E.g. "*cardinalis,*" which was Cencius' title since 1191/92, is always a difficult one to compose in the hexameter following classical rules: at least once, in 11:262, this is solved with the somewhat tedious placement of the enclitic copula *que* in between the word: "*cardiq(ue)nalis.*"

80 Kajanto, "Dating," 46.

81 Favreau, *Épigraphie médiévale*, 177–178; see also Blennow, *Latin Consecrative Inscriptions*, 253.

82 Indictions still lingered in full use in notarial dating formulae at the end of our period, as in one amusing example from 1500 in Alexis Gauvain, *Una storia dalla Roma del Quattrocento: quaderni di Ansuino di Anticoli, parroco in Roma e beneficiato vaticano (1468–1502)* (Città del Vaticano: Edizioni capitolo vaticano, 2014), 242 (I thank Holger Kaasik for the

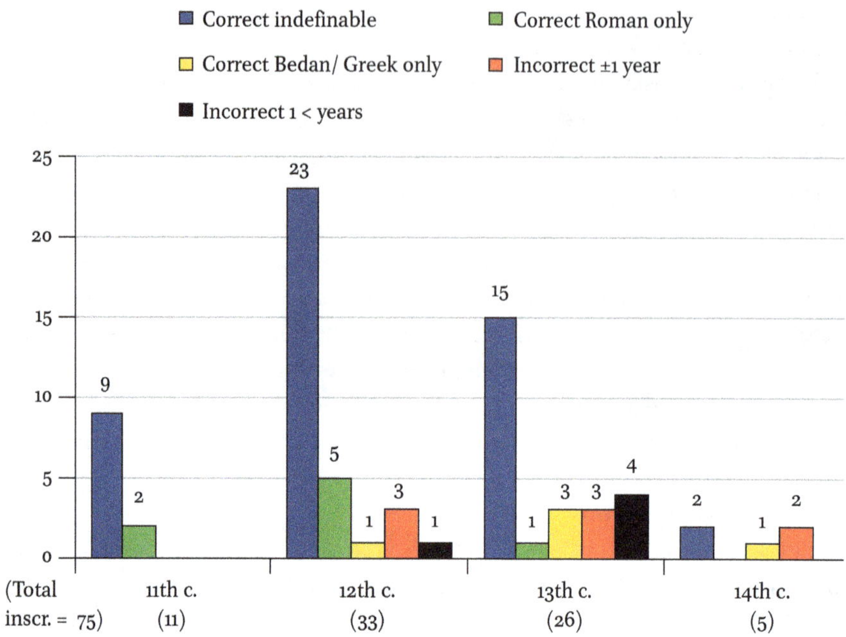

GRAPH 4.9 Occurrences of indiction in single inscriptions from the 11th to the 14th century.

Three indiction systems are known to have been in circulation in medieval Rome, differing in the date when the year changes: the oldest, the so called Greek indiction, begins on 1 September; the Bedan indiction on 24 September; and the Roman or pontifical indiction on 25 December.[83] As Blennow notes, the Bedan type is difficult to discern from the other two in a single document, but, since in our material all the accurately datable autumn cases range from 29 September to 21 December, the relevant distinction is only between the cases where the Roman indiction is the sole possibility, and those where it is impossible. However, in graph 4.9, an imprecise September date is put under the category "correct indefinable,"[84] which otherwise includes both those without a day-level date and those with a date from January to August. The table also presents the indictions and their correspondence with dates recorded

reference): "*In nomine Domini amen. Anno Domini Millesimo quingentesimo, pontificatus etc.*(!) *indictione III, mensis ianuarii die XVI.*"

83 Blennow, *Latin Consecrative Inscriptions*, 251–253 with further bibliography.

84 8:338. One other inscription, a lost one (1:3), bears a similar date, but the indiction 13 is four years too late for the year 1296.

in the inscriptions, even though four inscriptions were erected at a later point of time.[85]

Somewhat surprisingly, it seems that the indiction was in fact left aside earlier in non-funerary epigraphy, since the latest exemplar, a record of a *"venerabile emptum"* for Ospedale di S. Giacomo, is dated to 1397, whereas Kajanto records four 15th century epitaphs that mention an indiction.[86] It is peculiar that the 13th century shows a larger proportion of incorrect indictions, the majority of which are in lost inscriptions. However, it is difficult to postulate what this might indicate.

It is noteworthy that all but two out of 24 inscriptions (19 of them consecrative) that were commissioned by a pope or through the action of a pope date from before the 14th century. The two exceptions refer to 12th century dates, but were composed in a later period.[87] Therefore the indiction seems to have been retained longer in use by other commissioners than the pope. The last contemporarily erected papal exemplar is 9:999 in S. Urbano a Campo Carleo, in a monument which Urban IV ordered in 1263 *"ad p(re)ces d(omine) Iacobe filie Pet(r)i Bla(n)ci i(n) domo patris sui."* One other inscription, recording a papal consecration performed in 1256, was erected by a cardinal some 30 years later.[88]

2.1.7 The Christian Jubilee

The Christian jubilee year, celebrated for the first time in 1300, then at varying intervals, and from 1450 onwards every 25 years,[89] also deserves closer attention. Since Kajanto in his work does not discuss the use of the jubilee in dating, I have added the cases appearing in epitaphs to the adjoining graph 4.10.[90]

85 Blennow, *Latin Consecrative Inscriptions*, 15–19 no. 1: dated to 1046 but (re?)made c.1200–1250; *ibid.* p. 41: dated to 1108 but, according to Blennow, should be considerably later; *ibid.* 204–210: dated to 1110 but is rather a 14th or early 15th century falsification; *ibid.* 226–228: dated to 1256 but made around 1281–1287. For further details, see Blennow.

86 Kajanto, "Dating," 46; see also Kajanto, *Classical and Christian*, 18 noting that the last case, 2:932 from 1464, is still of a "fully medieval" style. Thus, the indiction was omitted systematically from humanist-style epitaphs.

87 Blennow, *Latin Consecrative Inscriptions*, 41, and 204–210: see above note 85.

88 Blennow, *Latin Consecrative Inscriptions*, 226–228: see above note 85.

89 For the jubilee, particularly that of 1300, see Jasmin Lukkari's article in the present volume.

90 1300 12:569 / 1450 2:268, 4:1247, 6:55 / 1475 4:180, 13:1265 / 1500 1:1252 & 1655–1657, 2:460 & 678, 3:533, 791–792 & 1066, 4:932, 5:73–74, 820 & 824, 6:110, 7:606, 11:286 / 1525 1:1280–1281, 2:1344, 3:29, 8:926. In addition, 3:558 has on L. 5 *"obiit anno iubilaei, die VIII calend(as) Octobr(is)"*; the jubilee referred to is uncertain, but Forcella places the inscription between others from 1525 and 1527.

From other inscriptions, the majority with this dating type records building projects, or other actions certainly or probably connected to the holy year, such as indulgence concessions. The cross-lined red sections denote the two inscriptions without any apparent connection to the jubilee.[91] Four inscriptions from a conventional year, but mentioning a jubilee, are not included.[92]

The jubilee year, propagated throughout the Latin Christian world, was from the start an event of considerable religious and economic importance. Nevertheless, as graph 4.10 presents, its impact on the epigraphic habit is minimal; it can be observed in the first jubilee of 1300, but then only appears again 150 years later. No references to the jubilees of 1350, 1390, 1400, or 1423 can be found in the record, although no other non-funerary inscriptions are known from these particular years.[93] This is somewhat striking, but is undoubtedly connected to the relatively small organisational effort put into jubilees[94] – and to the general (non-funerary) epigraphic recession in Rome during this period.

The non-funerary inscriptions mentioning the 1300 jubilee differ from all the others in that they make no use of the word jubilee itself, but rather utilise other expressions of multiple words.[95] The one epitaph in verse has the actual

---

91 8:354 and 8:361. Both are donation records from the Ospedale del SS. Salvatore al Laterano, now lost, but substituted later with vernacular versions. The former one's date was already lost at the time the transcription was made of it, but the vernacular substitute has the year 1500, and no reference to a jubilee. This, and a similar lack of a jubilee reference in 8:361 from 1525 leads me to regard such a reference as fairly improbable; nevertheless, the possibility cannot be excluded.

92 In the Vatican, the two popes' respective jubilees are referred to in Nicholas V's epitaph (6:59, L. 8 "*sancta iobelei tempora cum celebret*") and a posthumous monument for Sixtus IV (6:89 from 1494, L. 5 "*iubileo celebrato*"); the two other peculiar cases with Iulianus de Dathis will be discussed below. In addition, one inscribed church bell came to my knowledge only later and therefore has not been included in the material. It is on display in Ravenna, in the Museo Archivescovile (inv. no. 32) and it is stated there to have been "produced on the occasion of the first Jubilee of 1300 donated by Pope Boniface VIII to the Primaro Hospice located north of Ravenna." If this holds true, the bell may well originate from Rome. The inscription "*an(n)o D(omi)ni MCCC t(empor)e d(omi)ni Bonifacii p(a)p(ae) o(c)t(avi)*" reads from right to left, presumably in error.

93 Neither are there any known funerary inscriptions from 1423.

94 Cf. e.g. Arnold Esch, "L'economia nei giubilei del quattrocento," in *I giubilei nella storia della Chiesa. Atti del Congresso internazionale in collaborazione con l'École française de Rome sotto il patrocinio del Comitato centrale per il giubileo del 2000 (Roma, Istituto patristico Augustinianum, 23–26 giugno 1999)* (Città del Vaticano: Libreria Editrice Vaticana, 2001), 342–345 with further references.

95 1:4 (LL. 2–22) in verse L. 2 "*mille trece(n)tenis D(omi)ni currentibus annis,*" L. 16 "*maiori pena p(o)p(u)li pietate remissa*"; 1:5 "*ann(o) D(omi)ni MCCC, quo Rome fuit indulge(n)tia omniu(m) peccato(rum)*"; 6:17, which is the papal bull initiating the holy year.

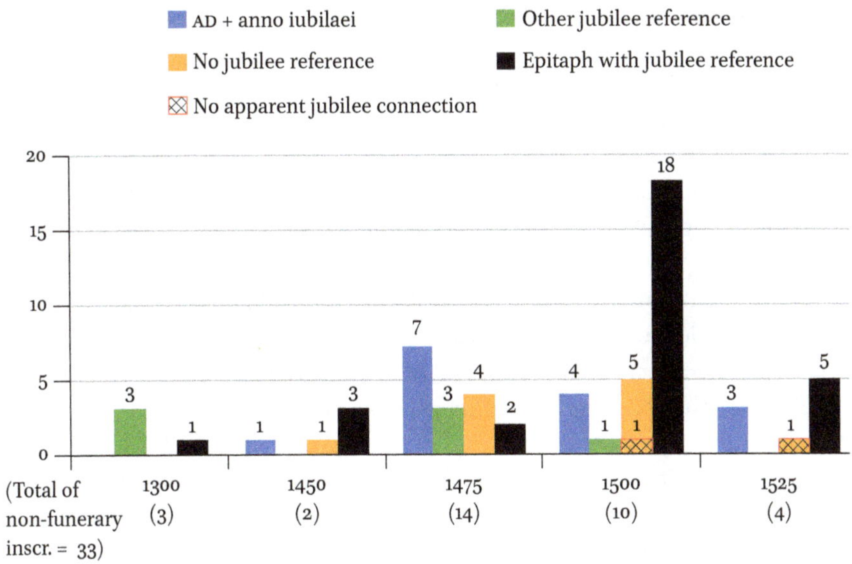

GRAPH 4.10 Jubilee references 1300–1525.

word, but the monument was likely only erected in the following year.[96] This accords with the fact that, in other contexts, the "jubilee" seems to have been employed in this sense only after the 1300 itself.[97]

From the later four "other jubilee references," all building inscriptions, three have the expression "*anno iubilaei*": one without other dating methods, and the other two mentioning a pope and another office-holder, who appears in the latter one in a temporal reference.[98] In the fourth inscription, commemorating the construction of the Ponte Sisto, no year is given but the place in time is evident from the expression "*Sixtus IIII pont(ifex) max(imus) | ad utilitatem p(opuli) Ro(mani) peregrinar(um)que multi|tudinis ad iubileum venturae pontem | (…) resti|tuit | (…).*"[99]

The 1475 jubilee of Sixtus IV is referred to more frequently in the building inscriptions than the other jubilees, which tells of the importance given to the preparations for this holy year. Regardless, the jubilee itself would gain only

96 12:569 LL. 5–7 "*anni transieru|nt tunc Christi mille tr|ecenti / anni vir fidei petiit veniam iubilei.*" According to Garms et al., *Mittelalterliche Grabmäler*, 96–97 no. XXI, 2, the man actually died on 9 January 1301.

97 See Jasmin Lukkari's article for the other sources. This applies to the church bell above in note 92 as well.

98 In the given order: 9:739 from 1500; 8:738 and 12:442 from 1475.

99 13:92.

modest success.[100] It is still noteworthy that four inscriptions out of 14 do not show any visible reference to the jubilee, even if Sixtus IV himself is mentioned, but here again the context seems to be decisive: two are indulgence concessions;[101] one relates to the construction of a bell tower connected to S. Croce in Gerusalemme, but only consists of the insignia of the responsible cardinal, Domenico Capranica, followed by "*MCCCCLXXV Alexandri Tiburt(ini),*" the name being that of the artisan;[102] and the remaining example is composed in verse.

The two non-funerary 1450 inscriptions follow the same pattern, as the one without a jubilee reference is again a metric composition. On the contrary, the four non-funerary cases of 1500 are all in prosaic building inscriptions, which implies that this year the jubilee marketing was not as pervasive as in 1475. The 1500 jubilee year produced other exceptions, too: in a dedicatory inscription of S. Pietro in Montorio, with a list of relics (13:1142), the jubilee has been included in the dating formula: "*anno salutis (Christ)ianae MD, sub iubilaeo et pontificatu | Alexandri VI, die IX Iunii, feria tertia post pentecostes.*" Furthermore, that year the number of epitaphs with jubilee references considerably surpass the respective numbers of the neighbouring jubilees, but the reason for this remains unclear to me.

Yet another interesting case is the one of Iulianus de Dathis, a clergyman who was the titular bishop of S. Leone (in Calabria) from 1518 until his death in 1524; he was buried in SS. Silvestro e Dorotea, in Trastevere.[103] For the holy year of 1500, he had two inscribed objects put up in the church mentioned, in which he was *poenitentiarus* and *rector:* a monument that recorded an angelic epiphany;[104] and a cippus serving as a base, at some point later, for a vessel of holy water.[105] Both inscriptions articulate the connection to the 1500 jubilee. This applies to two further monuments in the Lateran as well, that are

100 Esch, *L'economia nei Giubilei*, 348–358.

101 8:42, and 10:542 which is a vernacular translation(?).

102 8:500, identification of the persons by Forcella. For the cardinal, see his donation to the Ospedale del SS. Salvatore al Laterano: 8:432.

103 His epitaph is 9:743, now lost. Giovanna Curcio and Paola Farenga, "Dati, Giuliano," in *Dizionario Biografico degli Italiani* 33 (1987), 33 infer from the expression LL. 8–9 "*qui obiit IV kal(endas) Ianuar(ii) | MDXXIV*" that he died in 1523 (followed by Claussen and Senekovic, cf. below), but the year should be seen rather as separate from the day/month expression: 29 December 1524. For the record, Giuliano appears, in addition to the five inscriptions discussed here, in a sixth one not connected to a jubilee: a 1511 dedication of church and altars in S. Lucia in Gonfalone, 7:860.

104 9:740. According to Forcella (*ibid.* and the double publication 2:1057), this "apocryphal inscription" was transferred to S. Maria in Trastevere in 1737, and later removed.

105 9:739. Senekovic, *KSRM* 2, 387 note 150: "*diente zuerst vielleicht als Reliquienbehälter, später vermutlich als Weihwasserspender.*" Cf. also Forcella's comment at 9:739.

commemorations of the dead erected by the same man.[106] The latter two, however, share another interesting feature: in a contradictory manner to the given year dates, Iulianus appears as the bishop, which he became only in 1518. As other scholars have noted, this was intentional from Iulianus, who regarded the jubilee to be everlasting in the Lateran[107] – with this he seems to have conceived, as is implied by the epigraphic formulations, that the particular year 1500 lasted until the next holy year.

2.2 *Secondary Temporal References*

Among the inscriptions in category C, which show only a vague temporal reference, the most common method for doing this is by mentioning a pope. Here, the pope does not appear in a time-related context (marked with an expression "*tempore papae*" or similar), and neither is there any other clear method of defining a year or a longer period of time. However, these render themselves relevant through the fact that popes and papal years often appeared in dating, and were thus easily connected with the conceptualisation of time. Thus, it is likely that for anyone who had any knowledge of formal dating practices – probably most of those who were capable of reading an inscription – mentioning a (more or less contemporary) pope would, in a secondary function, immediately give a certain temporal framework for the main information.

The majority of these, 51 out of 60, are building inscriptions where, in many cases, the inscription consists of no more than the pope's name (often with his insignia). The remainder are: four 10th–11th century documentary inscriptions;[108] a poem on a pulpit;[109] two honorific monuments for Renaissance popes;[110] and three other commemorative inscriptions.[111] In one metric inscription we have a comparable imperial case on the Milanese *carroccio*,

106 One commemorates several deceased relatives of Iulianus, with "*anno iubilationis et | salutis MCCCCC*" (Claussen, *KSRM* 2, 318 with fig. 214), and the other was intended to be the grave monument for himself and some relatives, "*anno salutis et iubilei MD*" (Senekovic, *KSRM* 2, 387 note 145). Evidently, he was never buried in this monument (cf. 9:743 above) which was later lost, perhaps due to a falling out of papal favour after Leo X (Senekovic, *ibid.*).

107 Claussen, *KSRM* 2, 318 note 1324: "*die Datierung auf das Jubeljahr 1500 [muss] als eine gewollte Fiktion angesehen werden*"; he refers to Farenga and Curcio, *Dati, Giuliano*, 34 who take the dates on these inscriptions for "*omaggio al fatto che 'il giubileo è sempre in Santo Ianni,' come egli stesso [Giuliano Dati] scrive a nuova esplicita affermazione del suo atteggiamento verso la storia.*" Cf. also Senekovic, *KSRM* 2, 387.

108 2:5; De Rossi, *BCAM* 1872, 54; Favreau, *Épigraphie médiévale*, 33–39 no. 1; and Marini and Mai, *Inscriptiones Christianae*, 215 no. 1.

109 6:1060 (lost).

110 10:570 (lost), and Kajanto, *Papal Epigraphy*, 98–99.

111 8:18 (lost), 11:312, 11:313 (lost).

which Emperor Frederick II (1220–1250) took as a spoil and put on the Capitoline hill.[112]

One building inscription with a mention of a pope illustrates this usage further. In this lost record of the restoration of a chapel in the old Vatican Basilica by an Orsini lady (6:75), the previous history of the chapel is constructed with references first to Cardinal Napoleone Orsini Frangipani (1263–1342) and three other Orsini men, and then to Pope Pius II (1458–1464): "*Ursina d(omi)ni Iacobi de Ursinis cappellam hanc olim* | *in medio huius basilicae Martiali martiri dicatam per Neapolionem* | *r(everendissimum) d(iaconum) card(ina)lem* | *s(an)c(t)i Adriani et Matthaeum, Ursum Rai-*|$_5$*naldumq(ue) fr(atr)es de Ursinis, postmodum Pii* | *II su m(m)i pontificis iussu dirutam* (…)." In this deletion of the chapel, Pius II would certainly have served as a temporal landmark to a contemporary reader, since the inscription dates from 1472. Furthermore, the inscription makes us look at non-papal figures used in a similar manner, with examples of both a cardinal and relatives of an earlier generation.

Except for the previous case, 128 inscriptions of category C mention other contemporary persons other than a pope (or an emperor), of whom a similar use may be considered. Even though one may put one's own name on an object of a more permanent nature, with the intention of preserving the name for the future, this must not be taken automatically as an intended temporal reference. However, the succession of certain figures of an ecclesiastic or civic nature, such as cardinals or magistrates, or members of noble families, was a matter of interest to their respective communities and, accordingly, was recorded in one way or another for future reference – even if epigraphy does not seem to have been the medium for recording this in the area of Rome in our period.[113] Accordingly, for example, the inscription on the back of a cathedra in S. Clemente from around 1104–1125, "*Anastasius presbiter cardinalis huius tituli hoc opus cepit et perfecit,*"[114] helped at least the parochial sphere of Anastasius with a temporal landmark for the object in question.

112 1:1 "*Cesaris Augusti Friderici Roma(norum) secundi / dona tene currum p(er)pes in urbe decus.* | *hic Mediolani captus de strage triumphos / Cesaris, ut referat inclita preda, venit,* | *hostis in oppbrium(!) pendebit, in urbis honorem / mictitur, hunc urbis mictere iussit amor.*"

113 No Latin inscriptions of this kind are known. However, one Greek inscription (Guillou, *Recueil*, no. 112; date 1131/2) from the Greek monastery of Grottaferrata, south-east of Rome, constructs the past of the monastery with an intentional record of its leaders (ἡγούμενοι) from the founder up to the contemporary abbot. Conveniently, both ends are fixed with a year date, but for the abbots in between no years are recorded, thus only helping the reader with a relative chronology for them.

114 Claussen, *KSRM* 1, 303 note 17.

Apart from such high-ranking figures as cardinals, it is nevertheless a more cumbersome task to survey the possible individuals who may have served as implicit temporal anchors. Sculptors and other craftsmen, who often left their works signed,[115] form a significant group (36 inscriptions). It is difficult to assess how well-known these artisans were,[116] and it is far beyond the scope of this study to attempt to discern the ones illustrious enough – or those whose fame would have endured over time. Additionally, signatures were sometimes placed on spots of restricted visibility.

Six other inscriptions without a date mention men with non-ecclesiastic titles,[117] and in one further case without a title there is a famous *condottiero* Everso II degli Anguillara with his insignia.[118] These persons were probably generally known in their time, but little may be inferred from these scarce attestations. The remaining 31 cases, mainly dedications and building inscriptions, present one or more persons without a title; mostly men, but once a woman alone, and four times alongside a man. Some are clearly identifiable as members of the Roman nobility, such as in a vernacular dedication of a golden chalice in the Vatican: "*ques|to chali|cie fecie fare | Madonna |$_5$ Orsina | figlia | del magnifich|o signiore Iac|hopo Or|$_{10}$sino.*"[119] The temporal function is difficult to assess in these cases, even more so with persons not known from elsewhere.

2.3 *Day and Month in Dating*

After a lengthy discussion on long periods, it is time to look at the day/month-level. A year-level date emanates from the purpose to locate an inscription in a chronological framework,[120] and as such it occurs in various contexts: in a

115 Claussen, *Magistri doctissimi*, 1.

116 Claussen, *Magistri doctissimi*, 238–239.

117 12:6 (S. Paolo fuori le mura, 1070; metric) *consuli (A)malfigeno pro Pantaleone*; 13:89 (Ponte Cestio, 1193) *Benedictus almae urbis summ(us) senator*; Claussen, *KRSM* 1, 139–140 (S. Bartolomeo all'Isola, 12th century; metric) *Cencius (...) cancellarius Urbis*; Silvagni, *MEC* 1, tab. 40 no. 4 (S. Maria in Trivio, 12th century; metric) *patricius Vilisarius urbis amicus*; 8:339 (Ospedale del SS. Salvatore al Laterano, 1348) *Franciscus Vecchi not(arius) di Parione*; 9:245 (Ospedale di S. Giacomo, 1435) *Franciscus de Ursinis urbis prefectus*; 5:223 (S. Agostino, 16th century imprecise) *Marliani instauratori(!) Urbis s(anctae)*.

118 8:347(1) (Ospedale del SS. Salvatore al Laterano; after 1462) *Ever(s)o secundo* (degli Anguillara, died in 1464; cf. [unknown], "Anguillara, Everso," in *Dizionario Biografico degli Italiani* 3 (1961), 303).

119 6:76. Forcella does not inform us about its location but seems to consider it extant.

120 The need or desire to do so may, of course, have various reasons, ranging from juridical needs to historical self-consciousness, i.e. the desire to make one's name last.

simple building inscription, or in an artisan's signature, for instance. But more precise dates find a more restricted use in our material, because they are mostly recorded for one of two reasons: first, because of juridical requirements belonging to the practice of the papal chancery, or to testamentary or donatory matters; or second, for a commemorative purpose, where the calendar date was usually relevant for celebrating the commemorated object in the future.

The day/month indications divide into two categories: absolute and context-related datings. The absolute dating will be discussed first, and it follows one of three principles: either the ancient Roman dates oriented by *kalendae*, *nonae* and *idus* of a month; liturgical dates expressed with a Christian feast or *feria*; and "running" dates, that is with a current numbering given for days of a month, starting from one. The choice between Roman and running dates is primarily a stylistic one, even though the Roman date was prone to appear only in certain contexts in the late medieval period. But articulating, for example, "*festivit(ate) s(an)c(t)or(um) Gord(iani) et Epimachi*"[121] brings a distinctly Christian tone to the temporal construction. Thus, attention will be drawn to the use of Roman and modern datings in respect to each other, and to the circumstances where liturgical dating is used. For the record, days of the week are mentioned almost exclusively in liturgical dating as Sundays or *feriae*, whereas other "normal" names of days do not occur in the non-funerary material before the 16th century, and even then never referring to a particular date but to recurring days.[122]

The following graphs 4.11 and 4.12 show the general picture of their distribution: first, how diverse systems occur in general (sometimes more than one in the same inscription); and second, which one of the absolute datings was used uniformly in single inscriptions, or whether the usage was mixed. In the mixed cases, the day indications do not necessarily belong to the same dating formula and may refer to different days. Again, the somewhat meagre figures are insufficient for a statistically relevant analysis. However, some tendencies are discernible, in particular between medieval and Renaissance usage.

121 Blennow, *Latin Consecrative Inscriptions*, 120–123 no. 26, following "*men(se) Ma|dio, die* x."

122 The first precisely datable inscription, 8:363 from Ospedale del SS. Salvatore al Laterano from 1528, is slightly beyond the present time limit: LL. 4–6 "*cum onere singulis diebus Sabbati celebrandi unam missam.*" In addition, the three vaguely dated 16th century inscriptions are of little relevance here (7:245, 8:1230, 9:701). For the few funerary occurrences, see Kajanto, "Dating," 55–56. The Sundays and *feriae* are, of course, significant in indicating a single day in the liturgical dating – unlike the "normal" day names.

The general observations are as follows. Roman dating is very rarely used in epigraphy before the late 11th century,[123] and the first Roman date emerges between approximately 1073–1085, in an epigram in S. Pudenziana (11:262): L. 6 "*Augusti mensis septeno nempe kalendis,*" which makes use of *septeno* with *kalendis* in ablative instead of the classical accusative. Throughout our period of study, running datings retain their popularity, and in each examined period it is used in a uniform manner in the majority of cases. However, Roman and liturgical datings are attested during most periods – often together after the 13th century. In the 12th to 14th centuries, liturgical-only dating appears several times,[124] with its heyday in the 13th century; in the late 15th century, there is only one liturgical-only case, with nine others using mixed datings, eight of these with Roman dates.

For the figures in general, a comparison with funerary inscriptions presented by Iiro Kajanto[125] is justified here, although only after 1450 do both funerary and non-funerary material present easily comparable figures. The Roman type dating dropped in both categories between 1300 and 1450: it is found in ten out of 246 epitaphs – with nothing but one papal grave from 1350–1400[126] – and, in other genres, the four cases are confined to papal and liturgical contexts. However, the classicising Renaissance brought it back into epitaphs with a remarkable increase to 30 % of cases by 1450–1500, and up to 40 % by 1500–1527. In the group of non-funerary inscriptions, the changes in proportions in the same period are relatively small: the 11 with Roman-only dates from 1450–1527 are found in various genres, four of them flood inscriptions, whereas the mixed "Roman and liturgical" cases are almost exclusively of papal authorship

123 This applies to three cases known to me from the 10th century: Giovenale, *S. Maria in Cosmedin*, 72 no. 69 (902), Marini and Mai, *Inscriptiones Christianae*, 17 no. 2, and Silvagni, *MEC* 1, tab. 17 no. 2 (both 984).

124 Among the uncertain dates, a liturgical double indication is found in a forged inscription from S. Matteo in Merulana (Blennow, *Latin Consecrative Inscriptions*, 204–210): "*feria{e} sexta{e} 4 dominicae in quadragesima, die incarnationis eiusdem Domini nostri Iesu Christi*" (transl. Blennow: "in the sixth *feria* (Friday) after the fourth Sunday in Lent, in the day of the incarnation of our same Lord Jesus Christ"). This only seems to make sense if it refers to the Annunciation of the Lord, on 25 March.

125 Kajanto, "Dating," 48–49.

126 6:32 is the only grave monument from this period with Roman dating, but as it is for Pope Urban VI, Kajanto duly attributes the choice to the papal context. However, I do not comprehend why he has included 8:31 in the statistics as the other instance of Roman dating in the period in question, since it is not an epitaph but a papal bull: Kajanto, "Dating," 49.

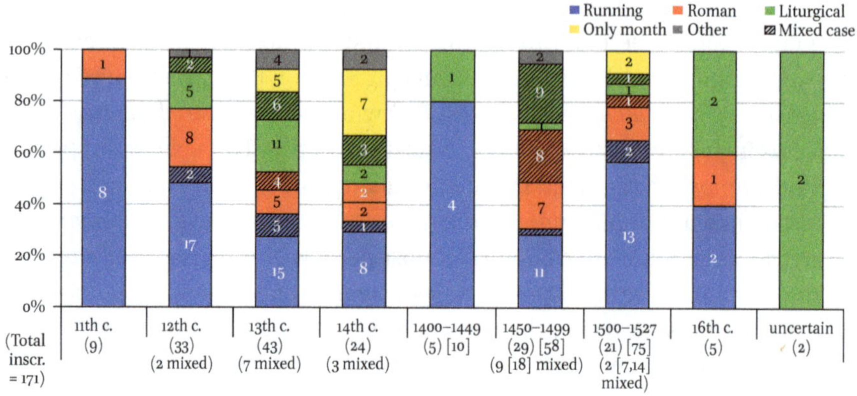

GRAPH 4.11 Occurrences of different day / month indications (one or more per inscription).

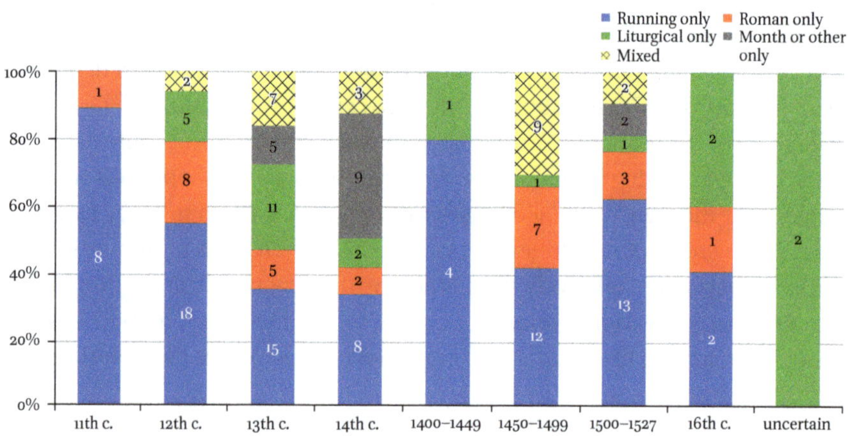

GRAPH 4.12 Uniformity of day / month indications in single inscriptions.

and concern concessions of indulgences.[127] The rise remained modest in the later 15th century – and diminished in the years 1500–1527, in contrast to epitaphs where its proportion kept growing. Regarding liturgical dating, Kajanto leaves it out of his statistical analysis, because it occurs in epitaphs only 12 times.[128]

Putting the liturgical dating method aside for now, the main point of interest in the distribution between Roman and running dating types lies within

127 All from the latter half of the 15th century, except for the double-dated 1514 flood inscription 13:436.

128 Kajanto, "Dating," 58–59: nine from between 1286–1428, and three from 1504–1527.

their different contexts. The general pattern is that running dating was used in most contexts, whereas Roman dating was confined to specific genres in the medieval non-funerary epigraphy and spread only modestly into other genres in the Renaissance period. In fact, there are roughly speaking only three genres where Roman medieval epigraphy made use of ancient Roman dating: church-related consecrations (11); epigraphs from papal documents (6); and inscriptions related to testaments or donations (3) and to altar renovation and relics (1);[129] in addition, there is one other exceptional case.[130] As popes were involved in many consecrations, papal presence seems to be indicated where Roman dating occurs. In the Renaissance period, Roman datings continued to be applied in papal epigraphs (8), but otherwise only in dedicatory or building inscriptions (7), in flood inscriptions (3), and in one other case.[131]

The question follows, then, whether running dating was also used in these inscription types or not. The papal chancery clung persistently to Roman dating, which is reflected in the lapidary versions or adaptations of its documents; the two exceptions to this, both late, are a brief concession record from 1494, which commemorates the concession but does not refer the decision, and a resume of a bull from 1521.[132] One 1518 copy of a 1394 indulgence concession has Roman, liturgical, and other day indications, which are somewhat unclear.[133] Apart from these, a running date is found in an inscription that records

129 Consecrations: 2:645, 11:262, 12:445; Blennow, *Latin Consecrative Inscriptions*, 43–48 no. 7, 46 note 95, 48–53 no. 8, 62–64 no. 12, 76–84 no. 16, 110–112 no. 23, 171–178 no. 37 (missed by Blennow in the table on p. 257). Papal epigraphs: 3:721, 6:17, 8:31, 11:375; Blennow 210–215(2); Pietro Luigi Galletti, *Inscriptiones Romanae infimi aevi Romae exstantes* (Romae: Typis J.G. Salomonj, 1760), 450–452 no. 2(1). Testaments or donations: 4:2, 6:5, 11:597. Finding relics in an altar, and renovation of the altar: 8:717(2) (cf. below note 154).

130 7:398, which has both running and Roman datings, and will be discussed below.

131 Papal epigraphs: 1:1196–1197, 1:1629, 3:507, 4:821, 7:398, 8:42, 8:780. Dedication or building: 1:41, 2:82, 2:581, 8:504(2), 12:245, 13:428, Kajanto, *Papal Epigraphy*, 85. Flood: 2:1080, 13:432–433. Papal commemoration of relic translation, with concession of indulgence 11:44.

132 8:352 Ospedale del SS. Salvatore in Laterano; 8:785 Ospedale della Consolazione. See also 10:542 (below note 134).

133 Blennow, *Latin Consecrative Inscriptions*, 210–215(2): LL. 14–16 "*i(n) die An(n)u(ntia)tionis | eiusd(em) de me(n)se Ma(r)tii et p(r)ima d(omi)nica Maii | co(n)cessit dictis duob(us), dieb(us) et q(u)olib(et) d(i)c(t)or(um) | duor(um) dier(um) illam et ea(n)de(m) i(n)dul(gen)t(ia)m*"; LL. 16–17 "*visita(n)tib(us)* (…) *illis VIII dieb(us) m(ensis) Aug(usti)*; LL. 17–18 *dat(um)* | (…) *XIIII k(a)l(end)as Nove(m)b(res)*." As far as I understand, the eight days of August have not been specified; the preceding part (1) of earlier origin is dated "*men(se) Aug(usto), d(ie) IX*," but I do not see any clear connection between these.

in the vernacular Sixtus IV's visit and indulgence concession to nuns at SS. Cosma e Damiano in Trastevere.[134]

Among other genres, the running date is considerably more common in consecrative inscriptions,[135] and in the remainder it surpasses the use of the Roman date by far. Furthermore, from the three cases with both running and Roman dating it can be inferred that the choices made show a similar pattern: one consecrative inscription from 1228 records a discovery of relics with a running date, and a subsequent church consecration uses Roman and liturgical dates.[136] Here the more traditional and Christian methods are implemented only with a sacred act, whereas the more casual running method suffices for dating a relic discovery. Another inscription has two separate records: that of *frater* Iohannes founding the Order of the Most Holy Trinity and of the Captives under the auspices of Innocent III, with a Roman date, and that of the burial of the same Iohannes with a running date.[137] A third case is in the already discussed flood inscription from 1514, where the same day is referred to with a Roman date in verse, and with a running date in prose.[138] In all three cases, the Roman dating is consistently reserved for the part invested with more prestige.

Let us return then to liturgical dating, which presents some interesting features. Apart from the funerary occurrences, its use was mostly limited to fundamentally ecclesiastic contexts: consecrative inscriptions constitute the majority; some late medieval donation records follow next; with mixed Roman and liturgical datings found only in inscriptions deriving from the pope, mostly those from the Renaissance period, mentioned above. Noticeably, in some cases, the liturgical date pinpoints the time when an indulgence is to be available, or when an action ordered in the donation is to take place; if the inscription itself is dated, running or Roman dating is used for that.[139] In one further case,

134 10:542, from 1475.

135 For the majority of the cases (17), see Blennow, *Latin Consecrative Inscriptions*, 256–257. The rest are: 10:540 + Guillou, *Recueil*, no. 117 (SS. Cosma e Damiano in Trastevere, 1065), 11:321 (SS. Salvatore in Primicerio, 1113), 2:1334 (S. Maria in Aquiro, 1295, lost), 2:1186 (S. Eustachio, 1336), 5:1120 (S. Lorenzo in Panisperna, 1383), 7:860 (S. Lucia del Gonfalone, 1511), 13:1142 (S. Pietro in Montorio, 1500).

136 Blennow, *Latin Consecrative Inscriptions*, 145–154 no. 32, L. 2 "*m(en)se Ianuar(io), d(ie) XVIII*; LL. 16–17 *XIII kalendas Ap(ri)|lis, in secunda feria post palmas.*" Note that the line divisions of this lost inscription are uncertain.

137 7:398, S. Tommaso in Formis, now lost. Forcella gives the year 1213 when Iohannes is said to have died, although the inscription may have been put up later as well.

138 13:436, above pp. 117 and 123.

139 These are, of course, mostly papal documents and donation records – and a single consecrative inscription 12:445, which however is in verse, and 1:60 discussed below. Note also

the actual dating formula of a consecration gives the day only in the running form, but a separate sentence at the end of the inscription states that it was the feast of St Thomas.[140]

Even the few notable exceptions all seem to have a contextual reason for having a liturgical date. An undated 16th century inscribed regulation in the Capitoline museums addresses a non-ecclesiastic public, but it concerns a religious procession.[141] An inscription on a column in S. Giacomo degli Spagnoli from 1503, now lost, brings St Jacob to the centre of the events: "*ista clavis est unius duarum | portarum Barbaricae civitati | Tripoli nu(n)cupatae, quam d(ie) | in festo s(ancti) Iacobi ab infedelib(us)* |$_{5}$ *in manibus Ferdinandi | fidelis Hisp(ani) cath(olici) regis dedit | et propterea hinc inde ex voto | debito posita fuit MDIII, | tempore Iulii p(a)p(ae) II.*"[142] Especially in this predominantly Hispanic context, the expression *ex voto debito* strongly suggests that the victory itself was attributed to the patron saint of Spain. Accordingly, the victorious day needed no other definition than being his feast.

In a comparable manner, another rare piece from 1327 from the Porta S. Sebastiano commemorates the victory of the Ghibelline faction over invading Guelph troops: "*anno D(omi)ni MCCC |XXVII, indictione | XI, mense "Septem|bris(!) die penultim*|$_{5}$*a, in festo s(an)c(t)i Micha|elis, intravit gens | foresteria in urb|e(!) et fuit debella|ta a populo Roma*|$_{10}$*no qui stante Ia|cobo de Pontia|nis capite regio|nis,*" with a depiction of the triumphant archangel next to the inscription.[143] This double reference to the celestial commander-in-chief testifies to his importance for the victory on the given day which, however, is indicated with the running method as well. Lastly, and again describing exceptional circumstances, two flood inscriptions from 1422 and 1495 bear a liturgical date, where its use perhaps interconnects with prayers directed to the respective

---

12:1495, a combination of epitaph and testamentary donation record, both of which refer to a liturgical date.

140 Blennow, *Latin Consecrative Inscriptions*, 100–104 no. 21 (S. Tommaso in Parione, 1139): L. 2 "*m(ense) Dec(embri), d(ie) XXI*"; LL. 22–30 "*h(aec) c(on)secra|tio in ei(us)dem | apostolice* |$_{25}$*festivi|tatis di|e fac|ta | es*|$_{30}$*t.*"

141 1:60. The regulation is concerned with the marching order of different lay officials and professions in the procession "*Dei genitricis Virginis festo | die, dum Christi salvatoris nostri mi|rabile simulacr(um) ex Laterano in Ex|quilias*(!) *ad Mariae Matris Maiorem aedem* | (...) *defertur*" (LL. 4–7, 9), and records a penalty for deviating conduct. Forcella does not mention any origin for the inscription, but a non-ecclesiastic authorship is suggested by the present location on the Capitoline hill, and by the fact that the clergy are absent from the discussion.

142 3:538.

143 13:3. Cf. Forcella's commentary 26–28.

saints to intervene against the high waters.[144] Nevertheless, neither the flood inscriptions nor commemorations of victory needed such a reference, as is proven by the majority of flood inscriptions – and by a graffito in Villa Lante al Gianicolo from the very end of the period discussed: "*a dì 6 de magio 1527 | fo la presa de Roma,*" that is deprived of any religious attributes.[145]

To sum up, there were three methods to produce a fixed date in an inscription, out of which running dating was used in almost every context; Christian feasts are used in prevalently ecclesiastic contexts, or indicate a special relationship to the saint of the day; and Roman dating, which had a certain traditional and prestigious quality about it, was mostly confined to the papal chancery – and was later revived in Renaissance funerary use. Perhaps in composing non-funerary epigraphic prose, emphasis was usually put on clarity, which was more readily achieved with running dating as the primary day/month style – something apparent in the Renaissance era as well. In contrast, in the epitaphs of the same period, style seems to have overshadowed practical considerations: the classical Roman date, and variations in language became fashionable alongside the humanist style. Moreover, imitation was greatly eased in this particular genre by the fact that it was easily comparable to classical models, which were already known in considerable numbers by the latter half of the 15th century.[146]

All this still leaves us with the 14 cases which record only a month but do not specify a day. This diverges from precisely-given dates in function: it only has a chronological purpose, supplementing a year date, but does not serve a juridical aim, nor provide a date for recurring commemoration. These sorts of inscriptions, mostly from the 13th and 14th centuries, form fairly distinguishable groups: firstly, six vernacular inscriptions in the Lateran from between 1345–1365, erected for the soul of a relative;[147] second, a 1353 epigram in the Vatican,

---

144 13:426, fixed onto the outer wall of S. Maria sopra Minerva: LL. 1–2 "*in die s(an)c(t)i | Andree.*" 13:431, Ospedale di S. Spirito, now lost: LL. 4–5 "*in festo s(an)c(t)e Barbare, IIIIo | mensis Decembris.*"

145 Tancredi Carunchio, "La Villa Lante al Gianicolo: L'architettura e i suoi restauri," in *Villa Lante al Gianicolo: storia della Fabbrica e cronaca degli abitatori*, eds, Tancredi Carunchio and Simo Örmä (Roma: Palombi editori / Institutum Romanum Finlandiae, 2005), 25.

146 Of course, building and honorific inscriptions, for instance, could draw from classical models, but many other inscription genres, especially those based largely on documents, were less flexible in this respect.

147 8:25, 8:27, 8:28, 8:29; Claussen, *KSRM* 2, 174 note 731(2); Paolo D'Achille, "Iscrizioni votive e sepolcrali in volgare dei secoli XIV–XVI," in *Il volgare nelle chiese di Roma: messaggi graffiti, dipinti e incisi dal IX al XVI secolo*, eds, Francesco Sabatini, Sergio Raffaelli, and Paolo D'Achille (Roma: Bonacci, 1987), 76 note 21. The last one is similar to the first inscription for the most, but left unfinished.

related to a bell or a bell-tower;[148] third, five consecrative or building inscriptions from between 1222–1297;[149] and fourth, the only two Renaissance period inscriptions, that happen to be from March 1518.[150] Consecrative inscriptions usually have a full date, which makes these few cases somewhat odd, particularly the example where the dating formula of the altar consecration includes only the month, "*me(n)se Martii.*" However, further below it is mentioned that relics were placed in the altar "*VIII k(a)l(endas) April(is).*"[151] In most cases, the consecration was presented as the important event that could be celebrated in anniversaries, not the placement of relics. Lastly, the omission of a day indication in the two 1518 inscriptions remains likewise inexplicable: one records the visit of Pope Leo X and a concession of an indulgence in S. Silvestro al Quirinale, but seems not to be fully modelled upon a papal document;[152] and the other is a building inscription of an altar in S. Annunziata all'Arco dei Pantani. In the former, the indulgence is to be acquired "*eodem die,*" but then the day has not been written out in full, and in the genre of the latter inscription, a full date or only the year could be expected. I cannot understand the reasons behind these half-way choices; in the first one, perhaps the day was omitted erroneously.[153]

2.4 *Indications of Specific Days with Relative or Unique Points of Reference*

The previously discussed day/month dating methods are fixed in a recurring calendrical year cycle, but dates could be related to unique events, too. These dates are mostly[154] found in inscriptions that mention a recurring commemoration of two kinds: 13 inscriptions of this sort related to the consecration of an altar or a church;[155] and 12 such inscriptions related to deceased (or whose

148 6:26.

149 1:3, 1:1175, 8:17, 10:901; Blennow, *Latin Consecrative Inscriptions*, 171–178 no. 37.

150 4:89; Blennow, *Latin Consecrative Inscriptions*, 210–15(3).

151 Blennow, *Latin Consecrative Inscriptions*, 171–178 no. 37, LL. 1–2 and 12.

152 4:89: "*ad posteri(tatis) memoriam sempiterna(m).* | *Leo X pont(ifex) max(imus), cum aedem hanc comitatus* | *unde triginta cardinal(-) subisset, cuicumq(ue)* | *imaginem hanc deiparae Virginis quo|5tiescumq(ue) etiamsi eodem die saepius* | *id fiat, visitanti quingentor(um) dier(um)* | *poenam condonavit, anno sal(utis) M|DXVIII, mense Martio.*"

153 However, this lost inscription has an extant later lapidary version, which does not indicate any date either: see Forcella at 4:89.

154 A relative date introducing an action posterior to a fixed date is possible, as in 8:717(2) (SS. Quattro Coronati, 1111), which records Pope Paschal II ordering relics to be excavated from below a broken altar on the Ides of June, and then *in tertia vero die* (L. 16) the altar to be restored.

155 12th century: Blennow, *Latin Consecrative Inscriptions*, 112–116 no. 24 (S. Maria in Aquiro; lost), 127–134 no. 28 (S. Eustachio); 13th century: 3:721 (S. Salvatore in Pensilis, 1285), 4:1144 (S. Francesco delle Stimmate), 11:375 (S. Lucia della Tinta), Blennow, *ibid.* 178–182 no. 38

passing away is expected) persons;[156] the latter often include a reference to a will. The recurring commemoration is usually annual, thus called *anniversarium*, and in the former cases it goes hand-in-hand with the concession of an indulgence for visitors, whereas in the latter cases it is a mass in memory, and for the soul of the deceased person, who had either donated it or someone else who had done so on their behalf.

Nevertheless, the consecrative inscriptions offer rather little to the present discussion, apart from the fact that they were directed particularly at the future, to serve as a visible reminder of the commemoration. The lapidary medium did indeed set the indulgence concessions authoritatively in stone, often in detail, where they could help to ascertain the correct conduct for devout (literate) pilgrims. But this was scarcely the primary means of bringing the visiting dates to general notice.

The situation is somewhat similar but slightly more interesting with personal commemorations, which seem to usually have two functions. Both are illustrated by the two lapidary donation records of Vannozza dei Cattanei (1442–1518), mistress of Rodrigo Borgia, the future Pope Alexander VI. The first function of these inscriptions lies rather in the public elevation of the donor than in serving as a practical reminder of celebrating the anniversary. The one in the Ospedale del SS. Salvatore in Laterano, put up posthumously in 1520, instructs three anniversaries to be celebrated annually for the donation – without further specifications.[157] These are explained only through the other inscription, from S. Maria della Consolazione, in which the anniversaries are said to belong to Vannozza and her first two husbands.[158] However, the

(S. Sabina), 184–189 no. 40 (S. Agnese fuori le Mura), 190–194 no. 41(1) (S. Costanza), 226–228 (SS. Marcellino e Pietro); 14th century: 2:1186 (S. Eustachio); early 16th century: 7:860 (S. Lucia del Gonfalone); uncertain date: Blennow, *ibid.* 204–210 (S. Matteo in Merulana; lost), 220–221 (S. Bibiana; lost).

156 11th century: Marini and Mai, *Inscriptiones Christianae*, 219 no. 1 (S. Nicola in Carcere); 13th century: 2:1494–1495 (S. Prassede), 4:2 (SS. Silvestro e Martino ai Monti; lost), 6:1061 (S. Giacomo alla Lungara; lost), Claussen, *KSRM* 3, 16 note 16 (S. Giorgio in Velabro); 14th century: 11:18 (S. Maria Maggiore), 12:457 (SS. Salvatore a Porta Paolo); late 15th century: 8:42 (S. Giovanni in Laterano), 8:363 (Ospedale del SS. Salvatore al Laterano; lost); early 16th century: 8:358 (Ospedale del SS. Salvatore al Laterano; lost), 8:1206 (S. Maria della Consolazione; vernacular); see also the ones of a vague 16th century date: 7:245 (SS. Venanzio ed Ansovino), 8:1231–1233 (S. Maria della Consolazione; all three vernacular).

157 8:358, LL. 6–8: "(…) *cum onere celebrandi tria | anniversaria singulis annis. | die I Aug(usti), MDXX.*" The last line records the date of the inscription, unrelated to the anniversaries (cf. the next footnote).

158 8:1206, LL. 6–17: "(…) *con pes(o) | di far celebrari 3 an(n)i|versarii l'anno: uno p(er) l'|anima di essa Van(n)ozza, |10 l'altro p(er) l'anima di | Giorgio Croce suo ma|rito primo, il terzo p(er) | l'anima di Carlo Cana|le suo 2° marito, come |15 p(er) l'atti d'Andrea Caro|so not(ari)o,*

anniversary dates have been left out of both inscriptions. A more practically oriented record with the actual dates of the anniversaries and other details was obviously kept elsewhere, for instance, in necrologies or obituaries – these, however, as a rule did not provide the year of the death of the deceased.[159] The second function is that these inscriptions, showing dates respectively of 1517 and 1520, furnished the anniversaries with a visible starting point, the piece of information usually lacking in obituaries. A year date is given in ten out of 12 cases, the two exceptions being the earliest example from the end of the 11th century, and a partially surviving 14th(?) century case from SS. Salvatore a Porta Paola.[160]

Other details were, in any case, not necessarily carved in stone, as is expressed in the last mentioned example: LL. 61–63 "*ut pleni(us) h(aec?) pate|nt in testam(en)to (et) co|dicello d(i)c(t)i d(omi)ni Iacobi.*" Furthermore, the laudatory function is evident in cases where the inscription has been erected while the individual to be remembered was still alive – here, naturally, the precise day could not yet have been indicated. Admittedly, instructions for a multi-pronged set of commemorative activities could be recorded in detail, as one Stephanus would have it in S. Giacomo alla Lungara in 1277,[161] but this does not alter the general interpretation of their primary function. To get the full picture, we must still take note of that fact that not all cases of commemoration were anniversaries: shorter intervals do occur as well, as in the previous example.

Lastly, historical narrations could refer to a day without dating it and return to it in the narration. This is fairly obvious, and as historical narrations will be discussed below, two examples here are sufficient. First, the already mentioned inscription erected in S. Croce in Gerusalemme in 1521–1523 by Bernardinus Lupi Carvaial informs us that the sign of Christ appeared in Rome the day when the Hispanic monarchs Ferdinand II and Isabella I gained a victory over

---

publico sotto | li 15 Gennaro, 1517." One may note the absence of Domenico Giannotti da Rignano: some have thought that he was her first husband before Giorgio Croce: Roberto Zapperi, "Cattanei, Vannozza," in *Dizionario Biografico degli Italiani* 22 (1979): 295.

159 Cf. Holger Kaasik, p. 246 in the present volume.

160 Marini and Mai, *Inscriptiones Christianae*, 219 no. 1; 12:457.

161 6:1061: "*an(no) D(omini) MCCLXXVII, indict(ione) VI, mensis Octobris | die XIII, memoria sit collegio huius | ecclesiae, quod tenentur firmiter | hoc agere, ut celebretur in ea |$_5$ semper ex nunc omni die missa | b(eatae) Mariae virg(inis) pro salute et anima | d(omini) Stephani filii nobilis viris d(omini) Petri | Stephani, et anniversarium etiam | ex die mortis eius celebrare singulis |$_{10}$ annis, et toto anno suae mortis | omni die missa defunctorum pro | anima eius, et deinde continue | singulis diebus missa beatae Vir|ginis in commemorationem animae |$_{15}$ suae perpetuo. haec autem | cauta sunt instrumento publico | scripto per Nicolaum Carbonis notarium.*"

the Muslims. Because of the sign, Pope Innocent VIII went over to the church in question to conduct a mass. Furthermore, an old chant recorded in a mosaic was sung that day, and the pope conceded a plenary indulgence to be obtained there on the same day every year.[162] In respect to the present context, the noteworthy feature is that the day has not been placed in the calendar year: not in terms of the historical discussion nor in terms of the conceded indulgence. We may only suppose that it was assumed this piece of information was known well enough. The other example presents the only case where a precise action in the past, the consecration of the Lateran church, gives the temporal framework, in this case for a miraculous epiphany: "(…) *sacru(m) vultu(m) Salvatoris (i)nteg(r)u(m) reponi fecit i(n) loco, ubi primo miraculose populo Romano apparuit, quando fuit ista eccl(esi)a consecrata.*"[163] Nevertheless, the focus is naturally on the coincidence of the consecration and the miracle; thus one would need external information in order to identify the particular day.

2.5 *Durations: Terms of Magistrates, Indulgences, and Eternity*

The last substantial temporal category to be discussed are the indications of duration. As we see in some previous examples, indulgences[164] were often obtainable annually during certain time windows, and masses for the dead could be instructed to be held for "eternity." Eternal references emerge also in plenary indulgences, and in the opening formulae of papal documents; on the other hand, historical narrations could speak of the distant past. In addition, in honorific inscriptions of the Renaissance period, lay magistrates could be recorded with their lengths of terms included.[165]

Unfortunately, there is not very much that can be gained from these types for the purposes of this study. As my aim here is not to try to get to the bottom of indulgences in general, I will only make the following note regarding their length: apart from the examples from between 1046–1263 and three others, already discussed by Anna Blennow in satisfactory detail,[166] the remaining 13th and 14th century occurrences of indulgences in non-funerary inscriptions show only one plenary example for the 1300 jubilee,[167] but from the 1450s

162 8:504(3).

163 Claussen, *KSRM* 2, 341, mosaic inscription in the apse, dating from 1291.

164 For the concept itself, see Jasmin Lukkari's article, p. 63 with note 16, with further references.

165 A record of posts held was new only in this context, honouring a person both living and lay. Epitaphs, especially papal ones, could well have already had such information included much earlier; for example, 8:1 for Silvester II who died in 1003.

166 Blennow, *Latin Consecrative Inscriptions*, nos. 11, 24–26, 28, 33–43, and pp. 41, 210–215, and 226–228; discussion on 268–273.

167 13th century 140 days 4:1144; one year 40 days 11:375; one year 40 days & two years 80 days 3:721; 7000 years(?) 11:228. – 14th century one year 6:24; one year 40 days and 40 days 2:1186;

onwards the indulgences conceded were either relatively small or, in most cases, plenary.[168] The limits of the availability of indulgences were usually presented with liturgical or particular local church-related dates, as already discussed.

Honorific monuments were, as the genre definition implies, erected to leave a memory of praiseworthy men and deeds for the future generations to admire,[169] although this was seldom stated *expressis verbis.*[170] From a more practical point of view, however, this genre presents one rare feature, in recording terms in office. In many cases, only the date when the monument was erected is given, and in one of the early cases, the dating method used leaves the length open: "*tempore Martini V (…) incepit officium anno 1425.*"[171] Out of ten cases with a recorded length appearing, it appears once as "*integru(m) biennium,*"[172] once only in days, shortly after the beginning of the term,[173] but in eight cases, somewhat exceptionally, in months, ranging from six to 21 – using varying expressions, such as plain "*men(sibus) XX,*" or with the multipliers

---

plenary 6:17; not indicated 2:645. Cf. Claussen, *KSRM* 1, 179 note 4, for 11:228, and for the reference to another similar inscription only known from a transcription, not in Forcella. The lines in question are 14–20: "*ibide(m) e(st) indulgenti|a maxi(m)a, p(ro)ut in cro|nitis veratit(er) enar|ratur, | s(un)t etia(m) secte(m) milia | ani(?), i(n) festo o(mn)iu(m) s(an)c(t)oru(m)* |20 *usq(ue) ad octava(m).*"

168 Late 15th century three days 4:821; plenary 1:1196, 1:1197, 1:1629, 3:507, 8:780, 11:44, 12:245; not indicated 10:542. Early 16th century 40 days 7:860; 50 days 4:89; plenary 8:504(3) (said to have been conceded by Innocent VIII, i.e. 1484–1492).

169 We can here give a brief summary of the exemplars of the genre in this material. The earliest three from 1425–1427 (1:11–13) still show features compatible with medieval style, and the surviving one (1:13) has, according to Forcella, gothic letter-forms. After a 40-year gap until the next known exemplar, the style has changed. Notably though, even if the language of the inscriptions was influenced by Renaissance humanism, most of these monuments (including the earliest three) were about the insignia of the honorand, often put up by them during or after their term in office, mainly as *senator* (1:11–15, 18–19, 21, 23–25, 29, 30, 33–34, 36, 48(1)). Otherwise, 1:20 (1481) is a poem on a statue base; 1:28, 31–32, and 35 do not refer to insignia; 1:17 has the insignia, but the honorand is presented in the dative case in a more classical style; and only from 1514 onwards can we see the emergence of honorific inscriptions that imitate classical models more closely (1:37, 40, 42, and in S. Prisca 11:313). 6:53 does not belong to this discussion, since the main issue is the restoration of Innocent VII's grave by Nicholas V. See also Kajanto, Papal E*pigraphy*, 99. Honorific monuments in churches for popes in the dative case are: Leo X 11:313, Clement VII 10:570; and honorific elements have been included in 1:3, 5:345, 13:853. See also 5:223 of uncertain nature, but the style does not seem funerary.

170 The rare instances are 1:15 (1468), 1:18 (1475).

171 1:12.

172 1:30 (1502), L. 2.

173 1:33 (1508), LL. 3–7 "*con|servatores senatus | officium quinque et | viginti diebus egre|gie exercentes*": an example of another office than that of *senator.*

"*bis senis me(n)s(ibus)*" or "*ter septem menses.*"[174] Nonetheless, the atypical choice to use months is logical, since it seems that the terms did not only last full years.[175] These expressions, which do not represent the simplest way to deliver numbers, should naturally be regarded as a stylistic feature of humanist epigraphy to embellish an otherwise straightforward statement.

The passing of time features in building inscriptions recording restoration works: the need for such works is manifested with expressions such as "*vetustate pene collapsam*" or "*antiquitate collaps{s}a*" or similar,[176] which remain vague in terms of time. In a case from 1504, the purpose to restore the "original" form is stated: "*temporum | iniuria confractum (…) primamq(ue) in formam | restituit*"; the interesting "injury of times" has been identified as a Ciceronian expression by Iiro Kajanto.[177] On one occasion the building time of a renovation project is mentioned, but as with numerous other unique cases, this occurs in a poem: "*t(em)p(or)e sub modico su(n)t plene cu(nc)ta p(er)acta.*"[178] The expression remains, of course, rather uninformative about the actual time spent. In a comparable manner to these cases, one inscription juxtaposes "ancient" and "modern" times: "*ad locu(m), quod vocaba|tur antiquo te(m)pore | Ursi pileati et mo|derno te(m)pore mon|asterium s(an)c(t)e Bibia|ne.*"[179]

In addition to these vague periods of time, eternity, being a crucial element in the Christian context, is naturally referred to in several inscriptions. The occurrences are mostly in connection with the transcendent future, i.e. the Apocalypse and the afterlife, whereas references to perpetuity in the immanent scope are rare. A discussion of transcendent cases is unnecessary here, but three remarks on immanent references are worth making. First, nearly

---

174 1:14, 1:15, 1:19, 1:20, 1:21, 1:25, 1:28, 1:32. In 1:14 LL. 2–4 (with a somewhat confused expression, but apparently the honorand was appointed senator twice) "*ut tempore Calisti semestrem dignitate(m) | senatoriam integre ita nunc Paulo II pon(tifice) max(imo) | eadem ann(o?) magna cum laude gessit*"; 1:15 LL. 2–3 "*bis denis mensibus de officio senatorio | benemerito decoratus*"; 1:19 (metric) LL. 5–6 "*sex me‹n›ses annumq(ue) | dedi bona iura{s} senator*"; 1:20 L. 9 "*III semestri*"; 1:21 (metric) LL. 1–2 "*ter septem menses Sixto regna(n)te senator | Romulidas rexit*"; 1:25 L. 2 "*bis senis me(n)s(ibus) senatorio munere funct(us)*"; 1:28 LL. 2–3 "*senator urbis regius | consiliarius sedit mensibus XIIII MCCCCLXXXIIII*" and L. 4 "*idem bis senator men(sibus) XX an(no) MCCCCXCVIII*"; 1:32 LL. 1–3 "*ur|bis senator Iulio II p(ontifice) max(imo) sedente an(n)o | D(omi)no(!) MDVII mensibus XVI.*"

175 Otherwise in 13:436, above 10 and 15, which is, however, explained by the metric composition.

176 4:1314 (S. Pantaleone, 1418), 4:1114 (S. Maria in Publicolis, the second half of the 15th century).

177 A vessel said to be found in Therms of Vespasian and Titus and brought to the Vatican: 6:122; Kajanto, Papal E*pigraphy*, 95–96 with note 18.

178 3:721 (S. Salvatore in Pensilis, 1285), L. 14. The poem is added after a papal epistle.

179 11:228 (S. Bibiana, 13th century), LL. 2–7.

all of these are in epigraphs corresponding to, or deriving from, papal documents that usually include in the ingress the self-explanatory phrase "*ad perpetuam rei memoriam.*"[180] Notable among these is the jubilee bull "*Antiquorum habet*" of Boniface VIII, the lapidary version of which has an ingress different from the documentary one: "*ad certitudine(m) presentiu(m) (et) memoriam | futuror(um).*"[181] Second, one restoration record from SS. Cosma e Damiano dei Barberi alludes to this style in "*divae Virginis et Cosmi Domianiq(ue)(!) martir(um) aedem vetustate collabentem ad perpetuam artis tonsor(um) curialium memoriam (…) restaurarunt.*"[182] And third, the metric part of a building inscription from 1296 bestows perpetual fame to the two senators responsible for the construction: "*Transtiberim gaude, quia cives sunt tibi, laude | hi duo solemni digni famaq(ue) perenni*";[183] a comparable case is known from 1427.[184]

2.6 *Other Illustrative Cases*

Lastly, I will briefly discuss the following points emerging from the material: historical references with special attention to family history; classical references; and church bells. These might well deserve closer attention, but cannot be examined here in as much detail as they might merit.

There is perhaps no need for an extensive discussion of the few examples of historical narrations here, because as texts they should be fairly similar to other written historical narrations, and thus any inquiry might produce rather inconsequential results.[185] In addition, sometimes the historical references are

180 Cf. Frenz, *Papsturkunden*, 26. In inscriptions: 1:1196–1197, 8:31, 8:785, 11:44; 4:89 "*ad posteritatis memoriam sempiternam*"; 8:42 "*ad futuram rei memoriam.*" See also an epistle, Marini and Mai, *Inscriptiones Christianae*, 211–212(1) "*per vos in eodem titulo in perpetuum.*"

181 6:17 LL. 1–2; cf. Jasmin Lukkari's article in the present volume, p. 92. The versions have some other discrepancies, too.

182 6:86, 1479, transported later to the Vatican grotto; LL. 1 and 4. Cf. 13:587 for another version of this recorded in a transcription: "*ad perpetuam barbitonsor(um) memoriam.*"

183 1:3, seen on the Capitoline hill, now lost; LL. 19–20.

184 5:345, S. Lorenzo in Lucina (now lost), L. 25: "*cui decus et meritae manent per secula laudes.*"

185 I would, of course, be happy to be proven wrong in regard to this point. The inscriptions are: 2:645 brief history of SS. XII Apostoli; 3:538 (above p. 143); 6:139–144 and 148–157 are labels on historical depictions in the Vatican library; 5:345 reference to unrest in Martin V's time and him putting an end to it; 7:624, a label on a depiction about the founding of a cult (see above notes 42 and 48); 8:18 the legend of Emperor Constantine and Pope Silvester I; 8:498 with Emperor Otto III and Popes Silvester II and Sergius III; removed later as "false history" (cf. Forcella's commentary, and above note 77); 8:504(2) 4th century history with a reference to temporal distance of "almost 1100 years" to the present time of the inscription; 9:740 an angelic epiphany (cf. above p. 134 with note 104); 11:4 a posthumous, non-funerary epigram for Cardinal Petrus Capoccinus († 1259), mentioning Emperor

very brief, such as in a verse comparing the flooded Tiber to the biblical deluge: "*territa Roma 'Noe redeunt nunc tempora' dixit.*"[186] Other types of instances where history is recalled may stimulate more interest, mainly those concerning familial presentation, which, in non-funerary epigraphy, occur mostly in the Renaissance period. One such is the new practice of mentioning a pope's family background in public inscriptions. There are two isolated earlier examples with popes of local origin: as supplementary information in a 1278 building inscription of Nicholas III,[187] and in a 1427 verse inscription of Martin V;[188] but this kind of record in prose becomes more common only at the end of the 15th century with Alexander VI, as in the following elaborate example: "*Alexander VI pont(ifex) max(imus) | Calisti III pont(ificis) max(imi) nepos, | natione Hispanus, patria | Valentinus, gente Borgia, |5 portas et propugnacula | a Vaticano ad Hadriani | molem vetustate confe|cta tutiora restituit | an(no) salutis MCCCCLXXXXII.*"[189] This is not at all to say that the feature was a completely novel one,[190] but showing a pope's family background in these circumstances

---

Frederick II († 1250); 11:312 history of the Aventine basilica of S. Sabina, going back to the mythical Evander; 11:375 a papal epistle with a narration on St Lucy; Claussen, *KSRM* 2, 342–343, verse inscription with Innocent III's dream and Nicholas IV's renovation; *ibid.* 343–345 reference to persecution of early Christians, and to Vespasian, Titus, and the Judaean war; Giovenale, *S. Maria in Cosmedin*, 72 no. 70 related to a past donation, fragmentary. See also commemorations of a recent victory 1:1 (above p. 135–136 with note 112), and 13:3 (above p. 143 note 143); inscriptions related to relics 10:867, Blennow, *Latin Consecrative Inscriptions*, 145–154 no. 32, Mondini, *KSRM* 3, 308 no. 111; an antique name for a place, next to the contemporary one, 11:228 (above p. 150 with note 179); a past miracle Claussen, *KSRM* 2, 341 (above p. 148 with note 163).

186 13:427 (S. Maria sopra Minerva, 1476), lost. Needless to say, the biblical events were considered historically valid.

187 6:12 (Vatican), LL. 5–7 "*fuit aut(em) p(rae)|d(i)c(tu)s sum(mus) po(n)tifex natio(n)e Roman(us), ex p(a)tre d(omi)ni Matth(ae)i Rubei | de domo Ursinorum.*"

188 5:345 (S. Lorenzo in Lucina, now lost), LL. 2–4 "*quo tempore pastor | Martinus quintus, prius Oddo a stirpe Columna | (…) regebat.*"

189 6:103; Alexander VI also in 6:115, 8:352, 13:248, 13:250 (only *Hisp(ano)* in 13:428, which was erected by a Spaniard and dedicated to St Jacob). Likewise Julius II in 6:130–131, 135; Leo X in 8:505 (cf. above note 76), 11:313, Kajanto, *Papal Epigraphy*, 98–99; Clement VII in 10:362. Not to be counted is 12:444 with Innocent II; see Appendix 1. See also 6:22 "*Benedictus p(a)p(a) XII Tholosanus*" already in 1341, and Paul II who carries the epithet "*Venetus*" in two laconic building inscriptions, 13:7–8; Eugene IV and Innocent VIII have been furnished with epithets in later references, 6:58 and 6:106 respectively. Notably, Kajanto pays no attention to any of these features in his *Papal Epigraphy*.

190 Familial backgrounds were, of course, recorded earlier as well, but in other contexts far less visible to the public gaze. See for instance, the 14th century papal obituaries in BAV, Arch. Cap. S. Pietro H.57, fol. 118r for Nicholas III, and fol. 144v for Boniface VIII (I thank Holger Kaasik for the references), or papal epitaphs: Kajanto, *Papal Epigraphy*, 20–21 for Boniface IX. In these, geographical origins had been already mentioned much earlier, e.g. 8:1 for Silvester II.

is telling of a change in attitude in terms of how popes could relate themselves to their worldly past in public monuments.

Humanist epigraphy, a sphere where one could boast of erudition, was imbued with classical or classicising references. On the level of language, these sometimes exhibited surprisingly un-Christian tones.[191] One obvious example in my material, which touches upon the presentation of a familial past, goes back again to the already discussed jubilee enthusiast Iulianus de Dathis. One of the two inscriptions he erected in 1500 in SS. Silvestro e Dorotea includes a curious dedication to "the ancestral spirits of forefathers and household," on side B: "*D(eo) o(ptimo) m(aximo), divo Silve|stro ac dive | Dorotheae, | manibus la|ribusq(ue) avi|tis sacrum* $|_5$ *an(no) iubilei.*"[192] This inscription seems to have survived, whereas the other inscription of his in the same church, recording an epiphany of an angel, was deemed apocryphal at some point and removed.[193] Another fascinating reference type is the invention of classical origins for one's family, but as these examples only occur in funerary material, they will not be discussed further here.[194]

Lastly, church bells, and the ringing of them in order to indicate time, deserve a brief note.[195] Bells were often inscribed, or they (or bell-towers) could be referred to in inscriptions. From those found in the present material,[196] two notable inscriptions emerge. The first, still to be found fixed to the front of

191 For quasi-pagan expressions in epitaphs, see Kajanto, *Classical and Christian*, 25, especially note 137 on the invocation in Greek αἰωνίῳ θεῶν αὐτοκράτορι ("to the eternal ruler of gods") in an otherwise Latin epitaph 6:111 (Vatican, 1501): "This particular dedication is only interpretable with difficulty in a Christian sense, suggestive as it is of polytheism."

192 9:739. Side A: "*Iulianus | de Dathis | ((insignia)) | penitentiarius | et rector.*"

193 9:740. For both inscriptions see above p. 134 with notes 103–104.

194 Cf. at least: 2:1052 de Quinque = Quinctii; 10:120 Caccia = Cassii; for Porcari = Porcii Catones, cf. Anna Modigliani, "La familia Porcari tra memorie repubblicane e curialismo," in *Un pontificato ed una città: Sisto IV (1471–1484): atti del convegno, Roma, 3–7 dicembre 1984*, eds, Massimo Miglio et al. (Città del Vaticano: Scuola vaticana di paleografia, diplomatica e archivistica, 1986), 317–354; for Santacroce = Valerii Publicolae, cf. Anna Esposito, "Famiglie aristocratiche e spazi sacri a Roma tra medioevo e prima età moderna," in *Ricerca come incontro. Archeologi, paleografi e storici per Paolo Delogu*, eds, Giulia Barone, Anna Esposito, and Carla Frova (Roma: Viella, 2013), 471–481, with further references.

195 Cf. Saku Pihko's article in the present volume, p. 40–41.

196 Bell inscriptions from our period in Forcella's corpus do not exceed four cases: 4:748–479, 6:14, 13:588; others known to me are Claussen, *KSRM* 1, 80; Claussen, *KSRM* 1, 349 note 6; the one in Ravenna, Museo Archivescovile, inv. no. 32, mentioned above in note 92; and the second example below. Also, the first example below and 1:1098 with "*nolae*" and "*nolarium*" (above p. 126) are relevant. See also 3:721 "*campanile*" mentioned on L. 14; 6:26 a lost verse inscription, which was related to a bell-tower in some way. Unfortunately, I have been unable to consult Sible de Blaauw's article "Campanae supra urbem – sull'uso delle campane nella Roma medievale," in *Rivista di Storia della Chiesa in Italia* 47 (1993), 367–414, for the purposes of this study.

S. Silvestro in Capite, includes a donation list, in which we find on line 10 the listed objects "[- - - *tu*]*rre*(*m*) *qua*(*m*) *ca*(*m*)*panile dicim*(*us*) I; *ca*(*m*)*panas* II; (…)."[197] The expression "tower that we call a bell-tower" shows that in 1060 this had not yet become a concept common enough for it to be called simply *campanile*. The other example, now lost, was a poem inscribed on a church bell put up by Honorius III in the Lateran.[198] In these verses, two and a half centuries later than the previous example, the bell itself could "speak" about its purpose to indicate the hours: "*ut pateant horae, me Salvatoris honore | sic de non esse precepit Honorius esse | trin*(*us*) *ap*(*osto*)*licus* (*Christi*)*q*(*ue*) *fid*(*e*)*lis amicus*."[199]

3 Summary

This study has examined temporal references in non-funerary epigraphy from Rome dating between 1000 and 1527. The material consists mostly of inscriptions on stone, usually set in public, accessible space, typically within churches, but other inscribed objects of durable material are included as well. Temporal references of various levels of preciseness and distance were studied, and since dates form a prominent part of the material, their formal qualities were also assessed.

For many of the points discussed, epigraphic practices changed over the course of time. Here the most relevant comparison may be made between medieval practices in the 12th–13th centuries and Renaissance forms from the mid-15th to early 16th century. This results primarily from the scarcity of available material from the 11th century, and again from the 14th to the first half of the 15th century. Even though the latter gap prevents us examining the latest medieval practices in detail, the material still leaves no doubt that the most important changes occurred with the emergence of Renaissance or humanist style epigraphy towards the second half of the 15th century. It is not surprising, then, that this is in full accordance with what Iiro Kajanto has presented regarding funerary epigraphy.

Other factors that influence the epigraphic output of temporal references are the author, genre and context of an inscription – in some particular cases these may be considered separately, but often the situation disallows any clear

197 Silvagni, *MEC* 1, tab. 19 no. 4; my transcription and conjecture. Note that the first R in [*tu*]*rre*(*m*) lies on the edge of the surviving surface and is not shown in Silvagni's photo. However, it is clearly visible on the stone (my examination on 15 May 2017).

198 Claussen, *KSRM* 2, 90 note 367.

199 My translation: "So that the hours would be known, the threefold apostolic Honorius, loyal friend of Christ, thus ordered me, in the Saviour's honour, from not-being to be."

distinction. The material originating from the papal chancery often imitated the chancery's traditions, such as a preference for Roman dating, and some features from there are also found in epigraphy that resulted from other ecclesiastical activities (consecrative and building inscriptions), such as liturgical dating. However, hardly any actual rules can be inferred from the evidence: even the papal material does not follow the documentary rules fully.

The genre and the context of the inscription also guide the needs, preciseness and style of the temporal references to some extent. However, the material suggests that the traditions behind different epigraphic genres were not strict in this respect, but rather left the ultimate decision to each author. In comparison to documentary material, this is understandable due to the monumental nature of epigraphs set in a public setting. Poetic compositions could be used in every epigraphic genre, and metric expressions of time often defy those found in their prosaic counterparts, beginning with the numeral format. Somewhat comparable to this is also the variation applied to stylised Renaissance prose, as one may observe in honorific inscriptions, for example.

Nevertheless, the most common epigraphic dating practices made use of elements which can be found in documentary material, and, when a day-level date was included, often roughly similar formulations, too. AD years and papal reign were the typical methods for indicating a year; indiction was used more amply up to the 13th century, but faded away over the course of the 14th century. With day-level dating, running and Roman date types coexisted, and, in contrast to the funerary epigraphy, their use saw no drastic changes in non-funerary contexts, not even in the Renaissance period.

Other methods were mostly utilised only in specific contexts. Apart from the already mentioned liturgical dating, persons of local importance – such as church officials – appear in some dating formulas, as do magistrates, especially from the 15th century onwards. A temporal reference could be oriented by another suitable figure, such as a saint thought to be of assistance in a dramatic event, or a distant ancestor in a renovation of a family chapel. Royal personages are only found in isolated instances, but the unfinished inscription from the time of the investiture controversy, which mentions the emperor in its dating formula, tells of a short-lived attempt to replace the pope in this function. The Christian jubilee years were also variably mentioned as temporal points of reference, especially in the case of Sixtus IV's jubilee of 1475, and in epitaphs erected in the year of the following one of 1500.

Lastly, whereas day-level dates were often used for the purpose of recurring commemorative acts, many types of inscriptions, such as those concerning building projects or smaller donations – and those inscribed on smaller objects in general – also bore either only an AD year or a secondary temporal reference. The AD year reflects a desire to mark an object for posterity with a memory of its commission (and usually the commissioner) fixed in time. Similar intentions may lie behind many secondary temporal references, too, for instance when these involve popes who appear in dating formulas per se, thus making references to the precise year unnecessary. However, it is also possible that these are only to record the person in question without a conscious implication of a temporal reference.

Bibliography

Sources and Literature

Blennow, Anna Holst. *The Latin Consecrative Inscriptions in Prose of Churches and Altars in Rome 1046–1263: Edition with Translation and a Commentary on Language and Palaeography.* Roma: Società romana di storia patria, 2011.

Carunchio, Tancredi. "La Villa Lante al Gianicolo: L'architettura e i suoi restauri." In *Villa Lante al Gianicolo: Storia della Fabbrica e cronaca degli abitatori.* Edited by Tancredi Carunchio and Simo Örmä, 23–75. Roma: Palombi editori / Institutum Romanum Finlandiae, 2005.

Cavallaro, Anna, Anna Esposito, and Andreas Rehberg, "Donne di pietra. Immagini, vicende, protagoniste delle sepolture romane del Rinascimento." *Mélanges de l'École française de Rome, Moyen age* 127, no. 1 (2015), 5–7.

Claussen, Peter Cornelius, Daniela Mondini, and Darko Senekovic. *Die Kirchen der Stadt Rom im Mittelalter: 3. G-L.* Stuttgart: Franz Steiner Verlag Stuttgart, 2010.

Claussen, Peter Cornelius. *Die Kirchen der Stadt Rom im Mittelalter: 1. A–F.* Stuttgart: Franz Steiner Verlag Stuttgart, 2002.

Claussen, Peter Cornelius. *Die Kirchen der Stadt Rom im Mittelalter: 2. S. Giovanni in Laterano, mit einem Beitrag von Darko Senekovic über S. Giovanni in Fonte.* Stuttgart: Franz Steiner Verlag Stuttgart, 2008.

Claussen, Peter Cornelius. *Magistri doctissimi romani: Die römischen Marmorkünstler des Mittelalters.* Stuttgart: Steiner-Verlag-Wiesbaden, 1987.

Curcio, Giovanna, and Paola Farenga. "Dati, Giuliano." In *Dizionario Biografico degli Italiani* 33 (1987), 31–35. http://www.treccani.it/enciclopedia/giuliano-dati_%28Dizionario-Biografico%29/.

D'Achille, Paolo. "Iscrizioni votive e sepolcrali in volgare dei secoli XIV–XVI." In *Il volgare nelle chiese di Roma: messaggi graffiti, dipinti e incisi dal IX al XVI secolo.* Edited by Francesco Sabatini, Sergio Raffaelli, and Paolo D'Achille, 67–107. Roma: Bonacci, 1987.

Di Stefano Manzella, Ivan. *Index inscriptionum Musei Vaticani. 1, Ambulacrum Iulianum sive "Galleria lapidaria."* Città del Vaticano: Ex Officina Libraria Pontificia, 1995.

Esch, Arnold. "L'economia nei giubilei del quattrocento." In *I giubilei nella storia della Chiesa. Atti del Congresso internazionale in collaborazione con l'École française de Rome sotto il patrocinio del Comitato centrale per il giubileo del 2000 (Roma, Istituto patristico Augustinianum, 23–26 giugno 1999)*, 341–358. Città del Vaticano: Libreria Editrice Vaticana, 2001.

Esposito, Anna. "Famiglie aristocratiche e spazi sacri a Roma tra medioevo e prima età moderna." In *Ricerca come incontro. Archeologi, paleografi e storici per Paolo Delogu*. Edited by Giulia Barone, Anna Esposito, and Carla Frova, 471–481. Roma: Viella, 2013.

Favreau, Robert. *Épigraphie médiévale*. Turnhout: Brepols, 1997.

Filippi, Giorgio. *Indice della raccolta epigrafica di San Paolo fuori le mura*. Città del Vaticano: Monumenti, Musei e Gallerie Pontificie, 1998.

Forcella, Vincenzo. *Iscrizioni delle chiese e d'altri edificii di Roma dal secolo XI fino ai giorni nostri*. Roma: (vol. I) Tipografia delle scienze matematiche e fisiche, 1869 / (voll. II–VII) Tipografia dei Fratelli Bencini, 1873–1876 / (voll. VIII–XIII) Coi tipi di Ludovico Cecchini, 1876–1879 / (vol. XIV) Tipografia Tiberina, 1884.

Frenz, Thomas. *Papsturkunden des Mittelalters und der Neuzeit*. Stuttgart: Franz Steiner Verlag, 2000^2.

Galletti, Pietro Luigi. *Inscriptiones Romanae infimi aevi Romae exstantes*. Romae: Typis J.G. Salomonj, 1760.

Garms, Jörg, Tassilo Blittersdorff, and Hanns Jäger-Sunstenau. *Die mittelalterlichen Grabmäler in Rom und Latium vom 13. bis zum 15. Jahrhundert Bd. 1. Die Grabplatten und Tafeln*. Wien: Österrreichische Akademie der Wissenschaften, 1980.

Gauvain, Alexis. *Una storia dalla Roma del Quattrocento: quaderni di Ansuino di Anticoli, parroco in Roma e beneficiato vaticano (1468–1502)*. Città del Vaticano: Edizioni capitolo vaticano, 2014.

Giovenale, Giovanni Battista. *La basilica di S. Maria in Cosmedin*. Roma: P. Sansaini, 1927.

Gray, Nicolette. "The Paleography of Latin Inscriptions in the Eighth, Ninth and Tenth Centuries in Italy." *Papers of the British School at Rome* 16 (1948), 38–162.

Guillou, André. *Recueil des inscriptions grecques médiévales d'Italie*. Rome: École française de Rome, 1996.

Hülsen, Christian. *Le chiese di Roma nel Medio Evo*. Firenze: Leo S. Olschki, 1927.

Kajanto, Iiro. "Dating in the Latin Inscriptions of Medieval and Renaissance Rome." *Arctos: Acta philologica Fennica* 11 (1977), 41–61.

Kajanto, Iiro. *Classical and Christian: Studies in the Latin Epitaphs of Medieval and Renaissance Rome*. Helsinki: Suomalainen tiedeakatemia, 1980.

Kajanto, Iiro. *Papal Epigraphy in Renaissance Rome*. Helsinki: Suomalainen tiedeakatemia, 1982.

Kendall, Calvin B. *The Allegory of the Church: Romanesque Portals and Their Verse Inscriptions*. Toronto, Buffalo: University of Toronto Press, 1998.

KSRM = *Die Kirchen der Stadt Rom im Mittelalter* (see above Claussen).

Marini, Luigi Gaetano, and Angelo Mai. "[Inscriptiones Christianae]." In *Scriptorum veterum nova collectio e Vaticanis codicibus edita* 5. Edited by Angelo Mai, 1–473. Rome: Typis Vaticanis, 1831.

Modigliani, Anna. "La familia Porcari tra memorie repubblicane e curialismo." In *Un pontificato ed una città: Sisto IV (1471–1484): atti del convegno, Roma, 3–7 dicembre 1984*. Edited by Massimo Miglio, Francesca Niutta, Diego Quaglioni and Concetta Ranieri, 317–354. Città del Vaticano: Scuola vaticana di paleografia, diplomatica e archivistica, 1986.

Mondini, Daniela. KSRM 3 (see above).

Senekovic, Darko. KSRM 2–3 (see above).

Silvagni, Angelo. "Intorno alla pubblicazione delle iscrizioni cristiane, antiche e medioevali, di Roma e dell'Italia." *Rivista di archeologia cristiana* 5 (1928), 135–142.

Silvagni, Angelo. *Monumenta epigraphica christiana saeculo XIII antiquiora quae in Italiae finibus adhuc exstant iussu Pii XII pontificis maximi, vol. I*. Città del Vaticano: Pontificium institutum archaeologiae christianae, 1943.

Zapperi, Roberto. "Cattanei, Vannozza." In *Dizionario Biografico degli Italiani* 22 (1979), 294–298. http://www.treccani.it/enciclopedia/vannozza-catanei_(Dizionario-Biografico).

[unknown].[200] "Anguillara, Everso." In *Dizionario Biografico degli Italiani* 3 (1961), 302–303. http://www.treccani.it/enciclopedia/everso-anguillara_(Dizionario-Biografico)/.

200 In the place where the author's name is expected, there is *, the meaning of which remains obscure to me.

CHAPTER 5

The Medieval Calendars of S. Pietro in Vaticano and S. Maria Maggiore in Rome

Holger Kaasik

1 The Medieval Calendars

Medieval calendars offer a unique perspective on liturgical life, commemorative practices, and conceptions of time in medieval communities. Although virtually all calendars note the most important annual celebrations observed all around Christendom, much of their content is in fact local in nature, featuring feasts of local saints, dedications of local churches and altars, and obits of benefactors and notable community members. Space was limited on medieval calendar pages, even when the standard format of one page per month, which was dominant in the 14th and 15th centuries, was not observed. As a consequence, calendars were much more exclusive in content than other related texts such as martyrologies, which could fit in dozens of saints per day when necessary. Furthermore, medieval calendars, unlike their contemporary counterparts, remained in use for years, even centuries.[1] As a result of this combination of longevity, exclusivity, and locality, medieval calendars were able to serve as repositories of communal memory spanning generations.

The utility of calendars for scholarship has been facilitated by a further three factors. Firstly, they are ubiquitous, numbering in their thousands in archives and libraries around the world; secondly, they are concise in content; and thirdly, they all follow the structure of the 365-day course of the year. These factors add up to a range of possibilities for various types of comparative study. Traditionally used for reconstructions of liturgical traditions deemed

1 E.g., calendars Città del Vaticano, Biblioteca Apostolica Vaticana, Arch. Cap. S. Pietro A.9 and E.14, the first dating from the early 14th century, the latter likely from the thirteenth. The 20 May feast of Bernardino of Siena (died in 1444, canonised in 1450) has been added to both calendars, which indicates they were still in use around the mid-15th century. To avoid clutter and repetition, I use the format "ACSP A9" when referring to the collection's manuscripts, the corresponding abbreviation for S. Maria Maggiore being SMM. When referring to the Biblioteca Apostolica Vaticana is necessary for clarity, I use the abbreviation BAV. The computational mechanisms used to adapt calendars to the weekly and lunar cycles of any given year are dealt with by Marko Halonen in the present volume.

 | DOI:10.1163/9789004436251_006

particularly significant or in examining contact between various religious centres and communities, medieval calendars have recently been increasingly used in studies dealing with communal memory and local identity.[2] Moreover, as calendars are routinely used to localise and date their host manuscripts, knowledge of calendars has implications for a wide range of fields from codicology, palaeography, and heraldry to the history of liturgy, music, art, book production, material culture, and everyday religious life.

The basilicas of S. Maria Maggiore and S. Pietro are among the most important religious centres in Rome, and indeed the whole of Latin Christendom. As the burial place of the Prince of the Apostles, S. Pietro functioned – as it still does – as one of the chief pilgrimage destinations in western Europe. For the same reason, the basilica has represented a physical argument for papal primacy, even as the see of the Bishop of Rome remains in the Lateran.[3] S. Maria Maggiore, in turn, is both the oldest and the most prestigious of the churches dedicated to Mary in Rome and a home to an array of significant relics, including the manger of the infant Christ. It was chief among the destinations for medieval papal processions, the Roman *stationes*, as it was the setting for both the Easter Sunday and Christmas processions.[4]

---

2 As examples of traditional approaches, see Stephen Joseph Peter van Dijk, *Sources of the Modern Roman Liturgy : The Ordinals by Haymo of Faversham and Related Documents (1243–1307), v. 1. Introduction. Description of Manuscripts*, Studia et Documenta Franciscana (Leiden: E.J. Brill, 1963); Stephen Joseph Peter van Dijk and Joan Hazelden Walker, *The Ordinal of the Papal Court from Innocent III to Boniface VIII and Related Documents* (Fribourg, Switzerland: The University Press, 1975); Elena Petterlini, "Il Santorale Marciano e Il Santorale Della Chiesa Di Verona," in *Da Bisanzio a San Marco. Musica e Liturgia*, ed. Giulio Cattin (Venezia: Fondazione Ugo e Olga Levi, 1997), 300–13; Anna Vildera, "Il Santorale Marciano e Il Santorale Della Chiesa Di Padova," in *Da Bisanzio a San Marco. Musica e Liturgia*, ed. Giulio Cattin (Venezia: Fondazione Ugo e Olga Levi, 1997), 271–300; for more recent takes, see Els Rose, "Liturgical Commemoration of the Saints in the 'Missale Gothicum' (Vat.Reg.Lat. 317). New Approaches to the Liturgy of Early Medieval Gaul," *Vigiliae Christianae* 58, no. 1 (2004), 75–97; Anna Welch, *Liturgy, Books and Franciscan Identity in Medieval Umbria* (Leiden and Boston: Brill, 2015).

3 During the long medieval contest for primacy between the Vatican and Lateran Basilicas, the former was favoured by some notable pontiffs, such as Innocent III (1198–1216) and Boniface VIII (1294–1303). The text on the apse mosaic commissioned for S. Pietro by the former is very revealing in this regard: "*Summa Petri sedes est haec sacra principis aedes / Mater cunctarum decor et decus ecclesiarum.*" See Sible de Blaauw, *Cultus et decor: liturgia e architettura nella Roma tardoantica e medievale: Basilica Salvatoris, Sanctae Mariae, Sancti Petri.* 2, Studi e testi 356 (Città del Vaticano: Biblioteca Apostolica Vaticana, 1994), 630; Richard Krautheimer, *Rome: Profile of a City, 312–1308* (Princeton, N.J: Princeton University Press, 2000), 203–06, 267.

4 Considering the Basilica's prominence, Victor Saxer has suggested it functioned as a "secondary" or "auxiliary" cathedral of Rome in the Middle Ages. Victor Saxer, "Il culto dei martiri romani durante il Medioevo centrale nelle basiliche Lateranense, Vaticana e Liberiana," in *Roma antica nel Medioevo: mito, rappresentazioni, sopravvivenze nella "Respublica Christiana"*

The situation regarding the manuscript collections of both basilicas is fortunate, as they have weathered Rome's turbulent history exceptionally well. Both collections are nowadays conserved in the Biblioteca Apostolica Vaticana; the *Archivio del Capitolo di S. Pietro* contains hundreds and the S. Maria Maggiore collection dozens of medieval manuscripts. There is a substantial corpus of calendars present in both cases, which is a rare luxury. Indeed, this is not the case for most of the multitude of churches of medieval Rome, among them the Lateran Cathedral, from which hardly a single calendar survives.[5]

My aim in this article is to bring these two invaluable collections of medieval calendars to the attention of wider scholarship, as despite their historical significance and extraordinary rate of survival, few of these calendars have previously been identified, and the information most readily available on those that have is in many cases inaccurate. I aim, then, to present a comprehensive overview of the calendars that are likely to have been produced for or used at one of the two basilicas, laying out the arguments for their localisations, and addressing some previously proposed attributions that seem doubtful when considered in the context of a broader corpus of material. In addition to addressing the calendars individually, I compare them to a compilation of 40 calendars transcribed for this study, as well as previously published material from Italy and north of the Alps.

2 Previous Research on the Calendars

Most of the calendars analysed in this study, dating primarily from the 14th and 15th centuries, have not yet been examined in much detail. Although Pierre Jounel's *Le Culte des saints dans les basiliques du Latran et du Vatican au douzième siècle* (École Française de Rome, 1977) far surpasses the chronological confines given in its title towards earlier centuries, its 12th-century *terminus* means the study only touches upon a single S. Pietro calendar, albeit a seminal one. Nevertheless, as Jounel has gone through the Herculean labour of providing a historico-liturgical commentary on every single feast in his primary sources, his study is an invaluable resource for any scholar of later Roman developments.[6]

dei secoli IX–XIII, atti della quattordicesima Settimana internazionale di studio, Mendola, 24–28 agosto 1998 (Vita e Pensiero, 2001), 151–52, 160.

5 See the Excursion at the end of the article.

6 Pierre Jounel, *Le Culte Des Saints Dans Les Basiliques Du Latran et Du Vatican Au Douzième Siècle*, Collection de l'Ecole Française de Rome 26 (Rome: École Française de Rome, 1977). On the manuscript, see the section on ACSP B79 below.

The calendars of S. Maria Maggiore, on the other hand, have been the subject of a more chronologically comprehensive study. Victor Saxer's *Sainte-Marie-Majeure: une basilique de Rome dans l'histoire de la ville et de son église, ve–xiiie siècle* (École française de Rome, 2001) is an impressive monograph, with its value significantly increased by the appended catalogue of the Basilica's library collection. Saxer devotes a great deal of attention to the cult of saints at S. Maria Maggiore, having gone through the Basilica's calendars and other liturgical materials. The fruits of these labours can be found in a further appendix to his monograph, which takes the form of two "synoptic tables." The first shows the contents of 15 calendars, and the other presents 17 *sanctorales*, both containing material up to the late 15th century.[7]

Unfortunately, Saxer's table of the S. Maria Maggiore calendars is not furnished with any notes on his principles of transcription, and the data is in a highly normalised and simplified format: a single calendar entry is represented only by the manuscript's signum under a normalised feast heading, common to all 15 calendars. Thus, there is no way of tracing the actual contents – including the titles of the saints, the colour of the ink, and possible feast grades – let alone the wording and spelling of individual entries. Additionally, the table contains a considerable number of errors; these factors all render Saxer's material practically unusable for the purposes of further scholarship.[8]

7 See Appendices III, IV, and VIII in Victor Saxer, *Sainte-Marie-Majeure : Une Basilique de Rome Dans l'histoire de La Ville et de Son Église, ve–xiiie Siècle*, Collection de l'École Française de Rome 283 (Rome: École française de Rome, 2001), 457–86, 572–674. In 2001, Saxer also published an article attempting to synthesise Jounel's work with his own. This contribution does not contain much in terms of new information with respect to the two monographs, however. Saxer, "Il culto dei martiri."

8 Considering just calendars SMM 119 and 120, I have identified the following 51 errors in Saxer's table: (1) 5 January, removal of entry not noted in 119; (2) 13 January, *Celsi* not noted in 119; (3–6) 1 February, two entries marked in incorrect columns for both calendars; (7–8) 5 February, *Agathe* not noted in either calendar; (9) 6 February, *Dorothee* not noted in 119; (10) 1 March *Herculani*, not noted in 120; (11–12) 9 March, entry marked in incorrect columns for both calendars; (13) 11 March, addition of *Festum pontificium* not noted in 120; (14) 20 March, *Plenaria Indulgentia ad sanctam Crucem in Ierusalem* not noted in 119; (15) 14 April, absence of *Maximi* not noted in 120; (16–19) 14 and 17 April, two entries marked in incorrect columns for both calendars; (20) 18 April, absence of *Naboris et Felicis* not noted in 119; (21–22) 26 April, *Cleti et Marcellini* not noted in either calendar; (23–26) 4 and 5 May, two entries marked in incorrect columns for both calendars; (27) 19 May, *Yuonis* not noted in 119; (28–29) 12 June, *Cyrini Naboris et Naçarii* not noted in 119, and an entry marked in an incorrect column for 120; (30) 13 June, *Antonii ordinis minorum* not noted in 119; (31) 16 July, *Viti Modesti et Crescentie* not noted in 120; (32–35) 19 July entries marked in incorrect columns for both calendars, the fact that *Epafre* is an addition in both is not noted; (36–37) 29 July, entries marked in incorrect columns for both calendars; (38–39) 1 September, entries marked in incorrect columns for both calendars; (40–41) 11 October, *Translatio Augustini prima* not noted in either

It is difficult to say how far the accumulation of such errors has distorted Saxer's analysis itself, but there is a real danger that his data may be perpetuated and built on in the future, particularly considering his promise of a *dépouillement systématique* of the Basilica's material.[9] Oddly enough, Saxer makes little use of the calendars elsewhere in his study, even in matters of localisation. While he devotes an entire section of his monograph to the Basilica's library collection, when it comes to the *destination et usage des livres liturgiques* and the *coordonnée topographique de ces livres*, Saxer limits the discussion to a few *incipits*.[10] Here he inexplicably relies on earlier catalogues, despite having the entire collection's calendar and *sanctorale* data at his disposal.[11]

There are three catalogues essential for studying the S. Pietro and S. Maria Maggiore manuscripts. The former have been covered in great detail by Cosimo Stornajolo in his *Inventarium*, composed in 1923 in two manuscript volumes.[12] Itself a part of the collection as manuscripts ACSP H99–100, the catalogue is also available at the Vatican Library in three unpublished typescript volumes from 1968.[13] Following the convention established in previous scholarship and other catalogues, my references are to the 1968 volumes, although the manuscript version is nowadays the more accessible one, via the Digital Vatican Library. The most recent catalogue of the S. Maria Maggiore collection is Saxer's aforementioned appendix to his monograph study of the Basilica's history. The volume also contains editions of some of the collection's earlier inventories, and as Saxer devotes special attention to liturgical manuscripts, the catalogue is an invaluable tool for the purposes of the study at hand.[14]

---

calendar; (42) 16 October, *Gallus* noted as an addition in 120, though the calendar shows nothing for this day; (43) 31 October, *Quintini civis romanus* not noted in 120; (44) 18 November, an entry marked in an incorrect column for 120; (45–46) 19 November, *Helisabeth* not noted in either calendar; (47) 2 December, *Bibiana* noted in 120, though the calendar shows nothing for this day; (48) 16 December, *Laçari* not noted in 120; (49–50) 20 December, entries marked in incorrect columns for both calendars; (51) 24 December, *Vigilia* not noted in 119.

9 Saxer, *Sainte-Marie-Majeure*, 399.

10 I use the terms *explicit* and *incipit* to refer to rubrics beginning with said expressions, not necessarily found in the very beginning and end of the manuscripts themselves.

11 Saxer, *Sainte-Marie-Majeure*, 258–60.

12 The full title of the catalogue reads: *Inventarium Codicum manuscriptorum latinorum Archivii Basilicae S. Petri in Vaticano maxima ex parte e recensione Cosimi Stornajolo depromptum*. I refer to the catalogue as "Stornajolo's catalogue," even though it carries additions and interpolations by several hands and is described as only *maxima ex parte* by Stornajolo.

13 Città del Vaticano, BAV, Sala Cons. Mss. 411 (1–3).

14 Saxer, *Sainte-Marie-Majeure*, 563–674.

Both collections are also covered in the five volumes of *Les manuscrits liturgiques latins de la Bibliothèque Vaticane* by Pierre Salmon (Biblioteca Apostolica Vaticana, 1968–1972), which constitutes the most recent take on the Vatican Library's liturgical material at large. Needless to say, this is an unsurpassed tool for scholars in a variety of fields. However, as a consequence of the quantity of material covered, descriptions of individual manuscripts are often brief and summarising. While Salmon's work predates Saxer's catalogue of the S. Maria Maggiore collection, he has taken into account Stornajolo's *Inventarium* of the S. Pietro manuscripts for volumes II through V of his series. For the first volume, covering many of the S. Pietro manuscripts discussed in this study, however, Salmon states not to have had access to Stornajolo's notes.[15] It has become evident while conducting research for this article that this disconnect has had unfortunate consequences for the current level of knowledge regarding the Vatican Basilica's material, as Salmon's widely accessible catalogue is probably the most influential work of its kind on many of the Vatican Library's collections. Indeed, Giacomo Baroffio's *Iter Liturgicum Italicum*, the most comprehensive inventory of Italian liturgical manuscripts ever compiled, seems largely to depend on Salmon for localisations of the S. Pietro manuscripts.

Iter Liturgicum Italicum is a massive inventory of over 30,000 Italian liturgical manuscripts, compiled by Italian liturgist and musicologist Giacomo Baroffio. As Baroffio's *Iter* is nowadays hosted online by the *Institut de recherche et d'histoire des textes* in France, it is a powerful tool for the study of Italian liturgical sources with vast potential for evolving with new research, both in terms of informing study and in terms of absorbing and disseminating its results.[16] As the greater part of the data in *Iter* has been derived from research and

15 Pierre Salmon, *Les manuscrits liturgiques latins de la Bibliothèque Vaticane. 1, Psautiers, antiphonaires, hymnaires, collectaires, bréviaires*, Studi e testi 251 (Città del Vaticano: Biblioteca apostolica vaticana, 1968), XIII: "*Je signale que je n'ai malheureusement pu voir les notes de Stornajolo sur les manuscrits de l'Archivio S. Pietro ... qu'après la rédaction des notices des manuscrits contenues dans le présent volume, sauf pour les manuscrits de la liste complémentaire.*"

16 Giacomo Baroffio and Laura Albiero, *Iter Liturgicum Italicum: Répertoire Des Manuscrits Liturgiques Italiens Établi Par Giacomo Baroffio*, ed. François Bougard (Institut de recherche et d'histoire des textes, 2016), http://liturgicum.irht.cnrs.fr. Originally published in 1999, the first edition of Barioffio's *Iter* was followed by a much expanded *Editio Maior* in 2011: Giacomo Baroffio, *Iter liturgicum Italicum* (Padova: CLEUP, 1999); Giacomo Baroffio, *Iter liturgicum Italicum: Editio Maior* (Stroncone (Terni): Associazione San Michele Arcangelo, 2011). In 2003, between the two editions of *Iter*, Baroffio published an extract on Italian calendars from his growing inventory. See Giacomo Baroffio, "Kalendaria Italica Inventario," *Aevum* 77, no. 2 (2003): 449–72. Unless otherwise indicated, all references to Baroffio's data in this study are to the most recent online edition.

catalogues from all over the world, the precision of the information varies according to the literature available on individual library collections.[17] It is likely, therefore, that the number of extant Italian calendars is greater than the 2168 manuscripts presently indicated in *Iter*. This is simply because not all catalogues display the contents of manuscripts with a sufficient level of detail to indicate the presence of a calendar. On the other hand, it means that many of the localisations given in *Iter* are still tentative at best, while some indicated calendars will turn out not to be calendars at all, due to variation in cataloguing practices and oversight in scholarship.[18]

In short, even though Baroffio's *Iter* provides a valuable tool and starting point for research on Italian liturgical sources, there is still likely to be a great deal of unidentified and uncontextualised material even in its current data, let alone in the material that remains off the grid altogether. With this in mind, the 70 calendars with Roman localisation presently listed in *Iter* provide the most credible estimate of the quantity of sources available for the study of medieval Roman calendars, and consequently for positioning the S. Pietro and S. Maria Maggiore material within a wider Roman context. Besides Baroffio, the only estimate of the number of medieval Roman calendars has been put forward by Marko Halonen, who in a 2017 article stated that there are 169 such calendars known to modern scholarship. As Halonen refers only to Baroffio, however, it is unclear what this assessment is based on.[19]

3 The Collections

The collection *Archivio del Capitolo di S. Pietro* was integrated into the Vatican Library as late as 1940, and its history is therefore intimately tied to that of the basilica itself. The collection boasts a significant number of donations, many from the personal library of Cardinal Giordano Orsini (died 1438), who

17 Baroffio, *Iter 1999*, V–VI. See also the description of the online inventory under the heading *Accueil* in Baroffio and Albiero, *Iter Online*.

18 For an example, see the Excursion at the end of the article.

19 Marko Halonen, "Kalendaria urbana et rustica: Keskiaikaisia kalentereita Roomassa sekä Suomen kaupungeissa ja maaseudulla," in *Leivoksia, kaupunkilaisia ja sivistysaatteita: Kävelyretkiä Euroopan historiaan*, ed. Sari Aalto, Samu Nyström, and Rose-Marie Peake (Helsinki: Kirjapaja, 2017), 131. Halonen's references are to Baroffio's personal website and 2003 article, where the number of Roman calendars identified is even smaller than the current figures in the online edition of *Iter*. Cf. Baroffio, "Kalendaria Italica." Upon enquiry, Halonen stated not to have recollection or records of the origins of the figure (email on 16 October 2017).

bequeathed the basilica an impressive 241 manuscripts.[20] The beginning of the 16th century, on the other hand, held a series of misfortunes: the sack of Rome in 1527, the construction of the new Vatican Basilica and the accompanying displacement of the library, and the selling of parts of the collection in 1535. Even though this all adds up to a rather heterogeneous collection, which has undoubtedly lost some of its core medieval material, it remains exceptionally substantial, consisting of some 500 items.[21]

20 Orsini's ecclesiastical career was extensive, leading him on travels around Europe. He was also an avid book collector and a humanist patron at a formative time for Italian libraries. On Orsini and his library, see Christopher S. Celenza, "The Will of Cardinal Giordano Orsini (Ob. 1438)," *Traditio* 51 (1996), 257–86; Christopher S. Celenza, "Orsini, Giordano," in *Dizionario biografico degli Italiani v. 79*, ed. Raffaele Romanelli (Roma: Istituto della Enciclopedia italiana, 2013); Erich König, *Kardinal Giordano Orsini († 1438) ; Ein Lebensbild Aus Der Zeit Der Großen Konzilien Und Des Humanismus*, Studien Und Darstellungen Aus Dem Gebiete Der Geschichte ; 5,1 (Freiburg i. Br.: Herder, 1906); Giuseppe Lombardi and Flavia Onofri, "La biblioteca di Giodano Orsini (c. 1360–1438)," in *Scrittura, biblioteche e stampa a Roma nel Quattrocento. Aspetti e problemi. Atti del Seminario 1–2 giugno 1979*, ed. C. Bianca et al., Littera Antiqua, 1,1 (Città del Vaticano: Scuola Vaticana di Paleografia, Diplomatica e Archivistica, 1980), 371–82. Charles L. Stinger has described Orsini's library as "the richest Roman library of classical and patristic manuscripts" during the pontificate of Eugene IV (1431–1447), while noting that S. Pietro's library, "a more important collection at this time than the papal library," was "unimpressive" in this regard. Charles L Stinger, *The Renaissance in Rome* (Bloomington: Indiana University Press, 1985), 283. As to why the cardinal showed the chapter such favour: in 1434 Orsini became the fifth of his family to hold the high position of Archpriest of S. Pietro. The basilica had become an Orsini stronghold in the power struggles of the Roman nobility already in the 13th century, and one *Ursus de domo Ursinorum* was long held to have been its legendary *primus Archipresbyter Cardinalis*. See Blaauw, *Cultus et decor*, 631; Dario Rezza and Mirko Stocchi, *Il capitolo di San Pietro in Vaticano dalle origini al XX secolo. Vol. 1: la storia e le persone* ([Città del Vaticano]: Edizioni Capitolo Vaticano, 2008), 151–53, 181–83.

21 For a general history of the Basilica's library, see Christine M. Grafinger, "Archivio del Capitolo di S. Pietro. [Manoscritti]," in *Guida ai fondi manoscritti, numismatici, a stampa della Biblioteca Vaticana, I. Dipartimento Manoscritti*, ed. Francesco D'Aiuto and Paolo Vian (Città del Vaticano: Biblioteca Apostolica Vaticana, 2011), 333–36; Giovanni Mercati, *Codici latini Pico Grimani Pio e di altra biblioteca ignota del secolo XVI esistenti nell' ottoboniana e i codici greci Pio di Modena con una digressione per la storia dei codici di S. Pietro in Vaticano*, Studi e testi 75 (Città del Vaticano: Biblioteca Apostolica Vaticana, 1938), 168. For early inventories of the collection, see Michael M. Gorman, "The Oldest Book List from St Peter's," in *Miscellanea Bibliothecae Apostolicae Vaticanae XI*, Studi e Testi 423 (Città del Vaticano: Biblioteca Apostolica Vaticana, 2004), 383–98; Giuseppe Lombardi, "Inventari di biblioteche romane del Quattrocento: un panorama," in *Libri, lettori e biblioteche dell'Italia medievale (secoli IX–XV). Fonti, testi, utilizzazione del libro. Atti della Tavola rotonda italo-francese (Roma, 7–8 marzo 1997)*, ed. Giuseppe Lombardi and Donatella

The manuscript collection of S. Maria Maggiore has a similar history to the collection of S. Pietro, as it was transferred to the Vatican Library relatively late, in 1931. This collection is smaller than its counterpart, comprising a little over 80 manuscripts and some incunabula. While there are no signs of single massive donations comparable to Orsini's, one can assume the collection has seen its share of both patronage and the adversities that have befallen Rome over the centuries.[22]

Indeed, despite the likely loss of material, the collections contain some 58 medieval calendars between them. Of these, I have been able to associate 17 with varying degrees of use at the two basilicas and several others with a broader Roman context. Keeping in mind the figure of 70 Roman calendars currently listed in *Iter*, it becomes clear that we are dealing with a substantial segment of the city's surviving patrimony. It should be reiterated, however, that the figures given in *Iter* are evolving: the total of 70 Roman calendars currently listed includes only eight of the S. Pietro and S. Maria Maggiore calendars I have identified. Further close inspection of individual library collections is likely to change the figures further, although the collections examined here are exceptional in their breadth and quantity of material, as well as their intimate connection to specific Roman basilicas. S. Pietro and S. Maria Maggiore, therefore, stand in stark contrast with the many hundreds of churches of medieval Rome whose patrimony has been entirely lost or is scattered in libraries and archives in Rome, Italy, and around the world, waiting to be identified.[23]

Nebbiai Dalla Guarda (Roma: Istituto centrale per il catalogo unico delle biblioteche italiane, 2000), 355, 361–64.

22 Piergiorgio Parodi, "S. Maria Maggiore," in *Guida ai fondi manoscritti, numismatici, a stampa della Biblioteca Vaticana, I. Dipartimento Manoscritti*, ed. Francesco D'Aiuto and Paolo Vian (Città del Vaticano: Biblioteca Apostolica Vaticana, 2011), 527–30. On the devastating effects of the 1527 sack of Rome on the city and its archival and library collections, see Jean-Claude Maire Vigueur, *The Forgotten Story: Rome in the Communal Period*, trans. David Fairservice (Roma: Viella, 2016), 7–8.

23 For a classic catalogue of medieval Roman churches, see Christian Hülsen, *Le chiese di Roma nel medio evo: Cataloghi ed appunti* (Firenze: Leo S. Olschki, 1927). Only a handful of Roman churches have one or at most a few calendars indicated in *Iter*. With no library collections of their own to examine, the likelihood of finding calendars to add to the material of such churches is slim. Moreover, existing localisations are tentative, and even when credible, they leave many questions open regarding the actual relationship between church and manuscript: namely, whether the calendars were actually used at the churches in question, or whether they were simply donations of virtually unused luxury manuscripts. Both groups are represented among the calendars previously attributed to S. Pietro, for instance, as will be demonstrated below.

4 Methodology

Localising calendars is largely based on identifying calendar entries that can be tied to a certain place. Although the core festivities and commemorations of the medieval liturgical year were shared, there were numerous local traditions and particularities. Indeed, even the dates of universal feasts could vary: the Apostle Bartholomew, for example, was celebrated in much of Latin Christendom on 24 August, as he still is today, but on 25 August in and around Rome.[24]

Moreover, many medieval calendars carry additional saints and celebrations particular to the individual communities that used them and to the surrounding regional contexts. The prevalence of such calendar traditions can to some extent be attributed to the decentralised nature of the production, dissemination, and procurement of liturgical books. In Rome, too, churches and religious communities relied on benefactors and their own resources to acquire the liturgical and religious literature they needed. The numerous coats of arms and other professions of private ownership and commission pertaining to wealthy individuals in the S. Pietro and S. Maria Maggiore manuscripts indicate that donations formed a considerable part of this process.[25] The calendars in these manuscripts could then be adapted according to the needs of their new users and surroundings.[26] As a result, a new local layer of entries can in some cases easily be identified, as the added entries stand out visibly among the original texts on calendar pages.

In localising material to S. Pietro and S. Maria Maggiore, my primary evidence is built around direct references to the physical structure of the basilicas themselves, mostly in mentions of their altars and chapels. I have also considered references to collectives, for example capitular coats of arms and specific phrasing in calendar entries, such as the specification "*cuius caput habemus,*"

24 Cf. calendars ACSP E4, fol. 4v, and SMM 94, fol. 4v. The first has Bartholomew's name written in rubrics on both days, with the description "*romani faciunt die sequenti*" on the first and "*prout Rome fit*" on the latter. The latter of the calendars, an incunable, has a similar note added to the margin of the August calendar page: "*Rome fit festum sancti Bartholomei XXV°, est vigilia XXIIII°.*"

25 See for example calendar manuscripts ACSP B63, B65, B66, B67, B68, B69, B72, B82, E4, F23; SMM 50, 119, 120, and the incunable SMM 94. See also the numerous mentions of donated books in the S. Pietro book of anniversaries ACSP H57, e.g. fols. 48r, 75r, 104r, 138r, 145r, 160r, 172r-v. Indications of production directly for Roman basilicas are rare in comparison, and even these cases seem to confirm the decentralised nature of the process: see sections on manuscripts ACSP A9 and SMM 119 and 120 below.

26 See the sections on the adopted S. Pietro and S. Maria Maggiore calendars below.

added to the name of a saint in calendar ACSP E14.[27] The rarity of such direct references indicates that they are relatively strong evidence when linking calendar to place, especially when considered in combination with other entries and evidence. Furthermore, I have expanded the selection of sources to include calendars that contain entries identified as indicative according to the previous criteria, but which nonetheless lack a direct reference to their surroundings, for example reading "*Translatio Ieronymi*" instead of "*Translatio Hieronymi ad sanctam Mariam maiorem.*"[28]

In assessing the indicative value of various calendar entries, especially those without direct references to their physical surroundings, it is useful to consider the frequency of their appearance in calendars outside Rome and Italy. I have therefore compared the 17 calendars I have identified with S. Pietro and S. Maria Maggiore with transcriptions I have prepared of a group of 24 additional calendars, mostly originating from central Italy in the 14th and 15th centuries.[29]

For a wider perspective, I have consulted the online calendar-resource *Calendoscope*, published by Denis Muzerelle and hosted by the *Institut de recherche et d'histoire des textes* of the *Centre national de la recherche scientifique* in France.[30] *Calendoscope* contains data from two distinct data sets: 176 calendars mainly from German dioceses and a further 339 calendars, mostly from France. This data has been refined into the current form by Muzerelle from two century-old sources: Hermann Grotefend's classic *Zeitrechnung des deutschen Mittelalters und der Neuzeit* and the Bollandists' collation of the hagiographical contents of over 350 Bibliothèque nationale de France manuscripts, appended to volume III of the *Catalogus codicum hagiographicorum latinorum*.[31]

This is heavily processed material, twice removed from the original sources, with already the initial stages of transcription and editing leaving much to be desired, although Muzerelle has taken editorial steps of his own to counter

27 See the section on manuscript ACSP E14 below.

28 See table 5.4 below.

29 Avignon, Bibliothèque Municipale, 100; BAV, ACSP B63, B64, B65, B66, B67, B68, B69, B71, B72, B81, B82, E1, E4, F28; SMM 41, 48, 106, 112, 113, 114; Stamp.ACSP A41; Vat. Lat. 4406; Roma, Biblioteca Angelica, 1098.

30 D. Muzerelle, *Calendoscope, Logiciel d'aide à l'identification Des Calendriers Liturgiques Médiévaux* (Institut de recherche et d'histoire des textes, 2013), http://calendoscope.irht.cnrs.fr/.

31 Grotefend, Hermann, *Zeitrechnung des deutschen Mittelalters und der Neuzeit. Bands 1–2* (Hannover: Hahn'sche Buchhandlung, 1891); Hagiographi Bollandiani, ed., *Catalogus codicum hagiographicorum latinorum antiquiorum saeculo XVI qui asservantur in Bibliotheca nationali Parisiensi, tom. III* (Bruxellis, 1893), 579–733.

some of these issues.[32] The data should, therefore, be consulted only for broad and unambiguous questions. Nevertheless, due to the great quantity of compiled sources, *Calendoscope* offers a useful comparison to the Roman material, especially when determining whether certain calendar entries should be considered particular or more or less universal. Furthermore, while both data sets have their geographical focus north of the Alps, there is some Italian material present in both, and the considerable number of mendicant calendars is a particularly useful point of comparison due to the notable presence of such calendars in Rome.[33]

Looking beyond the simple appearance of certain feasts, there are further liturgical indications in medieval calendars to consider when attempting to determine their origin and provenance.[34] Medieval scribes employed several methods for distinguishing feasts of greater importance from minor commemorations. Since most entries were written in black ink, the more prominent celebrations could be indicated by a change in colour. Indeed, the common use of red ink to this effect is the origin of the expression "red letter day." In some calendars, feasts are also ranked by grading systems that indicate the appropriate extent of liturgical services, although these varied through time and between religious orders and communities. Finally, additional services could be allocated to the most important feasts, signalled on calendar pages with the vigil, written on the date preceding the main celebration, and the octave, marked a week after. In practice, there was much variation in how these distinguishing features were used and what their exact implications were, and inconsistencies can be found even within individual calendars. Whatever the particularities, such nuances in emphasis can be revealing as to a calendar's history.

There are, however, some complications in localising medieval calendars. For one, it is not always a simple task to identify the name of a saint on a calendar page with a certain feast, as saints with similar names and attributes tend

32 For detailed caveats, see *Particularités des différents corpus* in Muzerelle, *Calendoscope*.

33 There are of course additional useful repertories to consult when studying medieval calendars, such as the works of S.J.P. van Dijk and the many volumes of *Acta Sanctorum*. For van Dijk, see e.g. Stephen Joseph Peter van Dijk, *Sources of the Modern Roman Liturgy: The Ordinals by Haymo of Faversham and Related Documents (1243–1307), v. 2. Texts*, Studia et Documenta Franciscana (Leiden: E.J. Brill, 1963); van Dijk and Walker, *The Ordinal of the Papal Court*.

34 With *origin*, I refer to the hypothesised context of the production or commission of manuscripts and calendars, whereas with *provenance* I indicate subsequent changes in ownership.

to be concentrated on the same dates.[35] What's more, medieval calendars typically display a high degree of blending of different traditions and strands of influence, with both famed and obscure saints from all around Christendom occupying the calendar pages. Indeed, entries could be entered and copied into the calendars from diverse sources and for a variety of reasons.[36] Although there must have been some logic to the act of writing in most cases, it is often impossible to say exactly what this may have been. Liturgical books in general, especially the breviary and the missal, are essentially syntheses of several earlier forms of books, collating various genres of texts, such as prayers, readings, sermons, rubrics, psalms, hymns, and antiphons. Copying and collating from several manuscripts was therefore a natural part of the development of liturgical books, and this was likely practiced even in the later Middle Ages.[37] Calendars are unlikely to be exempt from this phenomenon, and one should not expect to find "pure" calendar traditions or hold any single calendar to be too representative of a specific tradition.

Moreover, communities such as the chapters of S. Pietro and S. Maria Maggiore did not exist in a vacuum. The calendars of any such community are likely to reflect not only the customs of the community itself, but also those derived from their surroundings, from local to regional and even universal levels. There are also direct contacts between specific centres, even when geographically distant, to consider. These are especially prominent between centres belonging to monastic or mendicant orders but played their part in the

35 For the 9 February red-letter feast of a bishop and confessor *Savinus* in calendar ACSP B85, for example, *Acta Sanctorum* has three candidates to offer – all Italian, all bishops, all named Sabinus. See *Acta Sanctorum Februarii Tomus II* (Antverpiae: Apud Iacobum Meursium, 1658), 275.

36 Edward B. Garrison has suggested that calendar entries may occasionally have simply been "culled" from martyrologies, "either waywardly or for reasons of private piety that are likely to remain forever hidden from us." Edward B. Garrison, "Three Manuscripts for Lucchese Canons of S. Frediano in Rome," *Journal of the Warburg and Courtauld Institutes* 38 (1975), 10. This is certainly a sense one gets from some calendars where hardly a day has been left empty. Cf. the section on the S. Maria Maggiore incunable 94 below.

37 Giacomo Baroffio, "I codici liturgici: specchio della cultura italiana nel Medioevo. Punti fermi, appunti di lettura, spunti di ricerca," *Ecclesia orans* 9, no. 3 (1992), 264; Andrew Hughes, "The Scribe and the Late Medieval Liturgical Manuscript: Page Layout and Order of Work," in *The Centre and Its Compass: Studies in Medieval Literature in Honor of Professor John Leyerle*, ed. Robert A. Taylor, Studies in Medieval Culture 33 (Kalamazoo, Michigan: Medieval Institute Publications, 1993), 153–63; Andrew Hughes, *Medieval Manuscripts for Mass and Office: A Guide to Their Organization and Terminology* (Toronto: University of Toronto Press, 1995), 118–23.

lives of canons as well. S. Pietro, for example, was the centre of a vast network of subject churches, monasteries, and hospitals across Italy.[38]

A further problem to consider is the distinction between the communal and private use of calendars, as this has significant bearing on the conclusions one could hope to draw from the material. More specifically, what should one consider indicative of local communal continuity on the one hand, and what should one attribute to private preferences, potentially foreign to a Roman context, on the other? Fortunately, there are some clear cases that help illustrate these dynamics, such as the many lavishly decorated manuscripts adorned with coats of arms and other expressions of private ownership, which were donated or bequeathed to basilical chapters by wealthy individuals. Manuscripts ACSP B67 and B72 are good examples, as both were donated to S. Pietro close to their previous owners' deaths, and both have more or less pristine calendars. They are likely to reveal less about communal practices at the basilica than the personal tastes of the benefactors, even as these held positions in the basilical chapter themselves.[39]

Even manuscripts produced specifically for the use of a certain religious community may reflect more the particularities of the production process than the context for which they were destined. Manuscripts SMM 119 and 120 illustrate the kinds of factors that could influence the contents of a calendar. The manuscripts' long incipits state unambiguously and at great length that they were produced for the use of the chapter of canons at S. Maria Maggiore, and we can therefore confidently localise the calendars to the Basilica. The most prominent feature of the calendars, however, is their numerous Augustinian feasts, which are not found in the Basilica's other material. More often than not, such a case would remain shrouded in questions, but in this rare fortunate instance there is a likely answer: the scribe carrying out the work has identified himself as an Augustinian hermit.[40]

Whether the evidence is direct or more circumstantial, it is difficult to contextualise the calendars with much precision. They cannot, for example, be tied to specific altars or chapels. In fact, this is even true of churches, as some of the calendars I have identified with the two basilicas could very well have been used in one of their many subject churches, only subsequently making their way to the main basilica's library collection. Indeed, it must be noted that

38 See, for example, Mirko Stocchi, *La Basilica Vaticana e le sue chiese tra Medioevo ed Età Moderna: Volume I, Umbria* (Città del Vaticano: Edizioni Capitolo Vaticano, 2013); For examples of the effects various types of contacts between religious centres had on liturgical manuscripts, see Baroffio, "I codici liturgici."

39 See sections on manuscripts ACSP B67 and B72 below.

40 See section on manuscripts SMM 119 and 120 below.

among the 17 calendars discussed below, there is not a single identical pair, not even SMM 119 and 120, which are the results of the very same production process. Any attempts at localisation should therefore not be taken too literally but should rather be understood to suggest some form of contact between calendar and place. None of this differs much from the caveats that should always be kept in mind when contextualising medieval manuscripts, whether on palaeographical or art historical grounds, as someone painting in Bolognese style or writing in the French manner could very well have been operating in Rome or Naples.[41] Similarly, while calendar SMM 50 contains a range of original entries and later additions pertaining to Genoa and Pavia, mentioning these as indicative for the purposes of localisation does not imply that they were necessarily written within the confines of the cities in question.

On a final methodological note, calendars are often used not only to localise, but also to date manuscripts. While calendars can provide additional weight to a body of evidence that suggests a rough date or *terminus*, relying on calendars alone has several serious pitfalls. To give an example: Jacques Dubois and Jean-Loup Lemaitre propose in their hagiography textbook that one can determine a *terminus post quem* for manuscripts from calendar entries written in the original hand and a *terminus ante quem* from added entries.[42] The two underlying suggestions are that a feast of a saint will *not* be found in a calendar prior to the "adoption" of the feast, usually meaning the saint's canonisation, and that it *will* be found in calendars produced thereafter.[43]

However, the first suggestion does not take into account the fact that an existing cult and *fama sanctitatis* was usually the prerequisite for a formal recognition of a saint, especially in the case of canonisation. Thus, it is not out of order that Nicholas of Tolentino (died in 1305, canonised in 1446) is an original entry in calendar ACSP B85, which likely predates his canonisation, as the cult enjoyed considerable popularity and promotion already in the 14th century, and even the process of canonisation was started early, with enquiries

41 Cf. Baroffio, *Iter 1999*, VI: "*Spesso l'origine è stabilita unicamente in base a criteri iconografici e alle caratteristiche delle diverse scuole miniaturistiche, mentre è chiaro che come un Girolamo da Cremona, così tanti altri maestri e intere botteghe potevano operare distanti dalla sede operativa originaria.*"

42 Jacques Dubois and Jean-Loup Lemaitre, *Sources & méthodes de l'hagiographie médiévale* (Paris: Éditions du Cerf, 1993), 150: "*On rechercera donc les saints inscrits de première main susceptibles de fournir un terminus post quem, et ceux inscrits en addition, fournissant un terminus ante quem.*"

43 Granted, canonisations are not the only mechanism employed thus, but they are among the principal ones. Cf. for instance the chronological lists compiled by Victor Leroquais, also referred to by Dubois and Lemaitre. Victor Leroquais, *Les Bréviaires manuscrits des bibliothèques publiques de France* (Paris: Mâcon, Protat frères, 1934), XCVII–CXVIII.

conducted in 1325 and 1357. That the saint was credited with 301 miracles is in itself an indicator of the cult's success.[44] Indeed, even discounting all feasts that predate the institutionalisation of canonisation, there are plenty of saints in medieval calendars who were never canonised in the Middle Ages, such as the Franciscan Simon of Collazzone, found in calendar ACSP F23. It is therefore important to acknowledge that official recognitions of saints were just a specific mode of sanctification, and that calendars could, and often did, follow a different, more diverse, and locally oriented logic in their contents.

The suggestion that a feast can reliably be expected to be found in medieval calendars after some form of official recognition is at the very least equally problematic, as it relies on the modern-day assumption that something definitely ought to have been included in or excluded from a medieval calendar. Yet, entries were at times left out ostensibly by accident, especially as calendars could be worked on by scribes not fully versed in liturgy, or by people faithfully following inadequately composed instructions. Furthermore, local liturgical preferences and customs could change and run contrary to broader trends, and manuscripts could be copied from old exemplars. It is also possible that some information was simply considered too self-evident to merit later intervention into an already completed calendar. In sum, one must recognise that medieval practices varied enough to sometimes defy modern understanding. Why, for instance, has the entry specific to S. Maria Maggiore, "*Translatio sancti Hieronimi doctoris ad sanctam Mariam maiorem,*" been added to calendar SMM 97, while the feast of the Basilica's dedication, a much more common and broadly resonant feast, has been omitted? Whatever the reason, it is obvious that at various stages during a calendar's production and time in use, it was entirely possible to leave out entries, even locally significant ones and even during a period when such feasts were already theoretically in existence. Consequently, too much weight should not be given to the absence of something from a calendar when making positive statements about dating.

This is not to say that the absence or presence of certain key calendar entries could and should not be used in dating calendars and manuscripts at all. They can very well contribute additional circumstantial evidence to a dating that has simultaneously been reached through other avenues, such as palaeographical or art historical considerations. They should not, however, be relied

44 See André Vauchez, *Sainthood in the Later Middle Ages* (Cambridge: Cambridge University Press, 1997), 122–23, 254, 333, 503.

upon independently and certainly not as "often the best way to date" manuscripts, as suggested by Dubois and Lemaitre.[45] That being said, there are cases where the presence or absence of a feast can be particularly useful. Thus, the addition of Bernardino of Siena (died in 1444, canonised in 1450) to several S. Pietro calendars demonstrates that some of the calendars saw use even a century after their production.[46] Here the short distance in time between the death and the canonisation of the saint, as well as the fact that the canonisation took place in S. Pietro itself, renders the Franciscan saint's canonisation a reasonably reliable temporal marker. As the distance between death and canonisation increases, however, as in the case of Nicholas of Tolentino, one should consider the implications of the requirements of a pre-existing cult and *fama sanctitatis* for a successful canonisation.

5 The Calendars of S. Pietro in Vaticano

As of August 2020, there are five manuscript calendars in the Biblioteca Apostolica Vaticana localised to S. Pietro in Baroffio's *Iter*.[47] These are:

45 Dubois and Lemaitre, *Sources & méthodes de l'hagiographie médiévale*, 150: " … *le calendrier est souvent le meilleur moyen de dater le manuscrit qui le renferme.*"

46 Calendars ACSP A9, E14, E15, and F23.

47 Outside of the Vatican library, only one such manuscript is given: Veroli, Biblioteca Giovardiana, 16, a calendar-ceremonial dated to 1491–1500. I have been unable to consult this manuscript for the present study, and thus cannot confirm its relationship with S. Pietro. It should be noted, that while *Iter* lists the ceremonial on fols. 12–139 (*recto* and *verso* not indicated) as originating in Rome, the calendar on fols. 4–9 is specifically localised to *Roma, S. Pietro*. The link is not established in either of the collection's printed catalogues I have consulted, however: Lidia Buono, "Mns. 16," in *Catalogo dei più antichi manoscritti della Biblioteca Giovardiana di Veroli* (Roma: Gruppo editoriale internazionale, 1996), 69–71; Camillo Scaccia-Scarafoni, "Veroli – Biblioteca Giovardiana," in *Inventari dei manoscritti delle biblioteche d'Italia: vol. XXXIV*, ed. Albano Sorbelli (Firenze: Leo S. Olschki, 1926), 10. There is also a group of five calendars mentioned by van Dijk, which pertain to the pontificate of Nicholas III; one of these carries two added references to the Vatican Basilica and the adjacent papal palace. However, as the calendar (Paris, Bibliothèque nationale de France, Lat. 755) reflects a rather specific papal tradition, influenced by that of S. Pietro but not reflecting use at the basilica itself, I have not included it in this study. See Stephen Joseph Peter van Dijk and Joan Hazelden Walker, *The Origins of the Modern Roman Liturgy: The Liturgy of the Papal Court and the Franciscan Order in the Thirteenth Century* (Westminster, MD: The Newman Press, 1960), 407, fn. 2; S.A. Van Dijk, "Three Manuscripts of a Liturgical Reform by John Cajetan Orsini (Nicholas III)," *Scriptorium* 6, no. 2 (1952), 213–42; Stephen Joseph Peter van Dijk, "The Legend of 'the Missal of the Papal Chapel' and the Fact of Cardinal Orsini's Reform," *Sacris Erudiri* VIII, no. 1 (1956), 76–142; Dijk and Walker, *The Ordinal of the Papal Court*, XVIII–XIX, 61–85.

Città del Vaticano, Biblioteca Apostolica Vaticana,
- Arch. Cap. S. Pietro B.67
- Arch. Cap. S. Pietro B.72
- Arch. Cap. S. Pietro B.79
- Arch. Cap. S. Pietro B.87
- Arch. Cap. S. Pietro B.88

This reflects the contents of the S. Pietro library collection poorly, as five of its 42 medieval calendars are likely to originate at the basilica itself, and a further five are likely to have been modified there at a later point during their time in active use.[48] Of the former group, only B79 and B88 are currently indicated in *Iter*, and of the latter only B87 has been identified. Manuscripts B67 and B72, on the other hand, do not belong to either group, but are two of the numerous donations made to the basilica by its benefactors.

This state of affairs is partly due to confusion pertaining to individual manuscripts, and partly to the disconnect between the collection's detailed cataloguing by Stornajolo and the highly influential series on the Vatican library's liturgical manuscripts by Salmon. As Stornajolo's catalogue has up until recently only been available *in situ* at the library, it is easy to see how oversights by Salmon have been able to make their way into subsequent scholarship.

In identifying the calendars pertaining to S. Pietro, I have considered some general features of the manuscripts, such as incipits, coats of arms, and liturgical texts other than the calendars themselves. I am greatly indebted to previous scholarship, especially the many cataloguers who have written on the material, for locating many such points of interest in the manuscripts. The greater part of the evidence for localising the calendars, however, comes directly from the calendar pages themselves. For S. Pietro, the number of clear indicative entries is substantial, as can be seen in tables 5.1 and 5.2 below.

There is no need to give a detailed account of each entry here, although it should be noted that some are more straightforward to identify than others.[49]

48 The full list of the collection's 42 calendars consists of manuscripts: ACSP A9, B63, B64, B65, B66, B67, B68, B69, B70, B71, B72, B73, B79, B81, B82, B84, B85, B87, B88, D156, E1, E2, E4, E6, E7, E8, E9, E10, E11, E14, E15, F13, F19, F23, F26, F27, F28, F29, F45, and H28; and incunabula: Stamp.ACSP 31 and A41.

49 On the dedications of the main altar of S. Pietro and that of *Maria de cancellis*, see Jounel, *Le Culte Des Saints*, 230, 304. I have so far been unable to identify the altar *Salvatoris de pace*. The chapel *Sancti Salvatoris de abundantia* mentioned in the will of Cardinal Giordano Orsini could be a candidate, were it not for the fact that the two dedications are mentioned separately in the S. Pietro books of anniversaries. This makes the connection seem less likely, albeit not impossible. Cf. ACSP H57, fols. 132r, 163r; *Testamentum* in Celenza, "The Will of Cardinal Giordano Orsini (Ob. 1438)," 276. The entry appears exclusively in calendars containing other dedications of the Basilica, so it seems safe to assume a

TABLE 5.1 Indicative entries in calendars originating at San Pietro.

Date	A9	B79	B88	E15	F23
Mar. 25	Consecratio altaris beati Petri facta a papa Calisto secundo	Consecratio altaris sancti Petri		Consecracio al{taris} beati Petri facta a p{apa} Calisto secun{do}	
May 6	Dedicatio altaris [Boni]facii			Dedicatio altaris Bo{nifacii}	
May 7					Consecratio altaris capelle Bonifacii pape
Sep. 4	Sancti Bonifacii pape /Requi-escit ad altare pape Bonifacii\				Sancti Bonifacii pape et confessoris
Sep. 17	/Sancti Lamberti martyris in capella dominorum canonico-rum sancti Petri\		/Lamberti pape et martyris\		/Sancti Lamberti episcopi et martyris\
Nov. 6	Consecratio altaris sancte Marie de canzellis anno domini Mº CCº XVIº	Consecratio altaris sancte Marie de cacellis		Consecratio altaris sancte Marie de canzellis anno domini Mº CCº XVIº	Consecratio altaris sancte Marie de cancellis
Nov. 17	/Dedicatio altaris Salvatoris de pace\			Dedicatio altaris Salvatoris de pace	
Nov. 27	/Sancti Iacobi intercisi mar-tyris\ /cuius caput habemus\				/Sancti Iacobi interincisi Martyris\

Legend: /added entry\; [text illegible due to intervention]; {text illegible due to damage to manuscript}

TABLE 5.2 Indicative entries in calendars modified at San Pietro.

Date	B84	B85	B87	E14	F27
Jan. 23			/In isto die fuit consecratus dominus Bonifacius papa VIII^{us} et concessit in basilica principis apostolorum de urbe annuatim in isto die x dies indulgentie post alias indulgentias\		
Mar. 25	/Consacratio altaris maioris facta per papam Calistum\				/Consacratio maioris altaris nostre basilice\
Mar. 26		/Festum consecrationis maioris altaris basilice sancti Petri de urbe\			
May 6	/Consacratio altaris sancti Bonifatii\		/Consecratio altaris domini Bonifacii pape VIII\		
Sep. 4	/Sancti Bonifatii pape et confessoris, requiescit in cappella Bonifatii\				
Sep. 5					/Sancti Bonifatii pape et confessoris\
Nov. 6	/Consacratio sancte Marie de cancellis\		/Consacratio sancte Marie de cancellis\		/Consacratio altaris de cancellis\
Nov. 17					/Consacratio altaris sancti Salvatoris de pace\
Nov. 27				Iacobi intercisi martyris / cuius caput habemus\	

Legend: /added entry\

Both tables contain cases where my categorisation of the calendars in the two groups could be contested: calendar B88, for instance, contains little in the way of clear references to S. Pietro, while F23 carries a substantial number of entries that are foreign to Rome, written in original hand. On the other hand, one could argue that E14 should in fact be considered as one of the basilica's original calendars. As exact localisation of calendars is problematic, I do not propose that the divisions presented here should be considered final. However, I address some of these issues in discussions on individual calendars and manuscripts below.

I also discuss the two S. Pietro books of anniversaries, manuscripts ACSP H56 and H57, as they provide a useful point of comparison to the calendars with which they share some features. Furthermore, I address the issue of the two aforementioned calendars, manuscripts ACSP B67 and B72, which I consider to be mistakenly attributed to S. Pietro. Although both B67 and B72 were donated to S. Pietro by members of its chapter – the former by its archpriest, Cardinal Giordano Orsini (died 1438), the latter by one of its canons, Martinus de Roa (died 1475) – I propose that these reflect Orsini's activities as an avid book collector and de Roa's personal preferences, rather than the use of calendars at the basilica in general.

6 Original S. Pietro Calendars

6.1 *Manuscript A9*

Biblioteca Apostolica Vaticana, Arch. Cap. S. Pietro A.9
Dat. 1339; Legendar
Calendar on fols. Vr–Xv[50]

connection with S. Pietro in any case. St Lambert and the chapel of the basilica's canons are mentioned in Francesco Cancellieri, *De Secretariis Basilicae Vaticanae – Vol. IV: Sylloge Veterum Monumentorum* (Romae: ex officina Salvioniana ad lyceum Sapientiae, 1786), 1641–42, 1836, 1848, 1850. The head of *Iacobus intercisus* was donated to the basilica among other valuables by Cardinal Orsini in the 1430s. See Cancellieri, 1681–82; König, *Kardinal Giordano Orsini*, 81. See also the S. Pietro book of anniversaries, ACSP H57, fol. 75r; on the various entries pertaining to Boniface VIII, see Kaasik, *Calendars in use*, in the present volume.

50 Bibliography: Baroffio and Albiero, *Iter Online*; Baroffio, *Iter 1999*, 262; Baroffio, "Kalendaria Italica," 468; Baroffio, *Iter editio maior*, 488; José Ruysschaert, Albinia Catherine de la Mare, and Adriana Marucchi, eds, *I codici latini datati della Biblioteca apostolica Vaticana. 1, Nei fondi Archivio S. Pietro, Barberini, Boncompagni, Borghese, Borgia, Capponi, Chigi, Ferrajoli, Ottoboni* (Città del Vaticano: Biblioteca apostolica Vaticana, 1997), 3; Albert Poncelet,

The manuscript contains an exceptionally generous *explicit* on fols. 264v–265r, mentioning its commission by the "*prior et capitulum basilice apostolorum de urbe*" in the year 1339. In view of the plentiful calendar entries relating to S. Pietro, I concur with the hypothesis previously presented in *Codici Latini Datati*: there has for some reason been an omission of *principis* between the words *basilica* and *apostolorum*. The explicit, thus, refers to S. Pietro in Vaticano rather than the basilica of the Santi XII Apostoli in Rome.[51] Albert Poncelet has noted that the content of A9 has been extracted from various older Vatican Lectionaries, namely manuscripts ACSP A2 to A6, and that it served as the basis for a later copy, ACSP A8.[52] Manuscript A8, in turn, contains incipits with unambiguous references to S. Pietro, leaving no doubt of its Vatican origins.[53] Therefore, while A8 itself lacks a calendar, it adds to the evidence linking A9 to the basilica's tradition. The calendar in A9 seems to follow the contents of the manuscript proper closely, as testified by the foliation numbers added to feasts throughout the calendar, which refer to their respective readings in the

Catalogus codicum hagiographicorum latinorum bibliothecarum romanarum praeter quam Vaticanae (Bruxellis, 1909), 29–38; Pierre Salmon, *Les manuscrits liturgiques latins de la Bibliothèque Vaticane. 4, Les livres de lectures de l'office, les livres de l'office du chapitre, les livres d'heures*, Studi e testi 267 (Città del Vaticano: Biblioteca apostolica Vaticana, 1971), 5; Cosimo Stornajolo, *Inventarium Codicum Manuscriptorum Latinorum Archivii Basilicae S. Petri in Vaticano Maxima Ex Parte e Recensione Cosimi Stornajolo (Arch. S. Petri, H 99–100) Depromptum, TOMUS I (A-B)* (Biblioteca apostolica Vaticana: Sala Cons. Mss. 411 (1) (unpublished), 1968), 81–88. The typology of liturgical books is complex, and the cataloguers follow somewhat different standards. I refer to Salmon on the typologies of the manuscripts. A brief and useful summary on various types of liturgical manuscripts can be found in Hughes, *Medieval Manuscripts for Mass and Office*, 118–21. On typology, see also Giacomo Baroffio, "Catalogare manoscritti liturgici: tipologie semplici e complesse," in *La catalogazione dei manoscritti miniati come strumento di conoscenza: esperienze, metodologia, prospettive : atti del Convegno internazionale di studi, Viterbo, 4–5 marzo 2009*, ed. Silvia Maddalo and Michela Torquati (Roma: Istituto storico italiano per il Medio Evo, 2010), 115–26; John Harper, *The Forms and Orders of Western Liturgy from the Tenth to the Eighteenth Century: A Historical Introduction and Guide for Students and Musicians* (Oxford: Clarendon Press, 1993), 58–63.

51 Ruysschaert, Mare, and Marucchi, *I codici latini datati*, 3: "*Tra* basilice e apostolorum *è evidentemente omesso* principis … ". This is not the only instance where the head of the Basilica's chapter is styled a prior. Cf. Blaauw, *Cultus et decor*, 622.

52 Poncelet, *Catalogus codicum hagiographicorum*, 29.

53 See fols. 1r and 111r: "*Incipiunt legende sanctorum que in sacrosancta basilica principis apostolorum de urbe in matutinali officio leguntur.*"

manuscript. It seems, then, that in addition to being actively used as a calendar, it served effectively as a table of contents for its host manuscript.[54]

Although A9 is not listed as a S. Pietro calendar in *Iter*, it must be considered one of the principal witnesses of the Basilica's calendar tradition in the 14th century. This is not only because of the explicit, but also because of the fact that A9 is the only calendar to give the complete list of directly indicative S. Pietro entries identified in table 5.1.

6.2 *Manuscript B79*

Biblioteca Apostolica Vaticana, Arch. Cap. S. Pietro B.79
Dat. s. XII^{4}/^{4}; Antiphonal
Calendar on fols. 1r–3v[55]

Manuscript ACSP B79 contains the earliest known calendar from S. Pietro. The calendar has been thoroughly studied by Pierre Jounel, although Jounel does not refer directly to the manuscript, but uses a 1686 edition by Cardinal Giuseppe Maria Tommasi. This inevitably means that errors and misinterpretations from Tommasi are integrated into Jounel's analysis.[56] Moreover, neither

54 Another indication that the manuscript saw active use is the rather practical list on fol. Ir, detailing the feasts on which the canons were due four *denarii*, two for the vigil and two for mass.

55 Bibliography: Baroffio and Albiero, *Iter Online*; Baroffio, *Iter 1999*, 263; Baroffio, "Kalendaria Italica," 468; Baroffio, *Iter editio maior*, 488; Jounel, *Le Culte Des Saints*, 22–24, 416–21; Salmon, *Les manuscrits liturgiques 1*, 61; Stornajolo, *Inventarium 1*, 164; Giuseppe Maria Tomasi, "Responsoriale et antiphonarium Romanae Ecclesiae de circulo Anni iuxta veterem usum Canonicorum Basilicae Vaticanae Sci. Petri," in *Responsorialia et antiphonaria Romanae Ecclesiae a s. Gregorio Magno disposita* (Romae: Typis Iosephi Vannaccij, 1686), 1–215.

56 Tommasi's edition lacks the following entries in B79: 9 June, *Primi et Feliciani*; 11 June, *Barnabe*; 12 June, *Basilidis Cyrini Naboris et Nazarii*; 2 September, *Antonii*; 11 December, *Damasi*. This occasionally causes Jounel unnecessary complications. See for example Jounel, *Le Culte Des Saints*, 244: *"On explique donc difficilement l'absence de la fête des saints Prime et Félicien au Vatican."* Furthermore, on 9 September B79 carries the name of *Iordanus* as an addition. This is an extremely rare calendar entry shared by a few other S. Pietro calendars: A9, E15, and B84, the first two as an original entry, the last as an addition. Calendar F23 has the feast on the following date. 9 September, however, is also the date of the fairly common feast of Gorgonius, not found in calendar B79 but present in other S. Pietro calendars alongside Iordanus. While Tommasi has transcribed the name faithfully, he has added a marginal note indicating that Iordanus should be read as Gorgonius. Consequently, Jounel has only analysed the common feast of Gorgonius, and seems to be oblivious to the fact that the Vatican tradition differs here. See Jounel, 286–87; Tomasi,

Tommasi nor Jounel indicate additions, removals, nor variations in ink colour, which are all present in the calendar. Thus, any future analysis of this seminal Vatican calendar should be based on fresh transcriptions. Fortunately, B79 is one of the many S. Pietro manuscripts nowadays available through the Digital Vatican Library.

6.3 *Manuscript B88*

Biblioteca Apostolica Vaticana, Arch. Cap. S. Pietro B.88
Dat. s. XV^{2}/^{2}; Antiphonal
Calendar on fols. 1r–6v[57]

As shown in table 5.1, the calendar contains hardly any obvious indications of use at S. Pietro. Nevertheless, B88 is one of the few calendar manuscripts commonly localised to the Vatican Basilica, due to the basilica's coat of arms appearing on fol. 7r of the manuscript.[58] It occupies the opposite end of the chronological spectrum to B79 (discussed above), being the latest S. Pietro calendar examined here.[59] Among the five calendars of plausible Vatican origins, B88 has its closest match in ACSP F23, with which it shares a number of entries

"Responsorialia et antiphonaria," 11. Cf. ACSP B79 fol. 3r and Poncelet, *Catalogus codicum hagiographicorum*, 35–36.

57 For a detailed description of the manuscript's structure and contents, as well as art historical commentary, see Silvia Maddalo, "Breviario di San Pietro," in *Liturgia in figura: codici liturgici rinascimentali della Biblioteca Apostolica Vaticana*, ed. Giovanni Morello and Silvia Maddalo (Città del Vaticano ; Roma: Biblioteca apostolica vaticana ; De Luca, 1995), 159–62. Further bibliography: Baroffio and Albiero, *Iter Online*; Baroffio, *Iter 1999*, 263; Baroffio, "Kalendaria Italica," 468; Baroffio, *Iter editio maior*, 488; Salmon, *Les manuscrits liturgiques I*, 62; Stornajolo, *Inventarium I*, 179–80.

58 The arms are a distinct variant of the keys of St Peter: argent, two keys vertical gules. The first documented instances of their use in this form date from the 15th century. For a more detailed description and an overview of early use, see Giacomo Carlo Bascapé, *Sigillografia. Il sigillo nella diplomatica, nel diritto, nella storia, nell'arte – volume II: Sigillografia ecclesiastica*, Archivio della Fondazione italiana per la storia amministrativa 14 (Milano: Antonino Giuffrè, 1978), 72–73, 77–79. See also the illustrations in Giacomo C Bascapè, Marcello Piazzo del, and Luigi Borgia, *Insegne e simboli: araldica pubblica e privata medievale e moderna*, Pubblicazioni degli archivi di Stato: sussidi 11 (Roma: Ministero per i beni e le attività culturali, Ufficio centrale per i beni archivistici, 1999), 326–28.

59 The *terminus post quem* can be established conveniently by the presence of Bernardino of Siena in the calendar. It therefore cannot predate the Saint's 1444 death and is probably later than his 1450 canonisation. The only known potentially later calendar with possible S. Pietro ties is Veroli, Biblioteca Giovardiana, 16. See footnote 47.

rarely found in other Vatican calendars.[60] The likeness between the calendars extends even to phrasing and spelling: on 12 June both calendars read *"Basilidis, Cirini, Naboris, et Nazarii et Celsi,"* with two names at the end separated from the previous by *et*; on 16 September both calendars have the peculiar spelling *Germiniani*, instead of the usual *Geminiani*.[61] Moreover, both have the feast of Gorgonius marked on 10 September, a day later than usual. These types of variation would easily be dismissed as aberrations or even errors if found in isolation. However, when present in two calendars with joint provenance and many other shared particularities, they cannot be ignored.

6.4 *Manuscript E15*

Biblioteca Apostolica Vaticana, Arch. Cap. S. Pietro E.15
Dat. s. XIV–XV; Psalter
Calendar on fols. 3r–8v[62]

Little has been written on manuscript E15, localised to central Italy in Baroffio's *Iter*, in concordance with Salmon. Of the cataloguists, only Stornajolo gives a specific context, describing the manuscript as a *"Psalterium basilicae vaticanae cum hymnis et antiphonis notis musicalibus instructis."*[63] Looking at the calendar, a S. Pietro origin seems likely indeed, as not only is E15 one of the few calendars with several original entries referring directly to the basilica, but all the common entries between E15 and A9 are shared verbatim. Such a degree of similarity is rare and strongly suggests joint origin. Indeed, E15 is a perfect example of how the disconnect between Stornajolo's and Salmon's catalogues that originated in the 1960s has led to regression in the level of knowledge of the S. Pietro manuscripts, presently disseminated ever wider via the online edition of *Iter*.

---

60 27 January, *Iuliani*; 19 March, *Ioannis Pinarensis*; 11 April, *Ysaac*; 4 May, *Quiriaci*; 4 May, *Katerine de Senis*; 12 June, *Celsi*; 10 September, *Gorgonii*; 17 September, *Lamberti*; 7 November, *Prosdocimi*; Only two of these can be found in A9, and another two in F27.

61 In B88 the "r" in *Germinianus* is abbreviated. The meaning of the abbreviation is clear, however: cf. *Martiniani* on 2 July and *Cornelii* on 14 September.

62 Bibliography: Baroffio and Albiero, *Iter Online*; Baroffio, *Iter 1999*, 264; Baroffio, "Kalendaria Italica," 468; Baroffio, *Iter editio maior*, 489; Salmon, *Les manuscrits liturgiques I*, 6–7; Cosimo Stornajolo, *Inventarium Codicum Manuscriptorum Latinorum Archivii Basilicae S. Petri in Vaticano Maxima Ex Parte e Recensione Cosimi Stornajolo (Arch. S. Petri, H 99–100) Depromptum, TOMUS II (C-E)* (Biblioteca apostolica Vaticana: Sala Cons. Mss. 411 (2) (unpublished), 1968), 190–92.

63 Stornajolo, *Inventarium II*, 191.

6.5 *Manuscript F23*

Biblioteca Apostolica Vaticana, Arch. Cap. S. Pietro F.23
Dat. after 1296; Ordinal-Collectar-Antiphonal
Calendar on fols. 2r–7v[64]

Although Stornajolo dates the manuscript to "*saec. XIII–XIV*," the mention of Boniface's chapel, consecrated in 1296, provides a *terminus post quem* for the calendar. There is some danger of misattributing the manuscript with regard to its origins, as both Salmon and Stornajolo mention that the manuscript carries the coat of arms of Cardinal Giordano Orsini. Orsini's arms are not uncommon in the Vatican chapter's manuscripts, due to the cardinal's sizeable bequest of books, but the arms in the lower margin of the January calendar page of manuscript F23, fol. 2r, are nothing like those of the Orsini, nor do they show any cardinal's insignia. The confusion may result from the arms on F23 which appear to feature a bear, which the Orsini arms do not, despite their name.[65] Instead, the arms in manuscript F23 resemble those of Stephanus de Paccaronibus, a 13th-century prior of S. Pietro.[66] The question of the ownership of

64 Bibliography: Baroffio and Albiero, *Iter Online*; Baroffio, *Iter 1999*, 264; Baroffio, "Kalendaria Italica," 468; Baroffio, *Iter editio maior*, 490; Salmon, *Les manuscrits liturgiques I*, 75–76; Cosimo Stornajolo, *Inventarium Codicum Manuscriptorum Latinorum Archivii Basilicae S. Petri in Vaticano Maxima Ex Parte e Recensione Cosimi Stornajolo (Arch. S. Petri, H 99–100) Depromptum, TOMUS III (F-H)* (Biblioteca apostolica Vaticana: Sala Cons. Mss. 411 (3) (unpublished), 1968), 24–25.

65 For Orsini arms in the Vatican chapter's manuscripts, see for example: ACSP B66 fol. 7r; B67 fol. 10r; B82 fol. 71r. An adequate description of the Orsini family arms can be found in Michael Francis McCarthy, *Heraldica Collegii Cardinalium: A Roll of Arms of the College of Cardinals v. I. 1198–1799* (Sydney: Thylacine Press, 2002), 554: "Bendy gules and argent, on a chief argent sustained by a fillial or, a rose gules. By the 14th century the fillial is charged with a viper vert." It should be noted that this is a description of the family arms and not of any single variation adopted by individual family members. Unfortunately, this is not the only case where the apparent presence of a bear in a coat of arms in a liturgical manuscript has been confounded with the Orsini. As recently as 2006, Leonardo Magionami has attributed the arms found in Perugia, Biblioteca Capitolare Dominicini, 10, fol. 7r to the Roman family. The arms in question belong to the Vannucci family, who produced two consecutive bishops of Perugia from the mid-15th century onwards. See Leonardo Magionami, *I manoscritti del Capitolo di San Lorenzo di Perugia* (Roma: Jouvence, 2006), 62; Elvio Lunghi, "Le Miniature Nei Manoscritti Italiani Della Biblioteca Capitolare Di Perugia (Secoli XIII, XIV e XV)," in *Una Città e La Sua Cattedrale. Il Duomo Di Perugia* (Perugia, 1992), 259.

66 The arms can be seen on a seal mould in the Vatican Basilica's museum collection. An illustration of the mould has been reprinted several times over the centuries: cf. Bascapè, Piazzo, and Borgia, *Insegne e simboli*, 337; Ferdinando Ughelli and Nicola Coleti, *Italia sa-*

the arms is not of minor importance when considering the origin and provenance of the calendar – namely, whether its origins lie at S. Pietro or whether it came to the library via the bequest of a 15th-century bibliophile who amassed books from far and wide.

Based on two entries in the calendar, Salmon suggests the manuscript originates in Spoleto, a claim reiterated in *Iter*. The first, on 19 March, is the feast of an abbot and confessor Iohannes Pinarensis;[67] the second, written in rubrics on 11 April, refers to an abbot by the name of Ysaac, with the description: "*cuius corpus est prope Spoletum.*"[68] It is noteworthy, however, that no Spoletan origin has been attributed to manuscript B88 (discussed above), even though its calendar contains both of these feasts. Instead, B88 has been firmly localised to the Vatican Basilica, based on the capitular coat of arms in the manuscript. The presence of these Spoletan saints, not found elsewhere in the Roman material examined here, in the calendars serves to highlight the particular connection between F23 and the late 15th-century Vatican calendar.[69]

In addition to the feasts noted by Salmon, there are two further links to Spoleto in F23: Iohannes Spoletanus on 19 September and Pontianus, written in rubrics on 13 January. The former is a unique appearance among the calendars studied here, and it does not appear at all in the 515 calendars hosted on *Calendoscope*.[70] Furthermore, the saint is awarded the status of an archbishop in F23, which is a relatively uncommon title in calendars.[71] Pontianus, on the other hand, is probably the Ponziano still celebrated in Spoleto as the city's patron on 14 January. Considering the mounting evidence, proposing a Spoletan connection for F23 would seem to be well founded indeed.[72]

---

cra: sive de episcopis Italiae – Tomus secundus (Venetiis: apud Sebastianum Coleti, 1717), 701; Ferdinando Ughelli, *Italia sacra: sive de episcopis Italiae – Tomus secundus* (Romae: apud Bernardinum Tanum, 1647), 769. Note that Ughelli dates the term of Stephanus as prior of the Vatican Basilica to the time of Innocent IV (1243–1254), which is too early considering the *terminus post quem* of the calendar. It is not impossible, however, that other members of the family were present at the basilica into the 14th century.

67 See Salmon, *Les manuscrits liturgiques I*, 76: "*… il faut lire probablement Penariensis, actuellement Panaca, près de Spolète…*". Cf. *Bibliotheca Sanctorum*, vol. VI (Roma: Città nuova editrice, 1965), col. 866: " … *Parrano (Penara, Panaca), nella diocesi di Spoleto* … "; *Acta Sanctorum Martii Tomus III* (Antverpiae: Apud Iacobum Meursium, 1668), 30–31: "*Ioannes Abbas Paranensis, prope Spoletum in Vmbria.*"

68 Cf. *Bibliotheca Sanctorum*, vol. VII (Roma: Città nuova editrice, 1966), cols. 921–22.

69 See section on ACSP B88 above.

70 Muzerelle, *Calendoscope*.

71 With the notable exception of Thomas of Canterbury.

72 There are a couple of further Spoletan saints in F23 still: Gregorius on 23 December, found in some calendars as "Gregorius Spoletinus," and Savinus on 7 December, a martyr-bishop

Localising F23 is not quite as simple, however, as the calendar contains other notable entries that have not yet been mentioned, as shown in table 5.3 below. These have a near perfect match with known Franciscan commemorations, with a discrepancy of only two days regarding the July feast, usually marked on 21 July.[73] Additionally, a degree of prominence given to Anthony of Padua can be observed. While the popular Franciscan saint's feast itself is quite common, the appearance of its 20 June octave is much less frequent, let alone written in rubrics, as is the case in F23. Further still, the red-letter feast of a Simon on 24 April establishes a bridge between the Spoletan and Franciscan elements in the calendar. Known as Simon of Collazzone, the Franciscan was never canonised in the medieval period, despite a process started soon after his death in 1250. Rejected by the Papacy, he nevertheless enjoyed saintly status at a local level – in and around Spoleto.[74] As with Iohannes Spoletanus above, the feast in F23 is a unique occurrence in the material.

Before drawing conclusions on the manuscript's origin, however, the Vatican-related entries in the calendar need to be acknowledged. According to Stornajolo, saints celebrated at St Peter's have been added to the calendar by later hands. Indeed, there are quite a few additions, but important Vatican

TABLE 5.3 Notable Franciscan entries in F23.

July 19	*#Hic debet fieri officium pro defunctis fratribus et benefactoribus*
September 28	*Officium pro defunctis fratribus et benefactoribus*
November	*Officium pro animabus patrum et matrum fratrum fiat in ultimo die feriali ante adventum*

Legend: #entry in rubrics

of Spoleto. These feasts are not nearly as rare as the other examples given above, however, and consequently not as indicative of a calendar's origin.

73 Cf. van Dijk, *Sources: Texts*, 380–84; Hermann Grotefend, *Zeitrechnung des deutschen Mittelalters und der Neuzeit. Band 2. Abtheilung 2* (Hannover: Hahn'sche Buchhandlung, 1898), 39–41. Cf. also ACSP B84 and SMM 41.

74 Vauchez, *Sainthood*, 53, 117, 213, 254.

entries in original hand can be found as well, even when applying the narrow criteria used to compile table 5.1. Although slightly fewer in number than in calendars A9 and E15, these entries cannot be ignored when considering the origin of the manuscript.

In the end, then, a triangle of Spoletan, Vatican and Franciscan elements emerges, in the light of which the current hypothesis of Spoletan origin seems too simplistic. Considering the original calendar entries referring to the Vatican Basilica, the coat of arms possibly belonging to the family of one of its 13th-century priors, and the manuscript's location in the Basilica's library, it seems more likely than not that the blending of Spoletan, Franciscan, and Vatican elements took place at S. Pietro, whatever the precise mechanism behind the process. As the calendar has much in common with calendar B88, potentially the latest S. Pietro exemplar known to us, there is all the more reason to consider it one of the original S. Pietro calendars, despite its hybrid nature.

7 Adopted S. Pietro Calendars

7.1 *Manuscript B84*

Biblioteca Apostolica Vaticana, Arch. Cap. S. Pietro B.84
Dat. s. XIV; Breviary
Calendar on fols. 1r–6v[75]

Manuscript B84 forms a pair with ACSP B83, with the two complementing each other's liturgical cycles.[76] Stornajolo concludes the two manuscripts are most likely of Franciscan origin, having only later been adapted to use at S. Pietro. This is well supported by the fact that all Vatican-related entries in the calendar

75 Bibliography: Baroffio and Albiero, *Iter Online*; Baroffio, *Iter 1999*, 263; Baroffio, "Kalendaria Italica," 468; Baroffio, *Iter editio maior*, 488; Salmon, *Les manuscrits liturgiques I*, 100; Stornajolo, *Inventarium I*, 174–75.

76 Stornajolo considers both manuscripts to be written and illuminated by the same hands. Salmon makes no mention of B83 in his entry on B84, only taking note of the connection in the final volume of his catalogue, Pierre Salmon, *Les manuscrits liturgiques latins de la Bibliothèque Vaticane. 5, Liste complémentaire, tables gènérales*, Studi e testi 270 (Città del Vaticano: Biblioteca Apostolica Vaticana, 1972), 27.

are additions, while there are several Franciscan feasts in original hand.[77] Based on a number of rare entries, I would argue that the calendar can be localised to Perugia. The 22 April entry *Obitus fratris Egidii* in particular suggests it may originate from the convent of S. Francisco al Monte.[78]

7.2 *Manuscript B85*

Biblioteca Apostolica Vaticana, Arch. Cap. S. Pietro B.85
Dat. s. XIV–XV; Psalter-Hymnal
Calendar on fols. 1r–6v[79]

Little has been written on the manuscript's origin and provenance, although Stornajolo describes it as a *Psalterium basilicae vaticanae*. The added entry mentioning the basilica's main altar, seen in table 5.2, does imply some use at S. Pietro, but the evidence is scant. Both Stornajolo and Salmon point out hymns dedicated to a Berardus on fol. 131 of the manuscript. Judging by the text, it seems probable that the saint in question is Berardus, Bishop of Teramo in Abruzzo.[80] The calendar, too, carries a rare red-letter feast of a bishop and confessor of that name on 19 December. As there is a heavy concentration of central Italian saints in the calendar in addition to Berardus, the prominence awarded to this bishop is not without context.[81] While it may be difficult to

77 The 21 July office *pro benefactoribus et defunctis fratribus* and the octaves of Sts Francis and Anthony on 11 October and 20 June respectively. In calendar B84, red ink has not been used to differentiate between feasts of greater and lesser standing, leaving feast grades the only clues of these distinctions. It is noteworthy that the highest rank of a double feast is accorded not only to the feast of Francis proper, but also to its octave. From this perspective the feast would appear to outgrade those of Virgin Mary, among others. For entries pertaining to S. Pietro, see table 5.2 above.

78 The entry commemorates Giles of Assisi, one of the original companions of St Francis. See *Acta Sanctorum Aprilis Tomus III* (Antverpiae: Apud Michaelem Cnobarum, 1675), 3, 218–20. Other entries pertaining to a Perugian or central Italian context are: 29 January, *Constantii*; 1 February, *Severi*; 30 June, *Fortunati*; 3 July, *Mustiole*; 7 November, *Passio Erculani*. Note that calendar SMM 41 shares all of these entries, including the commemoration of Giles, with the exception of the feast of Mustiola.

79 Bibliography: Baroffio and Albiero, *Iter Online*; Baroffio, *Iter 1999*; Baroffio, "Kalendaria Italica," 263; Baroffio, *Iter editio maior*, 488; Salmon, *Les manuscrits liturgiques I*, 3; Salmon, *Les manuscrits liturgiques V*, 101; Stornajolo, *Inventarium I*, 175–76.

80 Fol. 131r: *"Aprutina plebs fidelis teranensis…"*; fol. 131v: *"Pontificali infula assumptus…"*; Cf. Ferdinando Ughelli and Nicola Coleti, *Italia sacra: sive de episcopis Italiae – Tomus primus* (Venetiis: apud Sebastianum Coleti, 1717), 355–57.

81 I have attempted to identify some of the rarer saints in B85: 18 May *Venantius*, rubricated, a martyr from Camerino. *Acta Sanctorum Maii Tomus IV* (Antverpiae: Apud Michaelem

determine where exactly the calendar originated – although Teramo seems to be an option – it is worth pointing out that it carries an exceptionally large number of entries that are foreign to a Roman context.[82]

7.3 *Manuscript B87*

Biblioteca Apostolica Vaticana, Arch. Cap. S. Pietro B.87
Dat. s. XIII–XIV; Antiphonal
Calendar on fols. 1r–6v[83]

B87 is one of the few calendars associated with the Vatican Basilica in Baroffio's *Iter*. Based on the additions made to the calendar, especially those listed in table 5.2, use at S. Pietro seems likely. The manuscript proper, however, places some emphasis on St Francis, with a full office beginning on fol. 223r and a large miniature on fol. 78r, which is probably what prompted Guglielmo Cavallo to describe the manuscript as an *Antifonario francescano*. The miniature is one of a kind in the manuscript, made to a scale far surpassing other decoration. Dominating the top half of the page and split in two, it depicts Francis receiving the stigmata on the left, and communicating with birds on the right. The calendar, on the other hand, lacks the Franciscan elements seen in B84 and F23, although it does feature other feasts associated with Francis himself. The 25 May feast of the saint's translation is fairly common and cannot, perhaps, be regarded as very indicative. The 16 July feast of Francis's canonisation,

Cnobarum, 1685), 133; 5 August *Emigdius*, martyr-bishop of Ascoli Piceno. *Acta Sanctorum Augusti Tomus II* (Antverpiae: Apud Bernardum Albertum vander Plassche, 1735), 1; 2 November *Amicus, Valentinus* and *Ylarius*, rubricated, the first a monk from Abruzzo, the latter two martyrs from Viterbo. *Acta Sanctorum Novembris Tomus I* (Parisiis: Apud Victorem Palme, 1887), 415, 420, 579–80. The calendar also carries the 7 December feast of Sabinus, most likely martyr-bishop of Spoleto. The entry is rubricated, although it may be due to the presence of St Ambrose on the same row in the calendar.

82 Some of these are common feasts falling on unusual days. There is a cluster of such cases towards the end of July, where the standard list of feasts starts two days ahead of the other calendars, and something similar is likely behind the heavy modifications visible on fol. 2v, the April page, where removals and additions both outnumber the remaining original entries. Such accumulation aside, the number of rare entries remains high.

83 Bibliography: Baroffio and Albiero, *Iter Online*; Baroffio, *Iter 1999*, 263; Baroffio, "Kalendaria Italica," 486; Baroffio, *Iter editio maior*, 488; Salmon, *Les manuscrits liturgiques I*, 61–62; Stornajolo, *Inventarium I*, 177–79. Guglielmo Cavallo, "Dallo 'scriptorium' senza biblioteca alla biblioteca senza 'scriptorium,'" in *Dall'eremo al cenobio: La civiltà monastica in Italia dalle origini all'età di Dante*, ed. Giovanni Pugliese Carratelli, Antica madre: collana di studi sull'Italia antica 10 (Milano: Scheiwiller, 1987), 399–412.

on the other hand, is very rare.[84] Thus, it is safe to say the calendar and the manuscript as a whole display something of a Franciscan character; it is, however, difficult to say much more on the question of its origins.

7.4 *Manuscript E14*

Biblioteca Apostolica Vaticana, Arch. Cap. S. Pietro E.14
Dat. s. XIII; Psalter–Hymnal
Calendar on fols. 1r–9r[85]

The calendar and the appended list of papal liturgical processions, the Roman *stationes*, contain numerous clues placing the manuscript firmly within a Roman context.[86] Moreover, the fortuitous specification *cuius caput habemus*, added to the 27 November feast of *Iacobus intercisus*, suggests the calendar saw some use at the Vatican Basilica.[87] Based on a prayer to St Martin, Stornajolo suggests that the manuscript may have belonged to S. Martino, one of the

84 In addition to B87, I have only encountered the feast in calendar ACSP E4. It can be found neither in Grotefend's collation of five Franciscan calendars nor in van Dijk's notes on 22 calendars of mostly Franciscan origin. Between the 339 Bibliothèque nationale de France calendars available through *Calendoscope*, 22 of which carry the label *Franciscain*, there is only one single occurrence of the feast. Van Dijk, *Sources: Texts*, 380; Grotefend, *Zeitrechnung Band 2*, 39; Muzerelle, *Calendoscope*.

85 Baroffio and Albiero, *Iter Online*; Baroffio, *Iter 1999*, 264; Baroffio, "Kalendaria Italica," 468; Baroffio, *Iter editio maior*, 489; Salmon, *Les manuscrits liturgiques I*, 6; Stornajolo, *Inventarium II*, 188–90.

86 In addition to the dedications of the most prominent Roman Basilicas – S. Maria Maggiore, the Lateran, and the basilicas of the apostles Peter and Paul – commonly, albeit not quite consistently, found in Italian calendars, the calendar carries rarities in both original and subsequent hands: 13 March *Dedicatio sancti Sebastiani martyris*; 20 March *Dedicatio basilice sancti Adriani martyris*; 15 November *Dedicatio sancte Marie in Transtiberi,* marked as an erasure with subscript dots; 13 May *Dedicatio sancte Marie rotunde ad martires*, addition; 4 June *Dedicatio sancta sanctorum in basilica sacrosancte lateranensis ecclesie*, addition; 24 October *Dedicatio ecclesie sancte Marie de vultu illa*, addition. The last of the added dedications likely refers to S. Maria della Mentorella, situated roughly in between Tivoli, Subiaco, and Palestrina. Property of the Benedictines of Subiaco for much of its history, the sanctuary had direct ties to Rome, being for a period a possession of the Benedictines of S. Gregorio al Celio. See Luchina Branciani, "Origine e Sviluppo Dell'eremitismo Nella Valle Sublacense," in *Le Valli Dei Monaci. Atti Del Convegno Internazionale Di Studio: Roma – Subiaco, 17–19 Maggio 2010*, ed. Letizia Ermini Pani (Spoleto: Fondazione Centro Italiano di Studi sull'Alto Medioevo, 2012), 597. Furthermore, some of the rarer feasts of saints in E14 are common in 14th-century Vatican calendars, and, finally, two corrections pertaining to Roman churches made to the list of *stationes* on fol. 9v suggest use in Rome.

87 See footnote 49.

Vatican monasteries involved in the canonical life of S. Pietro.[88] This hypothesis may be confirmed by a partly legible added note on fol. 170v of the manuscript, stating: "*Iste liber est ecclesie sancti Martinii...*". The calendar in its current form is the result of work and intervention by various scribes, making it a revealing testament to medieval practices regarding the use of calendars.[89]

7.5 *Manuscript F27*

Biblioteca Apostolica Vaticana, Arch. Cap. S. Pietro F.27
Dat. s. XIV–XV; Breviary
Calendar on fols. 415r–420v[90]

Both Stornajolo and Salmon have noted the altar dedications added to calendar F27, but only the former has localised the manuscript to S. Pietro. Baroffio's *Iter*, in turn, even casts doubt on the Italian character of the manuscript.[91] Judging by the entries listed in table 5.2, it seems clear that the calendar has been adapted to the S. Pietro tradition. There is not much that can be said of its origins. The calendar has an exceptional number of additions, as well as entries that are difficult to find anywhere else, let alone contextualise.[92] Although one can hardly expect to find normalised spelling in medieval calendars, even here F27 seems to display an exceptional degree of peculiarities.[93]

88 Jounel, *Le Culte Des Saints*, 309; Guy Ferrari, *Early Roman Monasteries: Notes for the History of the Monasteries and Convents at Rome from the V through the X Century* (Città del Vaticano: Pontificio Istituto di Archeologia Cristiana, 1957), 230–40; Rezza and Stocchi, *Il capitolo I*, 43–48; Alexis Gauvain, *Il capitolo di San Pietro in Vaticano dalle origini al XX secolo. Vol. II: il Patrimonio* ([Città del Vaticano]: Edizioni Capitolo Vaticano, 2011), 36–50.

89 See Kaasik, *Navigating the cycles of time*, in the present volume.

90 Bibliography: Baroffio and Albiero, *Iter Online*; Baroffio, *Iter 1999*, 264; Baroffio, "Kalendaria Italica," 468; Baroffio, *Iter editio maior*, 490; Salmon, *Les manuscrits liturgiques I*, 102; Stornajolo, *Inventarium III*, 36–37.

91 Baroffio and Albiero, *Iter Online*: "*Origine Italie (?)*."

92 Of the little over 300 entries in F27, around one-third are additions.

93 Thus, F27 has *Pantani* in the place of *Pancratii*, as written in most calendars on 12 May, and *Potrasii* instead of the usual *Protasii* on 19 June. Both examples are of entries written in original hand. A characteristically baffling added entry is the virgin-martyr *Theortea* on 6 February – seemingly halfway between Theodora and Dorothea, both found in calendars on this date. Additionally, occasional damage, especially to the August calendar page, makes it difficult to distinguish between faded entries and possible erasures. The calendar also features a number of apparent relocations of entries using a system of matching symbols. It is only on some occasions possible to understand without ambiguity the entries intended to be moved and, indeed, the direction of the movement. See especially the July and September pages. That being said, part of the peculiarity of the

7.6 *S. Pietro Books of Anniversaries, Manuscripts H56 and H57*

Biblioteca Apostolica Vaticana, Arch. Cap. S. Pietro H.56
Dat. s. XIII^{ex}; Martyrology–Book of anniversaries

Biblioteca Apostolica Vaticana, Arch. Cap. S. Pietro H.57
Dat. s. XV^{in}; Martyrology–Book of anniversaries[94]

Manuscripts ACSP H56 and H57 are two S. Pietro books of anniversaries that contain notices of the deaths and donations of the Basilica's benefactors, as well as details regarding their anniversary services.[95] H56 was in use from the close of the 13th century, being replaced in the early years of the 15th century by its copy, H57, which in turn remained in use all the way to the 17th century.[96] Both manuscripts follow a page-per-day structure, with martyrology entries at the top and plenty of space for obits and donations below. Between them, they carry obits dating from the 12th to the 17th centuries, while the martyrology entries are clearly related to the Basilica's calendars, featuring references to the dedications of altars seen in table 5.1 above, to give an example. Thus, these books provide valuable comparative material for any study of the S. Pietro calendars, as the two types of texts are closely related by genre and have partially overlapping contents.

calendar comes from feasts of common saints falling on unusual dates. However, only some of the aforementioned relocations of entries seem to serve to "normalise" such dates.

94 Bibliography: Marc Dykmans, "Les Obituaires Romains. Une Définition Suivie d'une Vue d'ensemble," *Studi Medievali, Serie Terza* 19, no. 2 (1978), 622–26; Pietro Egidi, *Necrologi e libri affini della Provincia romana* (Roma, Forzani e c., tipografi del Senato, 1908), 167–73; Baroffio and Albiero, *Iter Online*; Baroffio, *Iter 1999*, 264; Baroffio, *Iter editio maior*, 490; Salmon, *Les manuscrits liturgiques IV*, 77; Stornajolo, *Inventarium III*, 183–84. Unlike with other manuscripts discussed in this article, the typology for H56 and H57 given here is not from Salmon, who describes both manuscripts as *"Liber Officii capituli basilicae Vaticanae."*

95 Although most entries use the phrase *obiit*, in practice the focus is on donations and commemoration, as some benefactors have several obits spread throughout the manuscripts. For two examples of benefactors with 12 obits each, see Egidi, *Necrologi*, 172, fn. 3. Egidi, 174–281, has also edited the books' obituary notices up until 1501, although, like with most of the S. Pietro collection, both H56 and H57 can nowadays be viewed online through the Digital Vatican Library.

96 See H57, fol. 172r for a notice referring to the manuscript's commissioning. There appears to have been a third manuscript as well, somewhat earlier than H56, now sadly missing. On the dating of the manuscripts and the missing third manuscript, see Egidi, 168–70, 283–91. Note that Stornajolo dates both manuscripts to "*Saec. XIII ex.*", which must be an error. Stornajolo, *Inventarium III*, 183; Cf. ACSP H100 fols. 644–45.

8 Calendars Erroneously Attributed to S. Pietro

8.1 *Manuscript B67*

Biblioteca Apostolica Vaticana, Arch. Cap. S. Pietro B.67
Dat. s. XIV–XV?; Sacramentary
Calendar on fols. 2r–7v[97]

B67 is one of only five calendar manuscripts localised to S. Pietro in *Iter*. This view seems to have originated with Emmanuel Bourque, being subsequently reiterated in Salmon's influential catalogue and passed on in further research.[98] The localisation is most likely based on the manuscript's incipit on fol. 10r, which the catalogues give as *"Incipit Missale secundum ordinem camerae."* The mention of *camera*, then, has been taken to signify S. Pietro as the "apostolic chamber."[99]

97 Bibliography: Emmanuel Bourque, *Etude sur les sacramentaires Romains v. 2.2*, Studi di antichità cristiana 25 (Roma: Pontificio Istituto di Archeologia Cristiana, 1958), 304; Adalbert Ebner, *Quellen und Forschungen zur Geschichte und Kunstgeschichte des Missale Rromanum im Mittelalter : Iter italicum* (Freiburg im Breisgau: Herder, 1896), 174–75; Victor Saxer, "Trois Manuscrits Liturgiques de l'Archivio Di San Pietro, Dont Deux Datés, Aux Armes Du Cardinal Jordan Orsini.," *Rivista Di Storia Della Chiesa in Italia* 27, no. 2 (1973), 502–04; in addition, see the usual Baroffio and Albiero, *Iter Online*; Baroffio, *Iter 1999*, 263; Baroffio, "Kalendaria Italica," 468; Baroffio, *Iter editio maior*, 488; Pierre Salmon, *Les manuscrits liturgiques latins de la Bibliothèque Vaticane. 2, Sacramentaires, épistoliers, évangéliaires, graduels, missels*, Studi e testi 253 (Città del Vaticano: Biblioteca apostolica Vaticana, 1969), 3; Stornajolo, *Inventarium I*, 151–52. There is a discrepancy of a century in dating the manuscript, with older catalogues opting for an early 14th- and the more recent ones for an early 15th-century origin. The former dating is forwarded by Ebner, Bourque and Stornajolo; the latter by Salmon, Saxer and Baroffio. In his catalogue entry Salmon refers to all three of the earlier cataloguers, while Saxer uses Salmon as the basis of his analysis. Baroffio's dating, too, is likely derivative of either Salmon or Saxer or both. Therefore, Salmon can arguably be considered the institutor of the new dating, whether it came about by intent or accident.

98 Bourque laconically titles the manuscript *Sacramentaire de St-Pierre*. Salmon's view is probably derivative of Bourque, since neither Ebner nor Stornajolo, also referred to by Salmon, offer any localisation for the manuscript. Contrary to Bourque, Salmon's localisation is not presented without some doubt: pushed to the very end of the manuscript's description, it reads "Basilique de Saint-Pierre (?)." Salmon also notes that the manuscript's decoration seems to be of northern Italian character. Saxer, however refers directly to Salmon's assessment of the manuscript's origin, stating: "*Tenant compte sans doute du titre rubriqué du temporal ... Mgr Salmon suggère qu'il s'agit de la communauté monastique desservant la basilique Saint-Pierre. C'est pour la même raison que s'explique le titre que nous avons choisi pour le livre.*"

99 This reasoning is stated clearly by Saxer, who describes the manuscript as a *Sacramentaire monastique de la Chambre apostolique*.

The manuscript itself, however, does not give the word *camerae* at all – just an abbreviated *cam.* It is difficult to say when and where this highly unusual reading of the incipit originated, but it is already present in a 1598 inventory of the library's possessions.[100] Whatever its origins, it must be taken as erroneous, as the calendar strongly suggests that the manuscript is in fact Camaldolese.[101] A correct expansion of the incipit would thus read: *Incipit Missale s(ecun)d(u)m ordi(n)em cam(aldulensem)*. This would also explain the incipit's use of *ordo* instead of the broader *consuetudines*, *mos*, or *usus*, as this was indeed characteristic of Camaldolese incipits.[102]

Finally, one has to consider the coat of arms of Cardinal Giordano Orsini on fol. 10r of the manuscript, which is clearly superimposed on the original decoration. The arms present a further problem for the theory of Vatican origin, as they indicate the manuscript only made its way to S. Pietro in his sizeable bequest of books. As a solution, Saxer has suggested it may have been presented to Orsini as a gift at S. Pietro when the cardinal was made the basilica's archpriest, making a return among the larger part of Orsini's library a few years later. However, such speculations are only necessitated by the *camerae*-reading and the subsequent hypotheses regarding the manuscript's origin. Considering the manuscript's Camaldolese character, quite foreign to S. Pietro indeed, it

100 ACSP K4 fol. 49r.

101 Salmon and Saxer both describe the manuscript as monastic, and the latter points out rubrics pertaining to an abbot and a prior. See, for example fols. 73v–74v and 97v–98v. Moreover, many feasts in the calendar are assigned 12 readings, instead of the nine customary in secular churches. The calendar lacks any indications of use at the Vatican and seems foreign to Rome altogether, with the feast of Bartholomew not on its Roman date and the common 18 November dedicatory feast of S. Pietro and S. Paolo fuori le mura missing altogether. This would be odd indeed for a manuscript following the customs of the apostolic chamber. On the dating of Bartholomew's feast, see the section on methodology above. In contrast, the calendar has a number of entries not found in Roman calendars, among them the feast of St Romuald, the founder of the Camaldolese Order, written in rubrics. It shares other such rarities with known Camaldolese calendars as well: 1 June *Viriani et sociis eius* [sic]; 3 June *Laurentini et Pergentini*; 30 June *Martialis*; 25 August *Genesii et sociorum eius*; 23 November *Columbani*; 29 November *Crisanti et Da*[rie]; 9 December *Syri*. I have checked these entries against three Biblioteca Apostolica Vaticana manuscripts with Camaldolese incipits – Urb. Lat. 598, Urb.Lat. 599, and Vat. Lat. 10268 – as well as the "Missale ord. Camaldul. 1503 (London, British Museum)" presented in Grotefend, *Zeitrechnung Band* 2, 12–14. Although some of the rarer entries in B67 can be connected to Arezzo, these feasts had spread enough to lose their local character much earlier, already in the 12th century. See S.J.P. van Dijk, "The Lateran Missal," *Sacris Erudiri* VI (1954), 134–43.

102 See, for example Biblioteca Apostolica Vaticana, Urb. Lat. 598, fol. 123r: "*Incipit psalterium s(ecundu)m ordinem camaldule(n)sem*"; Urb.Lat. 599, fol. 1r: "*Incipit breviarium divini officii totius an(n)i s(ecundu)m ordine(m) et consuetudine(m) camaldulensem*"; Vat. Lat. 10268, fol. 231r: "*Incipit salterium s(ecundu)m ordi(n)em camaldulensem.*"

seems more plausible that the well-travelled book collector acquired the manuscript elsewhere, possibly through his contacts with Ambrogio Traversari, a fellow humanist and Prior General of the Camaldolese Order.

8.2 *Manuscript B72*

Biblioteca Apostolica Vaticana, Arch. Cap. S. Pietro B.72
Dat. s. $\text{XV}^{3/4}$; Missal
Calendar on fols. 3r–8v[103]

Similarly to the Camaldolese manuscript B67 discussed above, manuscript B72 is one of the few calendar manuscripts localised to S. Pietro in Baroffio's *Iter*. It was, in fact, donated to the basilica in 1475 by one of its canons, Martinus de Roa – a curial scribe and once the *cubicularius secretus* of Nicholas V.[104] The manuscript was likely a private commission intended for private use, and as the calendar displays no signs of use at S. Pietro nor any connection with its tradition, I do not count it among the basilica's calendars any more than most other donated calendar manuscripts in its library collection.[105]

---

103 Bibliography: Baroffio and Albiero, *Iter Online*; Baroffio, *Iter 1999*, 263; Baroffio, "Kalendaria Italica," 468; Baroffio, *Iter editio maior*, 488; Elisabetta Caldelli, *Copisti a Roma nel Quattrocento* (Roma: Viella, 2006), 157–58; Ruysschaert, Mare, and Marucchi, *I codici latini datati*, 209–10; Ebner, *Quellen und forschungen*, 176; Valérie Guéant, "Du nouveau sur la production des premières années à Rome du maître des missels della Rovere," *Rivista di storia della miniatura* 13 (2009), 145–46; Silvia Maddalo and Adalbert Roth, "Messale," in *Liturgia in figura: codici liturgici rinascimentali della Biblioteca Apostolica Vaticana*, ed. Giovanni Morello and Silvia Maddalo (Città del Vaticano ; Roma: Biblioteca apostolica vaticana ; De Luca, 1995), 155–59; Salmon, *Les manuscrits liturgiques II*, 100; Stornajolo, *Inventarium I*, 156–58.

104 De Roa, originally from Castile, spent much of his life and career in Rome and died shortly after his donation. The donation is detailed on fol. 423v of B72 and is stated to have taken place *"pro salute sui anime et corporis."* His death notice in the S. Pietro book of anniversaries, H57, fol. 104r, is dated just two days later. The latter notice also mentions a *"breviarium inceptum magnum completum sua impensa cum corrigiis fericis exclausuris argenteis deauratis"* among de Roa's donations to the Basilica. It is possible the breviary mentioned is ACSP B81, based on similarities of script and content between the calendars in B81 and B72. The wording *inceptum – completum* would also fit the layered character of B81. Unfortunately, B81 does not retain its original covers and it would take further research to confirm this hypothesis.

105 Adalbert Roth argues that the presence of musical liturgical elements without notation indicates the manuscript was produced for *missa privata* or *missa lecta*. The manuscript contains numerous professions of ownership and commission. See fols. 3r, 10r, 193v, 275r, 351r, and 423v. In contrast to the pristine calendar, however, added prayers mentioning Sixtus IV, Julius II, Leo X, and Clement VII on fol. 424r demonstrate the manuscript itself saw some use even in the 16th century. The calendar, in turn, has much in common with

9 The Calendars of S. Maria Maggiore

There are at present four calendar manuscripts localised to S. Maria Maggiore in *Iter*:

Città del Vaticano, Biblioteca Apostolica Vaticana,
- S. Maria Magg. 97
- S. Maria Magg. 99
- S. Maria Magg. 119
- S. Maria Magg. 120

Unlike in the case of S. Pietro, there are no misattributions here, although I have identified two additional manuscripts and one incunable with calendars pertaining to the basilica.[106] There are further differences compared to S. Pietro; for one, it is possible to attribute an origin close to S. Maria Maggiore only to two calendars – SMM 119 and 120 – and even for these the evidence is not derived from the calendars themselves. When it comes to calendars modified at the Basilica, the evidence is equally scarce, as shown in table 5.4 below.

As can be seen in table 5.4, S. Maria Maggiore calendars carry only a single entry that indicates use in the basilica. While there are further entries added to some of the basilica's adopted calendars, these are mostly Marian feasts, including that of St Anne and the dedication of the basilica itself, tied to the legend of Our Lady of the Snows.[107] As these appear frequently enough in calendars outside the basilica, Rome, and even Italy, they are of little indicative value for localisation.[108]

---

a group of mid to late 15th-century calendars: ACSP B81, E4, and E11. These share lengthy sequences of feasts concentrated in the spring months, sparsely filled in most calendars.

106 The full list of 16 S. Maria Maggiore calendars consists of manuscripts: SMM 41, 48, 49, 50, 51, 91, 97, 99, 106, 112, 113, 114, 119, 120; and incunabula: SMM 94 and 105.

107 For a detailed account on these feasts in the S. Maria Maggiore manuscripts, see Saxer, *Sainte-Marie-Majeure*, 318–28; Saxer, "Il culto dei martiri," 151–54.

108 The implications of the thin layer of distinctive S. Maria Maggiore entries, taking into account the Marian cult, for the use of calendars in practice are discussed in Kaasik, *Calendars in use*, in the present volume.

TABLE 5.4 Indicative entries in Santa Maria Maggiore calendars.

Date	50	91	94	97	99
May 9	/Traslatio sancti Ieronimi\	/Translatio sancti Ieronimi\		/Translatio sancti Hieronimi doctoris ad sanctam Mariam maiorem\	/Translatio divi Hieronimi de Betheleem ad sanctam Mariam maiorem, sed nunc celebratur in vigilia Ascensionis de mandato pape Pii II\
May			/Et nota quod vigilia ascensionis fit festum in sancta Maria ma(?) translationis sancti Ieronimi et in primis vesperis et ad missam vadiunt cardinales\		

Legend: /added entry\

10 Original S. Maria Maggiore Calendars

10.1 *Manuscripts 119 and 120*

Biblioteca Apostolica Vaticana, S. Maria Magg. 119
Dat. 1480; Psalter-Hymnal
Calendar on fols. 1r–6v

Biblioteca Apostolica Vaticana, S. Maria Magg. 120
Dat. 1478; Psalter-Hymnal
Calendar on fols. 2r–7v[109]

109 Bibliography: Baroffio and Albiero, *Iter Online*; Baroffio, *Iter 1999*, 277; Baroffio, "Kalendaria Italica," 470; Baroffio, *Iter editio maior*, 503; Caldelli, *Copisti a Roma nel Quattrocento*,

According to their incipits, these near-identical manuscripts were produced two years apart for the basilica of S. Maria Maggiore at the expense of a noblewoman of the Annibaldi family.[110] The work was carried out by a scribe identifying himself as *frater Dominicus* of the Augustinian Hermits, which explains the prominence of Augustinian feasts in both calendars.[111] The incipits' references to the Basilica's archpriest Guillaume d'Estouteville are also worth noting, as d'Estouteville was the cardinal protector of the Augustinians for 37 years, in addition to his long-standing patronage of S. Maria Maggiore.[112]

102–03, 173–74; Salmon, *Les manuscrits liturgiques I*, 27–28; Saxer, *Sainte-Marie-Majeure*, 673–74.

110 See SMM 119, fol. 10r–v: "*Ad honorem omnipotentis dei et gloriosissime virginis Marie matris eius. Incipit psalterium ad usum venerabilium dominorum canonicorum venerabilis basilice et ecclesie sancte Marie maioris de urbe, secundum consuetudinem clericorum Romane urbis eiusque districtus et cetera, quodque psalterium scriptum et illuminatum fuit sumptibus et expensis nobilissime domine domine Caredonne de Anibalis de urbe, tempore archypresbiteratus Reverendissimi domini domini Guillelmi de Estoutevilla, episcopi hostiensis cardinalis Rothomagensis, ac Sanctissimi domini nostri Sixti pape quarti camerarii dignissimi. Sub anno domini millesimo quadringentesimo octuagesimo, tempore sanctissimi domini domini Sixti pape quarti, anno vero eius nono, et cetera.*" Aside from the dating, the wording in SMM 120, fol. 11r–v is identical. It is likely the coats of arms on the manuscripts – SMM 119, fol. 11r; SMM 120, fol. 12r – featuring a scorpion belong to the Annibaldi benefactor. See *ANNIBALDI O ANNIBALDESCHI di Velletri* in G.B. di Crollalanza, *Dizionario storico-blasonico delle famiglie nobili e notabili italiane, estinte e fiorenti, vol. I* (Bologna: Arnaldo Forni, 1965), 48.

111 SMM 120 fol. 257r: "*Scriptum et notatum per me fratrem Dominicum de Roma ordinis heremitarum sancti Augustini. Anno 1478 et cetera.*"; fol. 297v: "*Frater Dominicus* [*de*] *Roma scripsit.*" The Augustinian entries found in both calendars are: the two translations of Augustine on 28 February and 11 October; the conversion of Augustine on 5 May; the translation of Monica on 9 April; the feast of Monica on 4 May; the feast of Nicholas of Tolentino on 10 September. It should be noted that the Annibaldis and the Augustinians shared a history, as one of the family's most outstanding members, the 13th-century cardinal Riccardo Annibaldi, had been instrumental in the Order's establishment and early success, also serving as its first Protector. Daniel Waley, "Annibaldi, Riccardo," in *Dizionario biografico degli Italiani v. 3*, ed. Alberto Maria Ghisalberti (Roma: Istituto della Enciclopedia italiana, 1961).

112 Even one of the Cardinal's children was named Agostino. D'Estouteville was immensely wealthy and his patronage involved renovations in S. Maria Maggiore and the construction of the church of Sant'Agostino in Rome. He also had an extensive library, from which he bequeathed nearly 200 books to the latter. The Cardinal's patronage was sought after even posthumously, as his funeral procession was interrupted by a row between the canons of Sant'Agostino and S. Maria Maggiore, during which swords were drawn. Gaspare Pontani, *Il Diario Romano Di Gaspare Pontani, Già Riferito al "Notaio Del Nantiporto"* [*30 Gennaio 1481 – 25 Luglio 1492*], ed. Diomede Toni, Rerum Italicarum Scriptores II, 3.2 (Città di Castello: Casa editrice S. Lapi, 1907), 23–24; Anna Esposito, "Estouteville, Guillaume d,'" in *Dizionario biografico degli Italiani v. 43*, ed. Fiorella Bartoccini and Mario Caravale (Roma: Istituto della Enciclopedia italiana, 1993); Anna Esposito Aliano, "Testamento e inventari per la riconstruzione della biblioteca del cardinale Guglielmo d'Estouteville," in

The calendars themselves are exceptionally similar: among the 279 calendar entries in SMM 120 and the 271 in SMM 119, there are only 16 entries that are present in one but not the other.[113]

11 Adopted S. Maria Maggiore Calendars

11.1 *Manuscript 50*

Biblioteca Apostolica Vaticana, S. Maria Magg. 50
Dat. s. XIV; Breviary
Calendar on fols. 3r–8v[114]

Manuscript 50 is of Franciscan origin, judging by both its incipit and its calendar.[115] The core of the calendar is markedly foreign to Rome, and contains original entries making special mention of Genoa[116] and an extensive series of additions pertaining to Pavia.[117] In the light of this, it seems strange that Saxer, in

Scrittura, biblioteche e stampa a Roma nel Quattrocento. Aspetti e problemi, ed. C. Bianca et al., Littera Antique, 1,1 (Città del Vaticano: Scuola Vaticana di Paleografia, Diplomatica e Archivistica, 1980), 311–19; Caldelli, *Copisti a Roma nel Quattrocento*, 103. Caldelli speculates that d'Estouteville is mentioned in the incipits of SMM 119 and SMM 120 because he was somehow involved in the manuscripts' realisation. It could, however, simply reflect reverence paid to the outstanding patron.

113 Additional differences between the calendars can be found in the use of red ink as well as the amount of other information provided: unlike the earlier SMM 120, SMM 119 lacks verses and markings on the ominous Egyptian days, as well as information on the year's length. Furthermore, it contains only half as many indications of feast grades. This is a good reminder that no matter how close in origin, no two calendars should be expected to be identical.

114 Bibliography: Baroffio and Albiero, *Iter Online*; Baroffio, *Iter 1999*, 277; Baroffio, "Kalendaria Italica," 470; Baroffio, *Iter editio maior*, 503; Salmon, *Les manuscrits liturgiques I*, 165; Saxer, *Sainte-Marie-Majeure*, 672.

115 See fol. 11r: "*In nomine Domini. Incipit ordo breviarii fratrum minorum secundum consuetudinem romane curie*"; Franciscan emphasis in the calendar: 25 May *Translatio Francisci*; 20 June *octava Antonii*; 12 July *Dedicatio ecclesie beati Francisci*; 12 August *Clare matris dominarum ordinis Damiani*; 11 October *octava beatissimi patris nostri Francisci*.

116 6 July *Siri*[i], with an added definition *Ianuensis*; 9 July *Felicis Ianuensis*; 13 October *Romoli Ianuensis*.

117 The Pavian additions I have been able to identify: 11 January *Translatio Honorate*; 21 January *Epiphanii*; 28 February *Translatio Augustini*; 17 May *Translatio Syri*; 20 May *Theodori*; 22 May *Bobonis*; 17 July *Enodii*; 22 July *Ieronimi*; 12 September *Adventus Syri et Yventii in Papiam*; 30 October *Crispini*; 9 December *Syri*; Furthermore, there are some additions pertaining to nearby centres, Milan and Verona: 4 April *Depositio Ambrosii*; 8 May *Victoris*; 18 July *Materni*; 8 December *Zenonis*; The 13 July entry *Hermagore et Fortunati* may also be counted among these additions, as the first Bishop of Pavia, Syrus, was held to have been

discussing the calendar, suggests that the manuscript is of Roman origin and has been adapted for use at S. Maria Maggiore at an early stage.[118] Be that as it may, the manuscript does show signs of modification at S. Maria Maggiore, as the office for the basilica's dedication has been added to the end of the manuscript, and the December calendar page features additions indicating the Roman *stationes* for the Christmas cycle. It should be noted that entries of this type are extremely rare in calendars and that S. Maria Maggiore held a prominent place among the stational churches of Rome. In SMM 50, for example, the additions for both the Nativity and its vigil read *statio ad sanctam Mariam maiorem*. The shifting of Bartholomew's feast from 24 to 25 August may indicate Roman use as well, although it must be said the Pavian layer of additions in the calendar far surpasses its Roman counterpart in volume.[119]

11.2 *Manuscript 91*

Biblioteca Apostolica Vaticana, S. Maria Magg. 91
Dat. s. XIV; Missal
Calendar on fols. 1r–6v[120]

Both Salmon and Saxer suggest the manuscript reflects the use of S. Maria Maggiore, further stating that the miniatures are typical of central Italy or Lazio. The calendar seems to be closely related to calendar SMM 99. Of the rarer entries in the two, *Constantie* on 18 February does hint towards a context near Rome, even if the evidence is too scant to attribute exact origins.[121] On the

a disciple of Hermagoras of Aquileia. The calendar contains further additions still, which I have not been able to identify for the present.

118 Saxer, *Sainte-Marie-Majeure*, 672: "*Calendrier (janv.-déc.), avec environ 50 additions de diverses mains. Celles du 7 mars: S. Thome conf., 9 mai.: Translatio S. Jeronimi, 5 août: Festum niuis ont été faites au XIVe s., voire pourraient être encore de 1e main. Le livre est romain par son origine et pourrait avoir été accommodé peu temps après à l'usage libérien.*" Saxer's suggestion that the three additions he has singled out may have been added by the original hand is equally strange – not only is the writing between the additions and the original calendar text different, but the last of the additions mentioned has clearly been written by a different hand than the first two.

119 The last leaves of the manuscript also carry an unidentified coat of arms, of which Saxer gives a mostly faithful description.

120 Bibliography: Baroffio and Albiero, *Iter Online*; Baroffio, *Iter 1999*, 277; Baroffio, "Kalendaria Italica," 470; Baroffio, *Iter editio maior*, 503; Salmon, *Les manuscrits liturgiques II*, 144; Saxer, *Sainte-Marie-Majeure*, 664–65.

121 Constantia appears with marked regularity in the Vatican Basilica's calendars in the 14th century: of the Basilica's original calendars, A9, E15, and F23 all carry the entry as an original. It is missing only in the chronological extremes, the late 12th-century B79 and the late

other hand, neither calendar shows any striking absences of important Roman feasts, nor too many oddities worth mentioning when compared to material more easily localised to Rome.[122] In sum, it is possible that they were adapted from an original context not too far removed from S. Maria Maggiore.

11.3 *Manuscript 97*

Biblioteca Apostolica Vaticana, S. Maria Magg. 97
Dat. s. XIII–XIV; Missal
Calendar on fols. 422r–427v[123]

15th-century B88. Of the adopted S. Pietro calendars, the feast appears as an original in E14, the only of its group likely to have originated in Rome. Calendars B84, B87, and F27 all carry the entry as additions, leaving B85, which lacks the feast, the odd one out. The entry can rarely be found in calendars outside of Rome, and of the 515 mostly French and German calendars listed on Calendoscope, it appears in just two, one of which is likely Roman in origin, the other seemingly a type of a compilation calendar where hardly a day has been left empty. See Muzerelle, *Calendoscope*; Dijk, "Three Manuscripts," 219–29. Garrison points out that the saint, daughter of Constantine, appears exclusively in Roman material, "with very few exceptions of the late thirteenth and of the fourteenth century." Garrison, "Three Manuscripts," 14. In 1256, Alexander IV (1254–1261) consecrated an altar in Constantia's mausoleum, already referred to as a church some centuries before, so there was evidently interest in the cult in Rome at the time. *Bibliotheca Sanctorum*, vol. IV (Roma: Società Grafica Romana, 1964), col. 259; Krautheimer, *Rome*, 25–26. This may explain why the feast appears in the Vatican calendars starting from the 13th-century E14 but is absent from the 12th-century B79. The feast seems to have fallen out of Roman calendars by the late 15th century, as it is not attested in the Roman productions of the period, namely calendars ACSP B88 and SMM 119 and 120.

122 To give an example, both calendars contain the November dedications of the Lateran and the two major apostolic basilicas in Rome, all in red letters. The few rarities the calendars share when compared to a broader corpus of calendars are, in calendrical order: 10 February *Guilielmi* or *Giulielmi*, not common in Roman calendars; 8 March *Perpetue et Felicitatis*, a common feast a day later than usual; 16 April *Aniceti*, a day earlier than typical, though this is not as rare an occurrence as with Perpetua and Felicitas; 1 September *VII fratrum*, as opposed to the more typical *XII fratrum*. This could be a mix-up between the two groups of martyrs or simply human error incurred when copying numerals. Either way, it is again a slightly more common find in the calendars examined for this study than the previous examples. Finally, there is the 16 February entry *Iuliane*, whose appearances are by no means restricted to Rome or even Italy, but which has been noted by Edward B. Garrison to "occur more or less often in … Roman books and calendars." See Garrison, "Three Manuscripts," 26, fn. 128; Cf. Muzerelle, *Calendoscope*.

123 Bibliography: Baroffio and Albiero, *Iter Online*; Baroffio, *Iter 1999*, 277; Baroffio, "Kalendaria Italica," 470; Baroffio, *Iter editio maior*, 503; Salmon, *Les manuscrits liturgiques II*, 144; Saxer, *Sainte-Marie-Majeure*, 665–66.

The manuscript contains a Franciscan incipit, although the words *fratrum minorum* have been erased.[124] In the calendar, Franciscan elements are limited, most prominently featuring the 25 May *Translatio Francisci* as a rubricated entry. While this feast is not confined solely to Franciscan calendars, it is far from universally attested.[125] Although the calendar lacks the feast of Clare of Assisi on 12 August, this does not necessarily imply it predates her death or canonisation, as it could just as well have been copied from such an exemplar. If the calendar saw no subsequent Franciscan use, the addition of the feast may not have been deemed necessary. The calendar is one of the four localised to S. Maria Maggiore in Baroffio's *Iter*, possibly due to Saxer's similar attribution. However, it is unlikely to originate in the Basilica, or even in Rome for that matter, as it lacks the dedications of the Lateran Basilica, the basilicas of Peter and Paul, and even S. Maria Maggiore itself. The calendar has been left largely unaltered despite this degree of foreignness to Rome, with only three additions present – although one of these does directly refer to S. Maria Maggiore.[126]

11.4 *Manuscript 99*

Biblioteca Apostolica Vaticana, S. Maria Magg. 99
Dat. s. XIV–XV; Missal
Calendar on fols. 2r–9v[127]

Both Salmon and Saxer state manuscript SMM 99 has been adapted for use in S. Maria Maggiore, which may be why it is thus localised in Baroffio's *Iter*. As mentioned above in the section on SMM 91, the two calendars are related, with some potential indications of origins in Rome or Lazio. In SMM 99, furthermore, the 11 June rubricated entry *Eunufrii* could point towards Rome, as a church on the Gianicolo was dedicated to the hermit saint in 1434.[128] An

124 Fol. 3r: "*Incipit ordo missalis* (*fratrum minorum* [erased]) *secundum consuetudinem Romane curie.*"

125 Of the 515 Calendoscope calendars, the feast appears in 31, the vast majority of which are indeed labelled Franciscan. Muzerelle, *Calendoscope*.

126 See table 5.4 above. The other added feasts are of Celestine V on 19 May and the Visitation of Mary on 2 July.

127 Bibliography: Baroffio and Albiero, *Iter Online*; Baroffio, *Iter 1999*, 277; Baroffio, "Kalendaria Italica," 470; Baroffio, *Iter editio maior*, 503; Salmon, *Les manuscrits liturgiques II*, 145; Saxer, *Sainte-Marie-Majeure*, 667.

128 Hülsen, *Le chiese di Roma*, 541, 47: S. Onuphrii. This entry can be found in several 15th-century calendars with ties to a Roman context: ACSP B72, B81, B88, and E4, as well as SMM 119 and 120. There are, however, appearances of the feast that likely predate the

addition mentioning the year 1508 indicates that the calendar still saw some use in the 16th century.[129]

11.5 *Incunable 94*

Biblioteca Apostolica Vaticana, S. Maria Magg. 94
[Genoa: Baptista Cavalus, 1476]; Breviary
Calendar on fols. 1r–6v[130]

Incunable SMM 94 is the only volume of its edition listed in the *Incunabula Short Title Catalogue*. The origin and dating given are by Thomas Accurti, based on circumstantial evidence pertaining to the characteristics of the paper and type in question. Saxer, on the other hand, makes no reference to Accurti, and while he offers no localisation for the edition, he dates it as late as the beginning of the 16th century. Perhaps this late dating is why Saxer ignores SMM 94 in his exposition and discussion of the calendars and *sanctorales* of the S. Maria Maggiore library collection. It is, nevertheless, strange that Saxer notes the numerous marginal additions in the psalter-hymnal sections of the book but does not mention that the calendar too is densely annotated. Indeed, the

founding of the church, such as in ACSP E14, F23, and F27, as well as Vat. Lat. 4406. It has furthermore been added to calendars ACSP A9 and E15, as well as SMM 41 and 48. Also predating the dedication of the church are the Vatican Basilica's two books of anniversaries, carrying the name of the hermit saint on fol. 81v in both cases. As many of the calendars listed above either originate or have been adapted for use in Rome, it is clear there was some interest in Onufrius in the city, both before and following the dedication of the Gianicolo church. Conversely, the feast seems to have been rare across the Alps, appearing only seven times in the 515 Calendoscope calendars. Muzerelle, *Calendoscope*.

129 21 November: "*Presentatio beate virginis, de mandato reverendissimi cardinalis domini Cesarinis arc*[margin lost] / *est missa canonicalis 1508.*" Saxer gives the name as "*Johannis Are,*" however, the cardinal in question is most likely Giuliano Cesarini, the archpriest of the basilica at the time. Consequently, it is likely the word cut out from the margin read *archipresbyteri*. Cf. Franca Petrucci, "Cesarini, Giuliano," in *Dizionario biografico degli Italiani v. 24*, ed. Alberto Maria Ghisalberti (Roma: Istituto della Enciclopedia italiana, 1980); Saxer, *Sainte-Marie-Majeure*, 667.

130 Bibliography: Thomas Accurti, *Aliae editiones saeculi XV pleraeque nondum descriptae* (Florentiae: Tipografia Giuntina, 1936), 2–3; Saxer, *Sainte-Marie-Majeure*, 587; British Library, *Incunabula Short Title Catalogue* (London: Consortium of European Research Libraries, 2016), n. ib01111400, https://data.cerl.org/istc/ib01111400; William J. Sheehan, ed., *Bibliothecae Apostolicae Vaticanae Incunabula – Volume 1: A-C*, Studi e Testi 380 (Città del Vaticano: Biblioteca Apostolica Vaticana, 1997), 288.

additions include several entries explicitly referring to a Roman context, and one cannot exclude the possibility of use at S. Maria Maggiore either.[131]

However, as the vast majority of the additions cannot be found in any other calendar examined in this study, Roman or otherwise, exactly how SMM 94 relates to a wider Roman context remains unclear. Admittedly, the calendar is a late exemplar, but perhaps not more so than SMM 119 and 120, produced for S. Maria Maggiore in 1480 and 1478.[132] The coat of arms on fol. 7r of the incunable may indicate private use, but this need not be the case.[133]

12 Conclusions

The material presented above constitutes a substantial expansion to the existing body of knowledge on the calendars and liturgical manuscripts used in the Roman basilicas of S. Pietro in Vaticano and S. Maria Maggiore in the Middle Ages. The broadest repertory of the most recently updated information on the manuscripts can be found in Giacomo Baroffio's *Iter Liturgicum Italicum*, the most ambitious inventory of Italian liturgical manuscripts compiled, which

131 See, for example: 20 March, added note: "*Rome: Hac die id est vespera sancti Benedicti est indulgentia perpetua ad Sanctam Crucem in Iherusalem tota die…*"; 19 May, added feast of St Ivo of Kermartin with the specification "*et fit festum in Sancto Petro per advocatos consistoriales et auditores, et conveniunt cardinales, in capite Vitecano, est festum palatii*"; August, added note: "*Rome fit festum sancti Bartholomei XXV°, est vigilia XXIIII°*"; 18 November, specification added to the dedications of S. Pietro and S. Paolo fuori le mura, "*sunt vespere papales et missa papalis in die*"; 30 November, specification added to the feast of St Andrew, "*est missa papalis et ostenditur capud sancti Andree.*" Furthermore, the additions include the feasts of Catherine of Siena and Monica on 5 May, which can be found in some Roman calendars, albeit a day earlier on 4 May. For Catherine of Siena, see ACSP A9, B88, and F23; for Monica, see ACSP A9, Stamp.ACSP A41, SMM 119 and 120. The mention of the translation of Jerome in a marginal note on the May calendar page may indicate the author of the additions was affiliated with S. Maria Maggiore. See table 5.4 above.

132 The November calendar page in SMM 94 contains a marginal note dated to 1480. While this does not necessarily mean the additions were written in the year in question, it does provide us with a *terminus post quem* for the insertions.

133 Cf. manuscripts SMM 119 and 120, which also carry coats of arms. Yet, the twin manuscripts were clearly not intended for private use, as evidenced by their incipits. See footnote 110 above. On the other hand, based on the additions listed in footnote 131, it could be proposed that SMM 94 may have belonged to somebody affiliated with the papal curia. See also the addition in table 5.4 above.

due to its conversion into an online format has vast potential for both helping to guide future research and evolving along with it. In the above sections, I have attempted to demonstrate how two of the five calendar manuscripts currently localised to S. Pietro in *Iter* should be considered misattributions, while a further seven manuscripts can be added to the corpus after a systematic examination of the Archivio del Capitolo di S. Pietro library collection in the Biblioteca Apostolica Vaticana. In the case of S. Maria Maggiore, three new calendars can be added to the four currently localised to the Basilica.

While some of the additional attributions are proposed here for the first time, many have appeared in literature and catalogue traditions before. Here I have strived to bridge the gaps that have occurred at critical points in previous scholarship and resulted in unnecessary confusion and a deteriorating level of general knowledge of the material.

The distinction between original and adopted calendars I have proposed allows for the examination of trends and consistencies within and between the groups of calendars in greater detail. To give a few examples, it can be suggested that after the oldest extant Vatican calendar ACSP B79, separated from its successors by more than a century, the Vatican tradition splits into two: the A9 and E15 line, which displays the greatest number of local Vatican entries without much in the way of obvious and radical external influence, and the F23 and B88 line, which carries a strand of Spoletan, and in the case of the former, Franciscan influence. Regarding S. Maria Maggiore calendars, a connection is evident between the two adopted calendars SMM 91 and 99, which seem to originate in a common context close to Rome.

Already at this rudimentary stage of categorising the evidence, a contrast can be observed between the material of the two basilicas, as the layer of obviously local entries is much greater in the calendars of S. Pietro than S. Maria Maggiore. Similarly, there is a greater quantity of calendars that can be considered to originate around S. Pietro than S. Maria Maggiore. Indeed, concerning the latter, the two original calendar manuscripts can only be thus localised due to the exceptionally generous incipits in the manuscripts, while the calendars themselves demonstrate no clear connection to the Basilica, lacking even the single indicative entry used to identify the basilica's adopted calendars.

What S. Pietro and S. Maria Maggiore have in common, however, is the high number and considerable geographical spread of adopted calendar manuscripts. Indeed, while ACSP E14 and SMM 91 and 99 seem to have originated in or around Rome, the contents of calendars ACSP B84 and SMM 50 point all the way to Perugia and Genoa–Pavia respectively. The two most common strands of "foreign" elements regarding S. Pietro and S. Maria Maggiore seem to pertain to central Italy on the one hand and the Franciscan order on the other.[134]

The mechanisms behind such diffusion and blending were varied, including elements such as donations and bequests of books, the potentially varied backgrounds of canons and beneficiaries, and the vast networks of subject churches possessed by basilicas of such prominence as S. Pietro and S. Maria Maggiore throughout Italy and beyond.[135] Although in the case of most individual manuscripts, it is impossible to trace the specifics of this process, questions relating to their function in their adoptive communities remain: how and why were calendars and liturgical manuscripts adopted with such seeming ease, and what function could such hybrids play in their new context? While these are questions easily identifiable through calendars, they have implications for a much broader spectrum of scholarship. To give an example, when placing the Vatican musical manuscript ACSP B80 in the context of "St. Peter's chant tradition," Christopher A. Reynolds identifies four of "the basilica's chant books" – among them B84, B85, and B88.[136] However, although calendars B84 and B85 indicate use and adaptation at S. Pietro, only B88 can be taken to originate at the basilica. More codicological and musicological research on the manuscripts should therefore be conducted in defining the Basilica's chant tradition, taking into account information relayed by their calendars.

In terms of further research on calendars, the proposed groupings of S. Pietro and S. Maria Maggiore material facilitate comparisons between calendar traditions and the use of calendars by different religious communities, starting with the two Basilicas themselves. What, beyond the indicative entries identified in tables 5.1, 5.2 and 5.4, do the calendars have in common within their respective groups? Do the answers differ between the basilicas and if so, why? Does such a comparison confirm the observation that the layer of locally significant entries seems to be much thinner in S. Maria Maggiore than S. Pietro? What implications does this have for the use of calendars, and consequently,

---

134 Van Dijk has noted such an interchange between the Franciscans and secular clergy became common from the mid-14th century onwards. Van Dijk, *Sources: Introduction & Description*, 132.

135 For a broad and concise overview, see Baroffio, "I codici liturgici." A revealing case is the 1259 decision of Alexander IV (1254–1261) to transfer the ownership of the possessions of the Franciscans of S. Maria Maggiore in Tivoli to the Chapter of S. Pietro in Vaticano, as the friars were not supposed to have possessions. It is difficult to say, however, whether this was a purely juridical move or whether it had some material consequences and if so, at what point. See Leonard Boyle, "The Date of the Consecration of the Basilica of San Clemente, Rome," *Archivum Fratrum Praedicatorum* XXX (1960), 425.

136 Christopher A. Reynolds, *Papal Patronage and the Music of St. Peter's, 1380–1513* (Berkeley: University of California Press, 1995), 83–84.

on our knowledge of the needs and practices of the respective communities? Answering these questions requires examining the full layer of local entries, comparing the consistency of contents among the Basilicas' original calendars, and considering the contents added to adopted calendars. As an alternative to a large-scale examination of groups of calendars, identifying an expanded corpus of material allows for more options when looking closely at specific manuscripts, and approaching the very same issues of the needs and practices of medieval religious communities and the functionality of medieval calendars from a different perspective. I employ both of these approaches in separate publications in the present volume.

13 Excursion: Searching for Lateran Calendars

The natural point of departure when searching for Lateran calendars is Pierre Jounel's *Le culte des saints dans les basiliques du Latran et du Vatican au douzième siècle*. As 12th-century and earlier calendars are rare, Jounel's chronological focus meant using just a single calendar, surviving only from March onwards, for his reconstruction of the Roman cathedral's tradition. The manuscript, Roma, Archivio di Stato di Roma, S. Salvatore 997, was stolen in 1977 and is missing to date.[137]

Consequently, the only evidence of the calendar's contents is to be found in Jounel's edition, which only gives a partial understanding of the manuscript. Although Jounel mentions that the calendar's entries were distinguished by ink colour, for instance, these are not indicated in his edition. Moreover, he states that additions by various hands demonstrate the manuscript remained in use over several centuries, but these are unfortunately not given in the edition.[138] How Jounel handled any removals of entries, which were more the rule than the exception in calendars in use over many centuries, is not documented. All of this means that while Jounel's edition is indeed of great value in preserving a reading of the calendar's original contents, its usefulness for examining how the use of calendars in the Lateran Basilica developed from the 12th century onwards is rather limited.

137 Robert Amiet, "Catalogue des livres liturgiques manuscrits conservés dans les archives et les bibliothèques de la Ville de Rome," *Studi medievali* ser. 3: 27, no. 2 (1986), 925, 930. Amiet mentions the theft of the manuscript along with two other codices, S. Salvatore 1001 and S. Spirito 73. At least part of manuscript S. Salvatore 1001 has recently been restored to the Archivio di Stato. All three manuscripts are indicated as extant in *Iter*, localising S. Salvatore 997 simply as *Lucca, S. Frediano*. Baroffio and Albiero, *Iter Online*.

138 Jounel, *Le Culte Des Saints*, 29–30, 84–94.

At least two further dead ends can be identified in previous scholarship. For one, there is the only notice of a Lateran calendar in *Iter*: manuscript Wien, Österreichische Nationalbibliothek, Cod. 1482, described as an Ordinal–Calendar. The manuscript does not contain a calendar, however. The confusion has its roots in Jounel's study on the Lateran calendar tradition, for which he extracted the saints mentioned in the Ordinal. He then attached dates to this list of extracted names, in order to compare the Ordinal's contents to other material pertaining to the Basilica. Thus, through *Iter*, Jounel's study has resulted in the creation of a brand-new medieval calendar, some 800 years after the composition of the original manuscript.[139]

There is also a mention of a Lateran calendar in a doctoral thesis by Costanza Rapone (2014), in conjunction with her analysis of a Crucifixion scene in an early 13th-century Italian New Testament manuscript, exhibited in 1970 at the Metropolitan Museum of Art in New York.[140] It seems evident, however, that Rapone herself has not studied the physical manuscript, but has based her analysis on the reproduction of the Crucifixion scene and the brief description of the manuscript provided in the exhibition catalogue.[141] Neither Rapone nor the catalogue provide clues as to the contents of the calendar or the subsequent whereabouts of the manuscript.[142] Therefore this is yet another promise of a Lateran calendar leading nowhere, at least for the time being.

139 Jounel, 26–27, 193. The Vienna manuscript has been edited in Ludwig Fischer, ed., *Bernhardi cardinalis et Lateranensis ecclesiae prioris ordo officiorum Ecclesiae Lateranensis* (München und Freising: Dr. F.P. Datterer & cie. (A. Sellier), 1916).

140 Costanza Rapone, "Tra Italia e Inghilterra nel Medioevo: Storie di codici e di uomini Giunti da Oltremanica Tra i Secoli XII e XIV" (doctoral thesis, Università degli studi della Tuscia di Viterbo, 2014), 342.

141 Rapone provides no information or analysis on the manuscript beyond what is in the catalogue and operates solely with the vague shelf mark, "Coll. H.P. Kraus," given therein. Furthermore, she provides no references to specific manuscript leaves, simply referring to "la *Crocifissione*." This differs from her treatment of other manuscripts, including those she uses for direct comparison with the Kraus manuscript. Cf. Rapone, 342–43, 470; Konrad Hoffmann, *The Year 1200: A Centennial Exhibition at the Metropolitan Museum of Art; February 12 through May 10, 1970. 1. Catalogue* (New York: The Metropolitan Museum of Art, 1970), 292–93.

142 The catalogue does forward the claim that "a sacramentary in the Vatican Library (Fondo S. Pietro F. 18) likewise combines a Sicilian style with a calendar of papal usage." This raises the question to what extent this unknown Lateran calendar is representative of the Basilica's, as opposed to papal, usage. Furthermore, it represents another instance of "calendar creation," as manuscript ACSP F18 does not contain a calendar. The claim is likely due to a misreading, on the part of the catalogue's compilers, of Garrison's analysis of the cycle of feasts in the Sacramentary. Cf. E.B. Garrison, "A Sacramentary for Roman Use with Sicilian Decoration (Vaticana, S. Pietro F.18)," in *Studies in the History of Mediaeval Italian Painting*, vol. IV:3 & 4 (Florence: L'Impronta, 1962), 411–15.

Considering existing manuscripts, the basilica's library itself offers no breakthroughs. Of the collection's six manuscript calendars, only one has been noted in previous scholarship to my knowledge.[143] Although some of these seem to have been used in Rome, none display any clear signs of use at the Basilica.[144]

143 Manuscript Roma, Archivio di San Giovanni in Laterano A65 was edited already in 1752 by Emmanuel de Azevedo under the heading *Vetus Missale Romanum Monasticum Lateranense*. It has been studied in detail by van Dijk, who has localised it to Città di Castello. See van Dijk, "The Lateran Missal"; Dijk and Walker, *Origins*, 402. The calendar survives in a fragmentary state, with parts of months legible from May onwards. Its contents are distant from S. Salvatore 997, edited by Jounel and discussed above.

144 The capitular archive of the Lateran Basilica shares its facilities with the *Archivio Storico Diocesano* but is an institutionally separate entity. As the Basilica's liturgical holdings are not well known, I briefly summarise my findings on the manuscripts currently not listed in *Iter*: A14, a small and probably personal 15th-century breviary with a densely filled calendar. The calendar's March page has three rubricated entries in original hand mentioning the arrival in Rome, the coronation, and the arrival in Naples of Emperor Frederick III; A15, another small Breviary. The calendar takes note of Nicholas of Tolentino and carries several entries relating to Augustine, giving it an Augustinian flavour. It lacks the November dedications of the Lateran and the basilicas of Peter and Paul in Rome, but marks the translation of St Mark on 31 January, an uncommon entry in Roman calendars. The feast of Bartholomew is marked on 24 August, as opposed to 25 August as customary in Rome; A16, a small 15th-century manuscript with computus tables and a calendar foreign to Rome; A38, a late 15th-century ordinal. The 17 May translation of Bernardino of Siena sets the *terminus post quem* at 1472. Although the calendar is densely filled, it bears no systematic similarity to that of MS A14 above. The papal insignia on fol. 1r suggest the manuscript may have been a papal commission or intended for the pope, but the space for the coat of arms below has been left empty. The ownership note of the Lateran Basilica that currently occupies the space is clearly a later addition. The calendar itself has been left unmodified and shows no signs of use; A66, a missal with a Neapolitan calendar, as attested by the multitude of saintly bishops of Naples recorded. The calendar may have seen some use in Rome, as it carries the November dedications of the major Roman basilicas as additions, as well as the 18 February feast of Constantia. Cf. footnote 121; Further liturgical manuscripts with no calendars and not noted in *Iter* are, as described in the Archive's typescript catalogue: A20 *Lectionarum ... piuttosto un Epistolario*; A73 *Salterio e antifone ... Inni*; A74 *Lezionario cum orationibus*; A77 *Homiliae SS. Patrum in Evangeliis adventus et Quadragesimae*; I was unable to check manuscript A64 for a calendar, as it could not be located upon my visits. It is described in the catalogue as "*Canon. Missae pro Episcopo cum Missa de Spiritu Sancto et in Alia pro omnibus Fidelibus Defunctis. In restauro.*"; According to the catalogue, the collection also hosts, under shelf mark B76, an exemplar of an incunable missal printed by Ulrich Hahn in Rome in 1476, currently not listed in the Incunabula Short Title Catalogue. This book was not located during my visits to the archive.

Finally, there is manuscript BAV Vat. Lat. 4406, which according to Edward B. Garrison contains a Lateran calendar.[145] Garrison dates the manuscript to 1200 and presents hagiological evidence to localise the calendar in Rome. He then establishes a connection with the Lucchese canons of S. Frediano – more specifically those of the Lateran Basilica. Even though the wealth of information Garrison employs makes his attribution seem plausible, there is some circularity to his arguments at key points.

The core evidence Garrison uses for his Lateran localisation consists of four calendar entries. Two of these, both written in rubrics, give the name Venantius, indicated as a martyr with companions on 18 May and as a bishop and confessor on 7 June. The conflicting attributions of martyr and confessor suggest these are in fact two distinct saints. The 18 May entry is not uncommon in central Italy, as it is the feast day of Venantius of Camerino, although the saint is not usually indicated as having companions. Nevertheless, Garrison insists that both entries pertain to Venantius, the Bishop of Salona, whose remains were transported from Dalmatia to the Lateran in the 7th century. Here Garrison uses the entries to support his Lateran localisation, while identifying the saint(s) as Venantius of Salona precisely because they appear in what he considers a Lateran calendar.[146]

Of the two remaining entries Garrison employs to further the Lateran localisation, yet another one involves a similar risk of circularity in argument. This is the 17 January entry *Dedicatio Pancratii*, which Garrison takes to refer to a certain church with such a dedication close to the Lateran. However, although Garrison points to several sources attesting to the ties between the two churches, nothing directly verifies the connection between the church of S. Pancrazio al Laterano and the calendar entry itself, such as an independent source confirming the date of the dedication. Instead, Garrison simply states: "This entry can, in the context, be taken without hesitation to refer to the dedication

145 Garrison, "Three Manuscripts," 12–32. Garrison also gives an edition of the calendar, collated with other relevant sources on 40–52. The level of detail Garrison has included in his transcription is commendable, although there are some errors where additions by later hands have been marked as rubricated entries: 23 February, *Romane*; 1 March, *Erculani*; 28 April, *Vitalis*; 26 July, *Anna mater Marie*. The last of these is both a rubricated entry and an addition. Some of these entries Garrison uses in discussing the calendar's provenance, treating later additions as highly indicative rubricated entries. Cf. ibid. 26–27.

146 Cf. Garrison, "Three Manuscripts," 21, fn. 97: "It is necessary to insist that in a S. Giovanni book, the presumption must be that it is this Dalmatian Venanzio who was being celebrated on both dates…". Garrison goes on to attribute the contradictory epithets of martyr and bishop to an anomaly. In Garrison's defence, confusion regarding saintly epithets is not uncommon in medieval calendars.

of the … church … ", the *context* apparently once again referring to the suggested Lateran localisation.[147]

The most significant entry Garrison draws attention to is the rubricated octave of the Lateran Basilica's dedicatory feast, which is, as he rightly points out, exceedingly rare. This does indeed provide some cover for Garrison's somewhat circular arguments detailed above. On the other hand, Garrison does not address the issue of the supposedly Lateran calendar omitting the double dedication of the Roman basilicas of the apostles Peter and Paul, which seems out of place for a Roman calendar of the time. After all, the feast was already present in the earlier Lateran calendar S. Salvatore 997, discussed above, at least according to Jounel's edition.[148]

To summarise, then, the situation regarding the survival and identification of medieval Lateran calendars is difficult. It may very well be that BAV Vat. Lat. 4406 is a Lateran calendar, as Garrison suggests, but even so, it would be the only one of its kind currently available to scholarship.[149] Apart from this, there is only Jounel's edition of the stolen S. Salvatore 997 to consult, omitting various layers of information present in the original manuscript, which, in turn, did not survive in its entirety to Jounel's day in any case. Thus, even though the Lateran calendar tradition may have been fundamental in the development of other Roman traditions, including that of the Curia and consequently Latin Christendom at large, the development of the use of calendars in the Roman Cathedral itself seem lost beyond recovery.

Bibliography

Sources

Acta Sanctorum Aprilis Tomus III. Antverpiae: Apud Michaelem Cnobarum, 1675.

Acta Sanctorum Augusti Tomus II. Antverpiae: Apud Bernardum Albertum vander Plassche, 1735.

Acta Sanctorum Februarii Tomus II. Antverpiae: Apud Iacobum Meursium, 1658.

Acta Sanctorum Maii Tomus IV. Antverpiae: Apud Michaelem Cnobarum, 1685.

Acta Sanctorum Martii Tomus III. Antverpiae: Apud Iacobum Meursium, 1668.

Acta Sanctorum Novembris Tomus I. Parisiis: Apud Victorem Palme, 1887.

147 Garrison, "Three Manuscripts," 20.

148 Cf. Jounel, *Le culte des saints*, 93, 379.

149 It is interesting to note that while Baroffio's 2003 calendar inventory characterises the manuscript's origin and provenance as "*Lucca S. Frediano* > *Roma S. Giovanni in Laterano*" the information currently given in *Iter* reads just "*Lucca, S. Frediano.*" Baroffio, "Kalendaria Italica," 470; Baroffio and Albiero, *Iter Online*.

Bibliotheca Sanctorum. Vol. IV. Roma: Società Grafica Romana, 1964.

Bibliotheca Sanctorum. Vol. VI. Roma: Città nuova editrice, 1965.

Bibliotheca Sanctorum. Vol. VII. Roma: Città nuova editrice, 1966.

British Library. *Incunabula Short Title Catalogue*. London: Consortium of European Research Libraries, 2016. https://data.cerl.org/istc/_search.

Hagiographi Bollandiani, ed. *Catalogus codicum hagiographicorum latinorum antiquiorum saeculo XVI qui asservantur in Bibliotheca nationali Parisiensi, tom. III*. Bruxellis, 1893.

Hülsen, Christian. *Le chiese di Roma nel medio evo: Cataloghi ed appunti*. Firenze: Leo S. Olschki, 1927.

Leroquais, Victor. *Les Bréviaires manuscrits des bibliothèques publiques de France*. 5 vols. Paris: Mâcon, Protat frères, 1934.

Poncelet, Albert. *Catalogus codicum hagiographicorum latinorum bibliothecarum romanarum praeter quam Vaticanae*. Bruxellis: Societas Bollandiana, 1909.

Ruysschaert, José, Albinia Catherine de la Mare, and Adriana Marucchi, eds, *I codici latini datati della Biblioteca apostolica Vaticana. 1, Nei fondi Archivio S. Pietro, Barberini, Boncompagni, Borghese, Borgia, Capponi, Chigi, Ferrajoli, Ottoboni*. Città del Vaticano: Biblioteca apostolica Vaticana, 1997.

Salmon, Pierre. *Les manuscrits liturgiques latins de la Bibliothèque Vaticane. 1, Psautiers, antiphonaires, hymnaires, collectaires, bréviaires*. Studi e testi 251. Città del Vaticano: Biblioteca apostolica vaticana, 1968.

Salmon, Pierre. *Les manuscrits liturgiques latins de la Bibliothèque Vaticane. 2, Sacramentaires, épistoliers, évangéliaires, graduels, missels*. Studi e testi 253. Città del Vaticano: Biblioteca apostolica Vaticana, 1969.

Salmon, Pierre. *Les manuscrits liturgiques latins de la Bibliothèque Vaticane. 4, Les livres de lectures de l'office, les livres de l'office du chapitre, les livres d'heures*. Studi e testi 267. Città del Vaticano: Biblioteca apostolica Vaticana, 1971.

Salmon, Pierre. *Les manuscrits liturgiques latins de la Bibliothèque Vaticane. 5, Liste complémentaire, tables gènérales*. Studi e testi 270. Città del Vaticano: Biblioteca Apostolica Vaticana, 1972.

Stornajolo, Cosimo. *Inventarium Codicum Manuscriptorum Latinorum Archivii Basilicae S. Petri in Vaticano Maxima Ex Parte e Recensione Cosimi Stornajolo (Arch. S. Petri, H 99–100) Depromptum, TOMUS I (A-B)*. Biblioteca apostolica Vaticana: Sala Cons. Mss. 411 (1) (unpublished), 1968.

Stornajolo, Cosimo. *Inventarium Codicum Manuscriptorum Latinorum Archivii Basilicae S. Petri in Vaticano Maxima Ex Parte e Recensione Cosimi Stornajolo (Arch. S. Petri, H 99–100) Depromptum, TOMUS II (C-E)*. Biblioteca apostolica Vaticana: Sala Cons. Mss. 411 (2) (unpublished), 1968.

Stornajolo, Cosimo. *Inventarium Codicum Manuscriptorum Latinorum Archivii Basilicae S. Petri in Vaticano Maxima Ex Parte e Recensione Cosimi Stornajolo (Arch. S.*

Petri, H 99–100) Depromptum, TOMUS III *(F-H)*. Biblioteca apostolica Vaticana: Sala Cons. Mss. 411 (3) (unpublished), 1968.

Literature

Accurti, Thomas. *Aliae editiones saeculi XV pleraeque nondum descriptae*. Florentiae: Tipografia Giuntina, 1936.

Amiet, Robert. "Catalogue des livres liturgiques manuscrits conservés dans les archives et les bibliothèques de la Ville de Rome." *Studi medievali* ser. 3: 27, no. 2 (1986), 925–97.

Baroffio, Giacomo. "Catalogare manoscritti liturgici: tipologie semplici e complesse." In *La catalogazione dei manoscritti miniati come strumento di conoscenza: esperienze, metodologia, prospettive : atti del Convegno internazionale di studi, Viterbo, 4–5 marzo 2009*. Edited by Silvia Maddalo and Michela Torquati, 115–26. Roma: Istituto storico italiano per il Medio Evo, 2010.

Baroffio, Giacomo. "I codici liturgici: specchio della cultura italiana nel Medioevo. Punti fermi, appunti di lettura, spunti di ricerca." *Ecclesia orans* 9, no. 3 (1992): 233–76.

Baroffio, Giacomo. *Iter liturgicum Italicum*. Padova: CLEUP, 1999.

Baroffio, Giacomo. *Iter liturgicum Italicum: Editio Maior*. Stroncone (Terni): Associazione San Michele Arcangelo, 2011.

Baroffio, Giacomo. "Kalendaria Italica Inventario." *Aevum* 77, no. 2 (2003), 449–72.

Baroffio, Giacomo, and Laura Albiero. *Iter Liturgicum Italicum: Répertoire des manuscrits liturgiques Italiens établi par Giacomo Baroffio*. Edited by François Bougard. Paris: Institut de recherche et d'histoire des textes, 2016. http://liturgicum.irht.cnrs.fr.

Bascapè, Giacomo C, Marcello Piazzo del, and Luigi Borgia. *Insegne e simboli: araldica pubblica e privata medievale e moderna*. Pubblicazioni degli archivi di Stato: sussidi 11. Roma: Ministero per i beni e le attività culturali, Ufficio centrale per i beni archivistici, 1999.

Bascapé, Giacomo Carlo. *Sigillografia. Il sigillo nella diplomatica, nel diritto, nella storia, nell'arte – volume II: Sigillografia ecclesiastica*. Archivio della Fondazione italiana per la storia amministrativa 14. Milano: Antonino Giuffrè, 1978.

Blaauw, Sible de. *Cultus et decor: liturgia e architettura nella Roma tardoantica e medievale: Basilica Salvatoris, Sanctae Mariae, Sancti Petri. 2*. Studi e testi 356. Città del Vaticano: Biblioteca Apostolica Vaticana, 1994.

Bourque, Emmanuel. *Etude sur les sacramentaires Romains v. 2.2*. Studi di antichità cristiana 25. Roma: Pontificio Istituto di Archeologia Cristiana, 1958.

Boyle, Leonard. "The Date of the Consecration of the Basilica of San Clemente, Rome." *Archivum Fratrum Praedicatorum* XXX (1960), 417–27.

Branciani, Luchina. "Origine e sviluppo dell'eremitismo nella Valle Sublacense." In *Le valli dei monaci. Atti del convegno internazionale di studio: Roma – Subiaco, 17–19*

Maggio 2010. Edited by Letizia Ermini Pani, 585–635. Spoleto: Fondazione Centro Italiano di Studi sull'Alto Medioevo, 2012.

Buono, Lidia. "Mns. 16." In *Catalogo dei più antichi manoscritti della Biblioteca Giovardiana di Veroli*, 69–71. Roma: Gruppo editoriale internazionale, 1996.

Caldelli, Elisabetta. *Copisti a Roma nel Quattrocento*. Roma: Viella, 2006.

Cancellieri, Francesco. *De Secretariis Basilicae Vaticanae – Vol. IV: Sylloge Veterum Monumentorum*. Romae: ex officina Salvioniana ad lyceum Sapientiae, 1786.

Cavallo, Guglielmo. "Dallo 'scriptorium' senza biblioteca alla biblioteca senza 'scriptorium.'" In *Dall'eremo al cenobio: La civiltà monastica in Italia dalle origini all'età di Dante*. Edited by Giovanni Pugliese Carratelli, 331–422. Antica madre: collana di studi sull'Italia antica 10. Milano: Scheiwiller, 1987.

Celenza, Christopher S. "Orsini, Giordano." In *Dizionario biografico degli Italiani v. 79*. Edited by Raffaele Romanelli. Roma: Istituto della Enciclopedia italiana, 2013.

Celenza, Christopher S. "The Will of Cardinal Giordano Orsini (Ob. 1438)." *Traditio* 51 (1996), 257–86.

Crollalanza, G.B. di. *Dizionario storico-blasonico delle famiglie nobili e notabili italiane, estinte e fiorenti, vol. I*. Bologna: Arnaldo Forni, 1965.

Dijk, S.A. Van. "Three Manuscripts of a Liturgical Reform by John Cajetan Orsini (Nicholas III)." *Scriptorium* 6, no. 2 (1952), 213–42.

Dijk, S.J.P. van. "The Lateran Missal." *Sacris Erudiri* VI (1954), 125–79.

Dijk, Stephen Joseph Peter van. *Sources of the Modern Roman Liturgy : The Ordinals by Haymo of Faversham and Related Documents (1243–1307), v. 1. Introduction. Description of Manuscripts*. Studia et Documenta Franciscana. Leiden: E.J. Brill, 1963.

Dijk, Stephen Joseph Peter van. *Sources of the Modern Roman Liturgy : The Ordinals by Haymo of Faversham and Related Documents (1243–1307), v. 2. Texts*. Studia et Documenta Franciscana. Leiden: E.J. Brill, 1963.

Dijk, Stephen Joseph Peter van. "The Legend of 'the Missal of the Papal Chapel' and the Fact of Cardinal Orsini's Reform." *Sacris Erudiri* VIII, no. 1 (1956), 76–142.

Dijk, Stephen Joseph Peter van, and Joan Hazelden Walker. *The Ordinal of the Papal Court from Innocent III to Boniface VIII and Related Documents*. Fribourg, Switzerland: The University Press, 1975.

Dijk, Stephen Joseph Peter van, and Joan Hazelden Walker. *The Origins of the Modern Roman Liturgy: The Liturgy of the Papal Court and the Franciscan Order in the Thirteenth Century*. Westminster, MD: The Newman Press, 1960.

Dubois, Jacques, and Jean-Loup Lemaitre. *Sources & méthodes de l'hagiographie médiévale*. Paris: Éditions du Cerf, 1993.

Dykmans, Marc. "Les obituaires Romains. Une définition suivie d'une vue d'ensemble." *Studi Medievali, Serie Terza* 19, no. 2 (1978), 591–652.

Ebner, Adalbert. *Quellen und Forschungen zur Geschichte und Kunstgeschichte des Missale Rromanum im Mittelalter: Iter italicum.* Freiburg im Breisgau: Herder, 1896.

Egidi, Pietro. *Necrologi e libri affini della Provincia romana.* 2 vols. Roma, Forzani e c., tipografi del Senato, 1908.

Esposito Aliano, Anna. "Testamento e inventari per la riconstruzione della biblioteca del cardinale Guglielmo d'Estouteville." In *Scrittura, biblioteche e stampa a Roma nel Quattrocento. Aspetti e problemi.* Edited by C. Bianca et al., 309–42. Littera Antique, 1,1. Città del Vaticano: Scuola Vaticana di Paleografia, Diplomatica e Archivistica, 1980.

Esposito, Anna. "Estouteville, Guillaume d'." In *Dizionario biografico degli Italiani* v. *43*. Edited by Fiorella Bartoccini and Mario Caravale. Roma: Istituto della Enciclopedia italiana, 1993.

Ferrari, Guy. *Early Roman Monasteries: Notes for the History of the Monasteries and Convents at Rome from the v through the x Century.* Città del Vaticano: Pontificio Istituto di Archeologia Cristiana, 1957.

Fischer, Ludwig, ed. *Bernhardi cardinalis et Lateranensis ecclesiae prioris ordo officiorum Ecclesiae Lateranensis.* München und Freising: Dr. F.P. Datterer & cie. (A. Sellier), 1916.

Garrison, E.B. "A Sacramentary for Roman Use with Sicilian Decoration (Vaticana, S. Pietro F.18)." In *Studies in the History of Mediaeval Italian Painting,* IV:3 & 4, 411–15. Florence: L'Impronta, 1962.

Garrison, Edward B. "Three Manuscripts for Lucchese Canons of S. Frediano in Rome." *Journal of the Warburg and Courtauld Institutes* 38 (1975), 1–52.

Gauvain, Alexis. *Il capitolo di San Pietro in Vaticano dalle origini al XX secolo. Vol. II: il Patrimonio.* [Città del Vaticano]: Edizioni Capitolo Vaticano, 2011.

Gorman, Michael M. "The Oldest Book List from St Peter's." In *Miscellanea Bibliothecae Apostolicae Vaticanae XI*, 383–98. Studi e Testi 423. Città del Vaticano: Biblioteca Apostolica Vaticana, 2004.

Grafinger, Christine M. "Archivio del Capitolo di S. Pietro. [Manoscritti]." In *Guida ai fondi manoscritti, numismatici, a stampa della Biblioteca Vaticana, I. Dipartimento Manoscritti.* Edited by Francesco D'Aiuto and Paolo Vian, 333–36. Città del Vaticano: Biblioteca Apostolica Vaticana, 2011.

Grotefend, Hermann. *Zeitrechnung des deutschen Mittelalters und der Neuzeit. Band 2. Abtheilung* 2. Hannover: Hahn'sche Buchhandlung, 1898.

Guéant, Valérie. "Du nouveau sur la production des premières années à Rome du maître des missels della Rovere." *Rivista di storia della miniatura* 13 (2009), 141–55.

Halonen, Marko. "Kalendaria urbana et rustica: Keskiaikaisia kalentereita Roomassa sekä Suomen kaupungeissa ja maaseudulla." In *Leivoksia, kaupunkilaisia ja sivisty-*

saatteita: Kävelyretkiä Euroopan historiaan. Edited by Sari Aalto, Samu Nyström, and Rose-Marie Peake, 126–48. Helsinki: Kirjapaja, 2017.

Harper, John. *The Forms and Orders of Western Liturgy from the Tenth to the Eighteenth Century: A Historical Introduction and Guide for Students and Musicians*. Oxford: Clarendon Press, 1993.

Hoffmann, Konrad. *The Year 1200: A Centennial Exhibition at the Metropolitan Museum of Art; February 12 through May 10, 1970. 1. Catalogue*. New York: The Metropolitan Museum of Art, 1970.

Hughes, Andrew. *Medieval Manuscripts for Mass and Office: A Guide to Their Organization and Terminology*. Toronto: University of Toronto Press, 1995.

Hughes, Andrew. "The Scribe and the Late Medieval Liturgical Manuscript: Page Layout and Order of Work." In *The Centre and Its Compass: Studies in Medieval Literature in Honor of Professor John Leyerle*. Edited by Robert A. Taylor, 151–224. Studies in Medieval Culture 33. Kalamazoo, Michigan: Medieval Institute Publications, 1993.

Jounel, Pierre. *Le culte des saints dans les basiliques du Latran et du Vatican au douzième siècle*. Collection de l'Ecole Française de Rome 26. Rome: École Française de Rome, 1977.

König, Erich. *Kardinal Giordano Orsini († 1438). Ein Lebensbild aus der Zeit der großen Konzilien und des Humanismus*. Studien und Darstellungen aus dem Gebiete der Geschichte 5,1. Freiburg i.Br.: Herder, 1906.

Krautheimer, Richard. *Rome: Profile of a City, 312–1308*. Princeton, N.J: Princeton University Press, 2000.

Lombardi, Giuseppe. "Inventari di biblioteche romane del Quattrocento: un panorama." In *Libri, lettori e biblioteche dell'Italia medievale (secoli IX–XV). Fonti, testi, utilizzazione del libro. Atti della Tavola rotonda italo-francese (Roma, 7–8 marzo 1997)*. Edited by Giuseppe Lombardi and Donatella Nebbiai Dalla Guarda, 349–72. Roma: Istituto centrale per il catalogo unico delle biblioteche italiane, 2000.

Lombardi, Giuseppe, and Flavia Onofri. "La biblioteca di Giodano Orsini (c. 1360–1438)." In *Scrittura, biblioteche e stampa a Roma nel Quattrocento. Aspetti e problemi. Atti del Seminario 1–2 giugno 1979*. Edited by C. Bianca, G. Lombardi, P. Farenga, A.G. Luciani, and M. Miglio, 371–82. Littera Antiqua, 1,1. Città del Vaticano: Scuola Vaticana di Paleografia, Diplomatica e Archivistica, 1980.

Lunghi, Elvio. "Le miniature nei manoscritti italiani della Biblioteca Capitolare di Perugia (Secoli XIII, XIV e XV)." In *Una città e la sua cattedrale. Il duomo di Perugia*, 249–76. Perugia, 1992.

Maddalo, Silvia. "Breviario di San Pietro." In *Liturgia in figura: codici liturgici rinascimentali della Biblioteca Apostolica Vaticana*. Edited by Giovanni Morello and Silvia Maddalo, 159–62. Città del Vaticano; Roma: Biblioteca Apostolica Vaticana, 1995.

Maddalo, Silvia, and Adalbert Roth. "Messale." In *Liturgia in figura: codici liturgici rinascimentali della Biblioteca Apostolica Vaticana*. Edited by Giovanni Morello and Silvia Maddalo, 155–59. Città del Vaticano ; Roma: Biblioteca Apostolica Vaticana, 1995.

Magionami, Leonardo. *I manoscritti del Capitolo di San Lorenzo di Perugia*. Roma: Jouvence, 2006.

Maire Vigueur, Jean-Claude. *The Forgotten Story: Rome in the Communal Period*. Translated by David Fairservice. Roma: Viella, 2016.

McCarthy, Michael Francis. *Heraldica Collegii Cardinalium: A Roll of Arms of the College of Cardinals v. 1. 1198–1799*. Sydney: Thylacine Press, 2002.

Mercati, Giovanni. *Codici latini Pico Grimani Pio e di altra biblioteca ignota del secolo XVI esistenti nell' ottoboniana e i codici greci Pio di Modena con una digressione per la storia dei codici di S. Pietro in Vaticano*. Studi e testi 75. Città del Vaticano: Biblioteca Apostolica Vaticana, 1938.

Muzerelle, D. *Calendoscope, Logiciel d'aide à l'identification Des Calendriers Liturgiques Médiévaux*. Paris: Institut de recherche et d'histoire des textes, 2013. http://calendoscope.irht.cnrs.fr/.

Parodi, Piergiorgio. "S. Maria Maggiore." In *Guida ai fondi manoscritti, numismatici, a stampa della Biblioteca Vaticana, I. Dipartimento Manoscritti*. Edited by Francesco D'Aiuto and Paolo Vian, 527–30. Città del Vaticano: Biblioteca Apostolica Vaticana, 2011.

Petrucci, Franca. "Cesarini, Giuliano." In *Dizionario biografico degli Italiani v. 24*. Edited by Alberto Maria Ghisalberti. Roma: Istituto della Enciclopedia italiana, 1980.

Petterlini, Elena. "Il Santorale Marciano e Il Santorale Della Chiesa Di Verona." In *Da Bisanzio a San Marco. Musica e Liturgia*. Edited by Giulio Cattin, 300–13. Venezia: Fondazione Ugo e Olga Levi, 1997.

Pontani, Gaspare. *Il Diario Romano Di Gaspare Pontani, Già Riferito al "Notaio Del Nantiporto" [30 Gennaio 1481 – 25 Luglio 1492]*. Edited by Diomede Toni. Rerum Italicarum Scriptores II, 3.2. Città di Castello: Casa editrice S. Lapi, 1907.

Rapone, Costanza. "Tra Italia e Inghilterra nel medioevo: Storie di codici e di uomini Giunti da Oltremanica tra i secoli XII e XIV." Doctoral diss., Università degli studi della Tuscia di Viterbo, 2014.

Reynolds, Christopher A. *Papal Patronage and the Music of St. Peter's, 1380–1513*. Berkeley: University of California Press, 1995.

Rezza, Dario, and Mirko Stocchi. *Il capitolo di San Pietro in Vaticano dalle origini al XX secolo. Vol. 1: la storia e le persone*. [Città del Vaticano]: Edizioni Capitolo Vaticano, 2008.

Rose, Els. "Liturgical Commemoration of the Saints in the 'Missale Gothicum' (Vat.Reg. Lat. 317). New Approaches to the Liturgy of Early Medieval Gaul." *Vigiliae Christianae* 58, no. 1 (2004), 75–97.

Saxer, Victor. "Il culto dei martiri romani durante il Medioevo centrale nelle basiliche Lateranense, Vaticana e Liberiana." In *Roma antica nel Medioevo: mito, rappresentazioni, sopravvivenze nella "Respublica Christiana" dei secoli IX–XIII, atti della quattordicesima Settimana internazionale di studio, Mendola, 24–28 agosto 1998*, 131–61. Milano: Vita e Pensiero, 2001.

Saxer, Victor. *Sainte-Marie-Majeure : Une Basilique de Rome Dans l'histoire de La Ville et de Son Église, Ve–XIIIe Siècle*. Collection de l'École Française de Rome 283. Rome: École française de Rome, 2001.

Saxer, Victor. "Trois Manuscrits Liturgiques de l'Archivio Di San Pietro, Dont Deux Datés, Aux Armes Du Cardinal Jordan Orsini." *Rivista Di Storia Della Chiesa in Italia* 27, no. 2 (1973), 501–05.

Scaccia-Scarafoni, Camillo. "Veroli – Biblioteca Giovardiana." In *Inventari dei manoscritti delle biblioteche d'Italia: vol. XXXIV*. Edited by Albano Sorbelli, 1–123. Firenze: Leo S. Olschki, 1926.

Sheehan, William J., ed. *Bibliothecae Apostolicae Vaticanae Incunabula – Volume I: A - C*. Studi e Testi 380. Città del Vaticano: Biblioteca Apostolica Vaticana, 1997.

Stinger, Charles L. *The Renaissance in Rome*. Bloomington: Indiana University Press, 1985.

Stocchi, Mirko. *La Basilica Vaticana e le sue chiese tra Medioevo ed Età Moderna: Volume I, Umbria*. Città del Vaticano: Edizioni Capitolo Vaticano, 2013.

Tomasi, Giuseppe Maria. "Responsoriale et antiphonarium Romanae Ecclesiae de circulo Anni iuxta veterem usum Canonicorum Basilicae Vaticanae Sci. Petri." In *Responsorialia et antiphonaria Romanae Ecclesiae a s. Gregorio Magno disposita*, 1–215. Romae: Typis Iosephi Vannaccij, 1686.

Ughelli, Ferdinando. *Italia sacra: sive de episcopis Italiae – Tomus secundus*. Romae: apud Bernardinum Tanum, 1647.

Ughelli, Ferdinando, and Nicola Coleti. *Italia sacra: sive de episcopis Italiae – Tomus primus*. Venetiis: apud Sebastianum Coleti, 1717.

Ughelli, Ferdinando, and Nicola Coleti. *Italia sacra: sive de episcopis Italiae – Tomus secundus*. Venetiis: apud Sebastianum Coleti, 1717.

Vauchez, André. *Sainthood in the Later Middle Ages*. Cambridge: Cambridge University Press, 1997.

Vildera, Anna. "Il Santorale Marciano e Il Santorale Della Chiesa Di Padova." In *Da Bisanzio a San Marco. Musica e Liturgia*. Edited by Giulio Cattin, 271–300. Venezia: Fondazione Ugo e Olga Levi, 1997.

Waley, Daniel. "Annibaldi, Riccardo." In *Dizionario biografico degli Italiani v. 3*. Edited by Alberto Maria Ghisalberti. Roma: Istituto della Enciclopedia italiana, 1961.

Welch, Anna. *Liturgy, Books and Franciscan Identity in Medieval Umbria*. Leiden and New York: Brill, 2015.

CHAPTER 6

Navigating the Cycles of Time: Calendar Dates and the Week in a 13th-Century Vatican Calendar

Holger Kaasik

The theory of how to read and use medieval calendars is well-known enough to be a staple of codicology, liturgy, and hagiography textbooks.[1] The use and functions of individual manuscript calendars as guides for navigating the cycles of time as used by individual communities, on the other hand, has rarely been touched on. There is, however, insight to be gained into medieval conceptions of time and practices of commemoration by examining in detail the hierarchies of temporal information on medieval calendar pages, as well as the choices and preferences of their users, manifested in the layers of intervention made to original texts of manuscript calendars. In this article, I examine closely such hierarchies and signs of intervention found in one of the Vatican Basilica's calendars, Città del Vaticano, Arch. Cap. S. Pietro, E.14 – henceforth ACSP E14.

In focusing on a single calendar to reach general arguments, I follow the example set by Matti Peikola in his 2009 article on the "instructional aspects of the calendar" examining the John Rylands University Library MS English 80.[2] Where Peikola discusses the calendar as a sort of paratext in guiding the reading of other texts in the manuscript, my focus is on the modalities with which calendar ACSP E14 was used to navigate and coordinate various layers of liturgical time. Thus, I examine what type of temporal information was prioritised by the people using the calendar and what seems to have been of secondary or negligible importance. In particular, I address the question, why medieval

1 See, for example, Raymond Clemens and Timothy Graham, *Introduction to Manuscript Studies* (Ithaca: Cornell University Press, 2007), 194–202; Andrew Hughes, *Medieval Manuscripts for Mass and Office: A Guide to Their Organization and Terminology* (Toronto: University of Toronto Press, 1995), 275–79; Jacques Dubois and Jean-Loup Lemaitre, *Sources & méthodes de l'hagiographie médiévale* (Paris: Éditions du Cerf, 1993), 140–43.

2 Matti Peikola, "Instructional Aspects of the Calendar in Later Medieval England, with Special Reference to The John Rylands University Library MS English 80," in *Instructional Writing in English: Studies in Honour of Risto Hiltunen*, ed. Matti Peikola, Janne Skaffari, and Sanna-Kaisa Tanskanen, Pragmatics & Beyond New Series 189 (Amsterdam: John Benjamins, 2009), 83–104.

 | DOI:10.1163/9789004436251_007

calendars, having been developed to facilitate the complex task of orienting between various overlapping but incommensurate temporal cycles, continued to use the less-than-convenient Roman dating system to represent dates, seemingly the calendars' most elementary building block.

The structure of the discussion is as follows: in the first section, I lay out the principles of the calendars' structure and functions, with particular regard to the difficulties caused by the continued use of Roman dates. In the second section, I analyse such problems and the medieval attempts of addressing them in detail as illustrated by the example of calendar ACSP E14. In order to reach conclusions on what the people using medieval calendars considered more and less relevant temporal information, in the third section I consider the importance of the week on the one hand and the relative insignificance of chronological information on the other. In the concluding section, I examine the calendars' capacity to accommodate change in the face of the advent and proliferation of print and return to the continued resilience of the Roman dating system.

1 The Practical, Sophisticated, and Cumbersome Structure of Medieval Calendars

The origin of Christian calendars relates to calculating Easter, a process necessitating awareness of the position of the sun within the course of the tropical year, the phases of the moon, as well as the cycle of the week. Thus, medieval calendars came to accommodate and reconcile a number of cyclical phenomena. On the calendar pages, the position of the sun and the phases of the moon in the sky run parallel to the cycles of the seven-day week and the months and dates of the Julian calendar. The calendars of the 12th to the 15th centuries typically indicate the position of the sun as solstices and equinoxes, while the moon has its own column of numerals, the *numeri aurei*, marking the date of the new moon according to a 19-year cycle.[3] The week, too, has its own column, filled with letters running from *a* to *g*, the so-called *litterae dominicales*. This ingenious layout worked well enough to allow medieval manuscript calendars decades or even centuries of use, as they were adaptable to the shifting lunar and weekly cycles of any given year.[4]

3 This so-called *metonic cycle* is roughly the interval it takes for solar and lunar cycles to concur.

4 For more detailed introductions, see Clemens and Graham, *Introduction to Manuscript Studies*, 199–207; Hughes, *Medieval Manuscripts for Mass and Office*, 277–79. There were, however,

It seems, then, that the structure of the medieval calendar developed to be practical and adaptable. There was one crucial exception, however: the system for indicating dates in calendars was still the same towards the end of the Middle Ages as it had been in Roman calendars of classical antiquity. To give a Roman example, let us compare two surviving fragmentary mural calendars in Rome: the first dated to the late 2nd to early 3rd centuries, now below the Basilica of S. Maria Maggiore;[5] the other dating from the mid-13th century, decorating the walls of a cardinal's palace adjoining the Basilica of Santi Quattro Coronati.[6] Although the two date from roughly a millennium apart, originate from vastly different cultural contexts, and have nothing in common in terms of contents, both calendars use the Roman dating system, revolving around the *Kalendae*, the *Nonae*, and the *Idus*. This convention was indeed to remain predominant in calendars and related texts, such as martyrologies and obituaries, to the close of the 15th century and beyond.[7]

---

some complications, most prominently the slowly accumulating dissonance between the Julian calendar and the tropical year, leading to the equinoxes and solstices gradually shifting out of place over the centuries. See Marko Halonen's article in the present volume.

5 M.R. Salzman, "New Evidence for the Dating of the Calendar at Santa Maria Maggiore in Rome," *Transactions of the American Philological Association (1974–)* 111 (1981), 215–27.

6 Antonio Muñoz, *Il restauro della chiesa e del chiostro dei SS. Quattro Coronati* (Roma: Danesi, 1914), 130–33; Theodor Klauser, "Ein Kirchenkalender aus der römischen Titelkirche der heiligen Vier Gekrönten," in *Scientia sacra: theologische Festgabe zugeeignet Seiner Eminenz dem Hochwürdigsten Herrn Karl Joseph Kardinal Schulte, Erzbischof von Köln, zum 25. Jahrestage der Bischofsweihe 19. März 1935*, ed. Carl Feckes (Köln; Düsseldorf: J.P. Bachem; L. Schwann, 1935), 11–40; Silvia Maddalo, "A proposito di un calendario dipinto a Roma, nel Duecento, e di una committenza cardinalizia," in *Immagine e Ideologia. Studi in onore di Arturo Carlo Quintavalle*, ed. Arturo Calzona, Roberto Campari, and Massimo Mussini (Milano: Electa, 2007), 302–12; Silvia Maddalo, "Rappresentare Il Tempo a Roma Nel Duecento: I Calendari Dipinti Tra Tradizione Laica e Riproposta Cristiana," in *Medioevo: La Chiesa e Il Palazzo. Atti Del Convegno Internazionale Di Studi Parma, 20–24 Settembre 2005*, ed. Arturo Carlo Quintavalle (Electa, 2007), 583–97.

7 There are a few exceptions, such as the late 15th-century calendar in manuscript Perugia, Biblioteca Comunale Augusta, 782 (L. 34), where "modern" or running dates have been used. The 16th century saw the growth of the practice where "modern" dates were added without leaving out Roman dating, resulting in doubly dated calendars. This, to give an example, is the case with the first ever calendar in Finnish, published in 1544 in the so-called prayer book of Mikael Agricola, a reformation era bishop of Turku. The same seems to be true for calendars in the authoritative Roman liturgical books issued by the post-tridentine Papacy. Cf. the facsimile calendar in Manlio Sodi and Achille Maria Triacca, eds, *Missale Romanum: Editio Princeps (1570)* (Città del Vaticano: Libreria Editrice Vaticana, 1998), 39–50.

Meanwhile, Roman dates were either supplanted or seriously challenged by running, or "modern"[8] dating in most other contexts.[9] In Rome, they seem to be hardly ever used by diarists in the 15th century, nor were they employed in any of the canonisation processes of Francesca Bussa dei Ponziani, better known as St Frances of Rome.[10] According to Iiro Kajanto, just 4.8 % of Roman epitaphs dated between 1301 and 1350 use Roman dating, the figure decreasing to 2.9 % for the period between 1351 and 1400 and rising to 5.2 % between 1401 and 1450.[11] Where there was no particular reason to retain the Roman system,

8 I use the term "modern" here to illustrate that what I refer to as *running dates* is indeed the dating system we use today. Besides that, there is nothing particularly modern about it, as it was in use already during the Middle Ages, and, further still, in antiquity. The Roman dating system is very much the oddity here.

9 For examples from a variety of sources, see F.K. Ginzel, *Handbuch der mathematischen und technischen Chronologie, das Zeitrechnungswesen der Völker*, vol. III (Leipzig: J.C. Hinrichs, 1914), 115–17. Ginzel, 121–22 also gives an overview of the *Consuetudo Bononiensis*, where dates were counted progressively towards the mid-point of the month, and then down towards the beginning of the next. Though this practice was in use in Italy from the 8th and 9th centuries onwards, it seems extremely rare in Roman sources of the 14th and 15th centuries. The only example I have encountered is an added calendar entry in Città del Vaticano, Biblioteca Apostolica Vaticana, S. Maria Magg. 99, on 25 October: "*Issta die VII[a] exeunte mense celebratur fesstum sancti Donati foresste.*"

10 The assessment is based on cursory reading of Roman diarists, with texts attributed to: Antonio di Pietro dello Schiavo (from 1404 to 1417), Paolo dello Mastro (from 1422 to 1482), Stefano Caffari (from 1423 to 1477), Cardinal Jacopo Ammanati-Piccolomini (from 1472 to 1479), Jacopo Gherardi da Volterra (from 1479 to 1484), Gaspare Pontani (from 1481 to 1492), Antonio de Vasco (from 1481 to 1492), and Sebastiano di Branca Tedallini (from 1485 to 1517). Of these, only Jacopo Gherardi da Volterra seems to occasionally resort to Roman dates. See Antonio di Pietro dello Schiavo, *Il diario romano di Antonio di Pietro dello Schiavo: dal 19 ottobre 1404 al 25 settembre 1417*, ed. Francesco Isoldi, Rerum Italicarum Scriptores II, 24.5 (Città di Castello: Casa editrice S. Lapi, 1916); Stefano Caffari, *Memorie di una famiglia della Roma del Quattrocento*, ed. Alba Ingletto and Stefania Santi, Miscellanea della Società romana di storia patria 54 (Roma: Presso la Societa alla Biblioteca Vallicelliana, 2009); Giacomo Gherardi, *Il diario romano di Jacopo Gherardi da Volterra dal VII settembre MCCCCLXXIX al XII agosto MCCCCLXXXIV*, ed. Enrico Carusi, Rerum Italicarum Scriptores II, 23.3 (Città di Castello: Casa editrice S. Lapi, 1904); Gaspare Pontani, *Il Diario Romano Di Gaspare Pontani, Già Riferito al "Notaio Del Nantiporto"* [*30 Gennaio 1481 – 25 Luglio 1492*], ed. Diomede Toni, Rerum Italicarum Scriptores II, 3.2 (Città di Castello: Casa editrice S. Lapi, 1907); The *diario* of dello Mastro can be found appended to Paolo di Lello Petrone, *La Mesticanza di Paolo di Lello Petrone*, ed. Francesco Isoldi, Rerum Italicarum Scriptores II, 24.2 (Città di Castello: Casa editrice S. Lapi, 1910); editions of Ammanati-Piccolomini, de Vasco, and di Branca Tedallini are appended to Gherardi, *Diario Romano*. For more on the *diario* genre and the canonisation process of St Frances of Rome, see Saku Pihko's article in the present volume.

11 Iiro Kajanto, "Dating in the Latin Inscriptions of Medieval and Renaissance Rome," *Arctos: Acta Philologica Fennica* XI (1977), 48–51. In absolute terms, the figures for the periods are

then, running numbering seems to have been preferred as the more practical choice. Indeed, running numbering could be opted for even in contexts closely related to calendars, such as the 8th-century lists of saints' days (*notitia natalicorum*) at S. Silvestro in Capite in Rome.[12] Similarly, although the early 11th-century martyrology in manuscript ACSP H58 gives both Roman and running dates up until 17 July, only the latter are indicated from thereon out.[13]

It is easy to see how Roman dating could have seemed unnecessarily cumbersome for use in most contexts, especially since there was a simpler alternative in use simultaneously. According to the Roman system, dates within a month do not increase cumulatively as the month progresses but are instead counted down towards whichever of the three named days – the *Kalendae*, the *Nonae*, and the *Idus* – is up next. What is more, as Roman counting was

---

three out of 62, two out of 69, and six out of 115 inscriptions respectively. Kajanto's data is derived from Vincenzo Forcella's *Iscrizioni delle chiese e d'altri edificii di Roma dal secolo XI fino ai giorni nostri*, published in 14 volumes between 1869 and 1884. The figure for 1351–1400 may be off by one, as one of the two inscriptions considered by Kajanto is not in fact an epitaph. See Urpo Kantola's article, fn. 126, in the present volume. For the period 1251–1300, the share of epitaphs with Roman dating in Kajanto's data is higher, 34.8 %, but the material is sparse, consisting of just 23 inscriptions in total – a problem which only intensifies when looking at earlier periods. From the mid-15th century onwards, Roman dating started gaining popularity, surging to 29.8 % in 1451–1500, and 40.2 % in 1501–1527, out of 382 and 321 epitaphs respectively. This development mirrors a renewed interest in classical ideals of epigraphy. Cf. Urpo Kantola in the present volume.

12 The lists are inscribed in two stone tables, one indicating female, the other male saints. For the dating, see Nicolette Gray, "The Paleography of Latin Inscriptions in the Eighth, Ninth and Tenth Centuries in Italy," *Papers of the British School at Rome* 16 (1948), 52–53, n. 10. Gray somewhat misleadingly describes the lists as "a calendar of the natal feasts of saints." However, as the lists do not give an overview of the whole year with saints inserted with various intervals, they are not really calendars in a strict sense, but rather simply lists of saints and their respective feast dates.

13 The martyrology, including appended computus elements, covers fols. 59r–80v of the manuscript. The transition from double dating to just running dates can be seen on fol. 69v, where 17 July is still given as "*D(ies) XVII, XVI k(a)l(endas) aug(ustas)*," while subsequent dates are given simply in the format "*D(ies) XVIII*." For more on the manuscript see Cosimo Stornajolo, *Inventarium Codicum Manuscriptorum Latinorum Archivii Basilicae S. Petri in Vaticano Maxima Ex Parte e Recensione Cosimi Stornajolo (Arch. S. Petri, H 99–100) Depromptum, TOMUS III (F-H)* (Biblioteca apostolica Vaticana: Sala Cons. Mss. 411 (3) (unpublished), 1968), 184–87; Supino Martini Paola, *Roma e l'area grafica romanesca (secoli X–XII)* (Alessandria: Edizioni dell'Orso, 1987), 72–75. Further examples closely related to the calendars of S. Pietro in Vaticano can be found in the Basilica's martyrology-based books of anniversaries – here, running dates have been given for some entries, despite the rubrics at the top of each page giving the Roman date for the day. As these entries and their dating apparatuses are often formulaic, there is reason to suspect they have been copied in from other documents. See, for example, ACSP H57 fols. 44r–v, 118r, 124v, 135v, 171r.

inclusive, a day prior to the next named day would actually be given as two days prior to said named day. Thus, the Roman equivalent of 30 October would be *ante diem III kalendas novembres*, or "three days before the first of November" – the first, or kalends, of November being the next named day after 30 October.[14] While using running numbering only requires one to know the number of days in a month, Roman dating necessitates awareness of the next named day and conducting an inclusive calculation. In sum, the latter is both less intuitively intelligible and more laborious to write and to decipher.[15]

It is therefore not surprising that Roman dating was prone to lead to errors and confusion. This is likely what Bede was hinting at in his discussion on the calendrical conventions of the ancients:

> Neither the Greeks nor the Egyptians ... observed the distinction of kalends, nones and ides in their months. Instead, they counted forward in the sequence of days from the beginning of the month until its end, simply and without error.[16]

The practice seems indeed to have been bothersome to medieval scribes, resulting in various kinds of confusion.[17] On calendar pages, too, errors often result from faulty dating apparatuses. The most typical case is the duplication or omission of a numeral marking a date, resulting in incorrect dating up until the next named day, at times covering the greater part of the month. While

14 This is the full formula of classical Roman dating. Medieval sources hardly ever use this, opting for various kinds of shortened alternatives and deviations. Ginzel, *Handbuch*, III:116; Kajanto, "Dating in the Latin Inscriptions," 51–55.

15 Space would be another concern, whether working with diary entries on parchment or public inscriptions in stone, but this would depend much on the exact formula and the extent of abbreviations used. Whatever the case, Roman dating requires more frequent changes to the dating formula, as the point of reference shifts from one named day to another.

16 Quoted here as translated by Faith Wallis in Bede, *The Reckoning of Time*, trans. Faith Wallis (Liverpool: Liverpool University Press, 1999), 51–52. The original can be found in Beda Venerabilis, *De Temporum Ratione Liber*, ed. Ch. W. Jones, Corpus Christianorum – Series Latina, CXXIII B (Turnholti: Brepols, 1977), 327: "*Sed et Graeci et Aegyptii, de quibus supra diximus, nullam in suis mensibus kalendarum, nonarum, iduum distinctionem obseruant, uerum ab incipiente cuiusque mensis exordio usque ad terminum eius, crescente simpliciter et inerrabiliter dierum concurrentium ordine computando, perueniunt.*"

17 For examples, see Ginzel, *Handbuch*, III:116. Indeed, interpreting Roman dating has caused chronological aberrations even in modern scholarship: see Kantola, fn. 103 in the present volume.

such mishaps were occasionally corrected, though not always successfully, other times they were simply left be.[18]

Consequently, there would seem to be a conflict embedded in the very core of the structure of medieval calendars, which on the one hand had evolved to be sophisticated devices simultaneously visualising and helping coordinate between various overlapping and incommensurate cycles of time, but on the other hand resisted simplifying the mechanism for answering questions of such elementary nature as "what day is it" and "what is the correct sequence of dates." A closer look at how errors relating to Roman dating were addressed at the level of individual manuscript calendars may provide clues as to why this appears to have been the case and whether it was in fact considered a problem at all by the people using the calendars.

2 The Case of ACSP E14

The 13th-century calendar ACSP E14 is firmly localisable to Rome and was likely used in or around the Vatican Basilica.[19] It boasts numerous additions by various hands, with a particularly large amount inserted by a second rubricator, probably contemporary to the original scribes. These, unlike most additions in medieval calendars, are carefully drawn, with initials often ornamented with double strokes. While finding additions in calendars is common, the level of intervention in ACSP E14 is exceptional. Even the range of additions is broad, including feasts of saints, some a part of the most common medieval *sanctorale*, the unlucky Egyptian Days, and the headers of months detailing the lengths of days and nights and giving versicles on the unlucky days of the month.[20] Moreover, the intervention is not limited to additions, but includes

18 For errors and occasional approaches to correction in the Vatican Basilica's calendars, see: ACSP B84, March; ACSP B85, June; ACSP E15, May; Further examples of errors and corrections, with varying degrees of success, in the Vatican Library's manuscripts: ACSP B81, October; ACSP E6, February, April, and May; S. Maria Magg. 41, July; S. Maria Magg. 113, October; Vat. Lat. 4406, January; Errors in Roman dating even found their way to incunabula. See, for example, Stamp.ACSP A41, April.

19 For a description of the manuscript, see Kaasik, *The Calendars of S. Pietro and S. Maria Maggiore* in the present volume.

20 Notable added feasts by various hands include octaves for the feasts of the Christmas cycle in the beginning of January, as well as the two feasts of Agnes towards the end of the month; The second rubricator has also appended extensive computational notes to the calendar on fol. 9r, while the list of Roman *stationes* on the following page is by yet another rubricator.

many removals and other rearranging of the original content.[21] In short, the calendar in its original form has been subjected to modification of a broad and substantial nature.

Judging by the added feast of St Bernardino of Siena (died in 1444, canonised in 1450), the use the calendar saw was extended as well as extensive. Yet, given all the attention it received from various scribes over at least a century and a half, it is remarkable how many notable errors originating from the original production process remain, in particular pertaining to the correct sequence of dates. In the month of September, for example, the omission of *a. d. x kal. oct.*, or 22 September, has resulted in erroneous dating up to 29 September, which, in turn, has been duplicated. The dates for the whole section in between have been left uncorrected, even though it contains a number of feasts and modifications, including both additions and removals.

A particularly revealing case, when it comes to medieval attitudes towards dates in calendars, however, is the section from mid- to late November, beginning from the top of fol. 8r of the manuscript. To understand the end result, it is necessary to be aware of the order in which different layers of information were added to the page. In table 6.1 below, I give a simplified step by step illustration of the process, leaving out any information on the calendar page not pertinent to visualising how the errors came about, including most feast entries on the manuscript page.

First, markings in black ink were written down. These include some feast entries, as well as the sequences of Dominical letters, marking the seven-day cycle of the week, except for the letter *a*, the first letter in each sequence, left for the rubricator. The Dominical letters run in a straight column down the entire left margin of the page and were consequently the primary point of reference the rubricator used when filling in entries in red ink. Next, the rubricator proceeded to fill in the gaps in the Dominical sequences and added parallel columns on either side of the Dominical column: the *numeri aurei* for the phases of the moon to the left, and the Roman dates to the right.

What the rubricator had not yet realised, however, was that the last Dominical letter on the previous page was *g*, while the first one to have been entered in black on the current page was *b*. Therefore, it would have been up to the rubricator to add the *a*, which, in turn, would have been the first indication of the existence of the first row, or day, on the current calendar page. Now, however, the rubricator was already in the process of filling in the dates, continuing

21 See, for example, the relocation of the feast of St Eustace in September, as well as the several removals of feasts via red subscript dots from the end of December. The latter page, fol. 8v, even contains removals of removals, as scratch marks and residue of red ink from subscript dots are visible under *Savini* and *Conceptio beatissime Marie virginis*.

TABLE 6.1 Simplified process of composition of the calendar in ACSP E14.

b		ritii epi & cfessoris
c		
d		
e		
f		
g		
b		
c		
d		
e		
f		
g		
b		
c		
d		
e		

		artini pp & mr
b	II	ritii epi & cfessoris
c	**idus**	
d	XVIII	
e	XVII	
f	XVI	
g	XV	
a	XIIII	
b	XIII	
c	XII	
d	XI	
e	X	
f	IX	
g	VIII	
a	VII	
b	VI	
c	V	
d	III	
e	II	

Note: Steps of Composition:

1. Dominical letters *b–g* and entries in black entered by original scribe.
2. Dominical letters a added by rubricator, erroneously ignoring the first row.
3. Date numerals added by rubricator, starting erroneously from the second row, as the first had not yet been indicated. This results in skipping numeral *IIII* towards the end of the month.
4. The rubricator adds the feast of Martin in its correct place before that of Britius. Due to the previous errors, however, the row remains without a date.

		Martini pp & mr
b	II	Britii epi & cfessoris
c	idus	
d	XVIII	
e	XVII	
f	XVI	
g	XV	
a	XIIII	
b	XIII	
c	XII	
d	XI	
e	X	
f	IX	
g	VIII	
a	VII	
b	VI	
c	V	
d	III	
e	II	

5. Initials added to feast entries by decorator.

a		Martini pp & mr
b	II	Britii epi & cfessoris
c	idus	
d	XVIII	
e	XVII	
f	XVI	
g	XV	
a	XIIII	
b	XIII	
c	XII	
d	XI	
e	X	
f	IX	
g	VIII	
a	VII	
b	VI	
c	V	
d	III	
e	II	

6. Missing Dominical letter a added on first row by second rubricator.

Result: feasts with correct sequence and intervals; feasts aligned correctly with Dominical letters; erroneous dating for the larger part of November; the feast of Martin on a "dateless" day; *a. d. IIII kal. dec.* (28 November) skipped.

the sequence from the previous page, not from where the missing *a* should have been, but in line with the *b*, on what should have been only the second row of the page. Thus, the dates and weeks came to run out of sync for the larger part of the month. The rubricator likely realised the mistake only when approaching the end of November, as the month was running out one step ahead of the dates. This prompted the omission of *a. d. IIII kal. dec.* (28 November) altogether. Further complications arose when the rubricator realised the need to enter a red-letter feast on the first row of the page. As the second row of the page, still being the top row at this stage, was already taken up by a feast in black letters, the rubricator entered the red-letter feast above, essentially creating a day without a date. Finally, the missing Dominical *a* was added to this row by the second rubricator.[22]

The result is the correct sequence of feasts with correct intervals, perfectly aligned with the cycle of the week. The dates, however, run incorrectly all the way from 12 November to 28 November, the former date being preceded by an undated day, and the latter being skipped entirely. Although this extensive section, again, contains numerous feasts, removals, and additions by several scribes, no effort has been made to amend the calendar dates in the column between the feasts and the Dominical letters. It would seem, then, that erroneous dating did not bother the people using the calendar, but primary importance was given instead to the correct alignment of the week with the sequence and intervals of feasts.

This relative indifference towards calendar dates is confirmed by one final example in E14. In the month of July, date numerals run correctly throughout, so the error in feast dates here was not the rubricator's fault. However, the

22 Further clues as to the order of work can be found towards the end of January, where the names Papias and Maurus have been inserted twice, with the latter entry, written in black, missing the decorated initial characteristic of the original entries in the calendar. This entry has been ruled over by the rubricator, who has entered and underlined the entry on the previous date, with the same names in red, this time completed with the initial *P*, drawn in blue and red. This indicates that the markings in black were indeed the first to have been written down, followed by rubrication and decoration. The calendar would have gone through a fairly standard order of work, then, were it not for the numerous modifications by the second rubricator. Cf. for example Clemens and Graham, *Introduction to Manuscript Studies*, 20–22. Considering the original production followed this common pattern, it may seem a little odd that the decorator ended up drawing a *B* in front of the 5 August entry "*edicatio basilice sancte marie ad presepe,*" spelling *Bedicatio*. It is possible the decorator was just following instructions and paying little mind to the actual text, however. This is suggested also by the 8 May feast of St Michael, where the original rubricator forgot to omit the first letter, which was to be added by the decorator, resulting in its eventual duplication as "*Aapparitio sancti Michaelis archangeli.*"

sequence of feasts between 8 and 16 July, written in black ink by the original scribe, was entered to begin two rows too early. This error was addressed twice, in both cases likely during the initial production process of the calendar or shortly thereafter. The first solution was a simple and revealing one: two Dominical letters were skipped at the appropriate point, with the *e* on 6 July followed directly by a rubricated *a* on 7 July. Though the letters were later reworked, the residues of this first correction, the *a* in particular, can still easily be made out. What is remarkable is that the solution only addressed the alignment of the feasts with the cycle of the week, leaving, once again, the incorrect dates as they were.

Subsequently, however, the sequence of Dominical letters was normalised. The discord was now solved in a two-fold manner: first, by red lines connecting the feasts to their appropriate Dominical letters, going out of their way not to attach themselves to the dates marked in between; secondly, by a duplication of the Dominical letters in red, crammed in between the feasts and the dates on the appropriate rows. While this finally brought the calendar entries in sync with both the week and the dates, that this was achieved by doubly connecting the feasts to the days of the week, while the red lines connecting the two navigate carefully around the dates, demonstrates the precedence the former enjoyed over the latter.[23] Indeed, one gets the impression that the July dates were corrected almost by accident, as a by-product of rectifying the anomaly in the Dominical sequence. Considering that this is the only instance, out of the three discussed thus far, where incorrect dates spanning over an extended section of a month have been remedied, the impression is only reinforced.

Finally, factoring in the many cases where an error at one point has been mitigated by casually inserting or omitting a date elsewhere, it seems that the people who produced and made use of calendar E14 did not care all that much about whether or not the calendar displayed the correct dates for specific days and feasts.[24] This observation can to some extent be generalised to apply to

---

23 A similar emphasis on the Dominical letters can, perhaps, be observed in the incunable calendar S. Maria Magg. 105, where horizontal strokes have been added, beginning from the Dominical letters and proceeding through the dates on each row. Where feasts are present, these strokes extend all the way from the Dominical letters to the initials of the feasts. The function of the strokes seems to be to act as a visual aid, helping to connect the days of the week with the sequence and intervals of feasts, whether a feast is actually present on a given row or not.

24 In addition to the examples already given, the calendar contains errors in dating that did not have such far-reaching consequences. This is the case with the first of January, which in calendar E14 has been marked with the numeral "IIII," actually the date of the following day. As the numeral is duplicated on 2 January, the error is confined, and, though left unfixed, provides no additional cause for confusion. Similarly, in October, where *a. d.*

other calendars used in liturgical contexts, as errors in dating are indeed common, and in many cases prevail despite attempts at correction.[25] A particularly striking example similar to E14 is Perugia, Biblioteca Capitolare Dominicini, Ms. 5, which carries additions suggesting use in Rome around 1301.[26] Although the calendar has been modified at several stages, at no point has it been deemed necessary to fill in the dates for the latter half of May, which stop suddenly and for no apparent reason on the 16th of the month. Again, as with calendar E14, this is despite the section containing a number of feasts, both in original and subsequent hands.[27]

3 The Primacy of the Week and the Sequences of Feasts

Considering how errors in dating were addressed and ignored in medieval manuscript calendars, it is necessary to concede that the question of (correct) dates seems to have been of relatively small importance to the people using them. Indeed, it is not at all difficult to find medieval calendars without dates entirely.[28] This may seem somewhat counterintuitive from a modern

X kal. nov. is missing, the error only results in the duplication of *a. d. IX kal. nov.* as *a. d. VIIII kal. nov*. The only instance of an immediate correction to such an error is 16 June, where what appears originally to have been written as *XIII* has been modified into something resembling *XUI*, seemingly by the original rubricator.

25 See footnote 18 above.

26 Perugia, Biblioteca Capitolare Dominicini, Ms. 5, fol. 2v: "*Obitus fratris Mathei Gaudri de Tr(??), prioris ecclesie sanctorum Aquile et Prisce de urbe*"; fol. 5r: "*Anno domini M° CCCI° pontificatus domini Bonifacii pape VIII in die dominica sancte Trinitatis frater Robertus Cla(?)mart, prior ecclesie sancte Prisce, fecit consecrari maius altare dicte ecclesie per manus domini fratris Nicolai episcopi Tortibulensis in quo posite sunt reliquie infrascripte, videlicet de calciamento beati Petri apostoli, item de cathedra eiusdem, item de sepulcro beate Virginis, Item de capite beate Prisce, et dedit supradicto altari dictus episcopus indulgentiam XL dierum.*" The reading given by Leonardo Magionami in the library collection's latest catalogue contains some errors with respect to the names and relics, as well as the diocese mentioned. Cf. Leonardo Magionami, *I manoscritti del Capitolo di San Lorenzo di Perugia* (Roma: Jouvence, 2006), 44.

27 Perugia, Biblioteca Capitolare Dominicini, Ms. 5, fol. 2r.

28 This seems to be especially true for books of hours. See, for example, Berlin, Staatsbibliothek zu Berlin – Preußischer Kulturbesitz, Mss. theol. lat. oct. 79 and 97; for well-known manuscripts, see the "London Rothschild Hours": London, British Library, Add Ms. 35313; or the "Rohan Hours": Paris, Bibliothèque Nationale, Ms. Latin 9471. The latter is particularly curious, as, although this vernacular calendar does not display dates, it does have a column running throughout the calendar that details the next named day for each date. To give an illustration, instead of providing "IIII" and "KL" for *a. d. IIII kal. feb.* (29 January), the calendar just displays "KL" – an entirely redundant point of reference on its own.

perspective, given how large a role dates play in how we use our calendars today. Since the vast majority of surviving medieval calendars are found within manuscripts pertaining to liturgical and other devotional contexts, however, this is not necessarily all that surprising.

In examining calendar ACSP E14, we observed how the rubricator had used the Dominical letters as a point of reference when filling in the incorrect sequence of dates to the month of November. Moreover, the only case of extensive correction to erroneous dates in the calendar resulted from amending the sequence of Dominical letters in July. Indeed, while the presence of dates in medieval calendars cannot be taken for granted at all, the Dominical letters seem to be something hardly any calendar could do without.[29] As even today, liturgical life in medieval Christendom was inseparably attached to the seven-day cycle of the week: from the weekly observation of the Lord's Day, reading cycles of psalms and holding special offices, to calculating Easter and the many *temporale* feasts attached to it, often focused on Sundays.[30] The

---

This is baffling, given the amount of effort that must have gone into the design and execution of these fabulous illuminated calendar pages; as examples of other types of manuscripts, see Perugia, Biblioteca Comunale Augusta, Mss. 780 (L. 32) and 1121 (N. 73). Note that the former has running dates in Hindu-Arabic numerals added by a later hand.

29 Very few exceptions seem to be known. Richard Pfaff has noted two 8th-century calendar fragments that lack Dominical letters, but their early dating means they originate from a context very different than the 12th- to 15th-century material examined here. Richard W. Pfaff, "Why Do Medieval Psalters Have Calendars?," in *Liturgical Calendars, Saints, and Services in Medieval England*, Variorum (Aldershot: Ashgate, 1998), 14. Though more such cases no doubt exist among the thousands of extant medieval calendars, I have yet to encounter one. All the above examples of calendars lacking dates consistently display Dominical letters. It is also worth noting, that while the late 13th-century S. Pietro martyrology–book of anniversaries, ACSP H56, originally lacked Dominical letters, these have been added to the manuscript by a later hand. In its early 15th-century copy, ACSP H57, the Dominical letters are rubricated. Clearly the week was considered too important a temporal cycle to be omitted.

30 Hughes, *Medieval Manuscripts for Mass and Office*, 3–20; John Harper, *The Forms and Orders of Western Liturgy from the Tenth to the Eighteenth Century: A Historical Introduction and Guide for Students and Musicians* (Oxford: Clarendon Press, 1993), 45–51; Clemens and Graham, *Introduction to Manuscript Studies*, 213; Bryan D. Spinks, "The Growth of Liturgy and the Church Year," in *The Cambridge History of Christianity, Vol. 2: Constantine to c. 600* (Cambridge University Press, 2007), 601, 613–14. Iiro Kajanto has observed that in antiquity, the days of the week, although expressed using their pagan names, were much more commonly indicated in Christian epitaphs than their pagan counterparts. Conversely, when Renaissance epigraphy moved to imitate classical ideals, weekdays all but disappear from the record. Kajanto, "Dating in the Latin Inscriptions," 55–58. As for non-funerary inscriptions from medieval Rome, references to the week seem to appear exclusively in liturgical contexts. See Urpo Kantola's article in the present volume. While the deponents of the 15th-century canonisation processes of St Frances of Rome made scant

interplay of these smaller and grander cycles with the fixed feast dates of saints was what constituted the core problematic of the medieval liturgical year, reflected on calendar pages by the Dominical letters and, occasionally, by specifications on which *sanctorale* feasts were to give way to Sundays and *temporale* feasts of greater import, and which were to stand their ground.[31] Indeed, the rubrics of the various sections in manuscript ACSP E14 are full of instructions regarding feasts, weekdays, and Sundays of varying importance.[32]

The calendar functioned here as a coordinative device between the overlapping cycles of liturgical time, and to do so adequately only two fundamental criteria had to be fulfilled: firstly, the cycle of the week had to be indicated correctly; secondly, the liturgical feasts had to be inserted in their proper order and with correct intervals. Questions regarding systems of dating, whether or not dates were indicated correctly, or, indeed, at all, were of secondary importance. In stark contrast to the week, with all its spiritual connotations and practical consequences to devotional life, in the sphere of the liturgical, dates were simply abstract numerals void of any intrinsic meaning and practical consequence.[33] While this may seem somewhat at odds with the concept of the anniversary – at the core of most calendar entries, be they feasts of saints or death notices – what mattered in practice was not the date itself, but rather

references to the week, denoting a period of eight days seems to have been commonplace. See Saku Pihko's article in the present volume. This has its parallel in the liturgical sphere in the ubiquitous concept of the octave, which is used to denote the interval of a week as eight days, including both the start and end days.

31 For a useful summary of the underlying problematic, see Harper, *Forms and Orders*, 49–53. The instructions on calendars are usually in the form of a specification "*non transfertur*" that indicates the celebration of a certain feast should not be postponed due to it coinciding with another feast or falling on a Sunday, for instance. To give a few examples, see calendars ACSP B88, Stamp.ACSP A41, and SMM 94, 119 and 120. Rubrics on such dynamics are common in liturgical books in general. See, for example, SMM 94, fol. 317r: "*De festivitatibus quae infra octavam sancti Francisci veniunt nihil tunc agitur, sed post octavam celebrantur.*"

32 See, for example, fols. 47v, 52v, 53r, 65r, 139v, and 152r–159v.

33 This does not mean dates were not occasionally mystified. 25 March is an apt example, as it was considered the date of both the conception and the crucifixion of Christ, as well as the creation of Adam, "the fall of Lucifer, the passing of Israel through the Red Sea and the immolation of Isaac." See Charles G. Herbermann et al., eds, *The Catholic Encyclopedia ; an International Work of Reference on the Constitution, Doctrine, Discipline and History of the Catholic Church, Volume 1* (New York: Robert Appleton Company, 1907), 542. Needless to say, such a level of emphasis is exceptional and can most likely be attributed to the marked significance of the Feast of the Annunciation. Moreover, this type of calendrical mystification had little impact on day-to-day liturgical practicalities and, consequently, the use of liturgical calendars.

the annual recurrence and repetition of acts of commemoration.[34] For this, all that was needed was a calendar showing the correct order and intervals of commemorations.

Here we come to the essential differences between the medieval calendar in its fully developed form and other types of related texts, such as the 8th-century lists of saints' days or *notitia nataliciorum* at S. Silvestro in Capite in Rome.[35] The latter presents the reader only with a list of saints and the dates of their feasts. There can be no question, therefore, as to whether the dates are of the highest importance here, given that they are the only means of temporal orientation available. The calendar, in turn, has this information encoded in its very structure, whether or not the dates are actually written out, and in coordinating various layers of time, it serves a variety of purposes for which simple lists and litanies would be useless.[36] On the other hand, given the laxity with which dates are recorded in some calendars, it seems unlikely that they were ever used to look up dates for feasts, let alone to coordinate events and activities outside the sphere of the liturgical and the commemorative.[37]

It is worth considering that calendar dates are not the only type of temporal information consigned to a position of relative unimportance in medieval calendars. Indeed, the emphasis on the practicalities of liturgy can also be observed in the nearly complete absence of a chronological dimension on calendar pages. Although most calendar entries refer to events in the past, information on years is very rarely given. This holds true for feasts of saints, obits of the recently deceased, and biblical events alike. Consider, for instance, calendar ACSP H28, where the only temporal distance between the Ascension of

34 This is also evident in the S. Pietro books of anniversaries, where up to 12 obits have been granted to a single benefactor. See Pietro Egidi, *Necrologi e libri affini della Provincia romana* (Roma, Forzani e c., tipografi del Senato, 1908), 172, n. 3. A similar phenomenon can be observed in the obituary calendar painted onto the walls of S. Giovenale in Orvieto, with entries from January to April reading: *"Primum anniversarium Bonutii domini Petri," "Secundum anniversarium Bonutii," "Tertium Anniversarium Bonutii," "Quartum offitium Bonutii."* Incidentally, this calendar does not indicate dates, but only lists Dominical letters as a means of temporal orientation within a month.

35 See footnote 12 above.

36 These are characteristics that calendars have in common with some martyrologies, e.g. ACSP H57. This is also what Pfaff calls *practical calendars*, that is calendars "capable of being used as a guide to the sanctorale cycle in the liturgical year," referring to the function of coordinating between the *sanctorale* and *temporale* cycles of liturgical time. Pfaff, "Medieval Psalters," 5.

37 Note that the observations here pertain primarily to calendars found in liturgical contexts, the most common context by far for medieval calendars. How well these conclusions can be generalised to other types of calendars remains a subject for future research.

Christ into Heaven, St John the Evangelist surviving immersion in boiling oil in Rome, the election of Matthias as an Apostle, and Noah disembarking from the Ark is that the first took place one calendar date before the others.[38]

The lack of chronological points of reference also characterises the vast majority of obits recorded in the S. Pietro books of anniversaries, ACSP H56 and H57.[39] Here, the process of turning the past into a flattened surface must have been to some degree conscious, considering how the obits were layered and copied between the two books over several centuries.[40] Indeed, in some entries detailing donations to the Basilica, information on what year it was has been omitted even as reference is made to it having been a jubilee year, which makes the choice seem deliberate.[41] This is a sharp contrast to medieval Roman epitaphs, which, according to Iiro Kajanto, give the year of death with very few exceptions.[42] Perhaps the discrepancy can be attributed to a difference in perspective: after all, calendars and related texts were focused on the whole commemorative year of a community, rather than the remembrance of a deceased individual. As such, the calendrical and the cyclical superseded the chronological to the extent that the latter was mostly excluded from calendars and related texts.

Thus, we may observe that various types of temporal information seemingly fundamental to our conception of anniversaries were neglected or ignored in the commemorative context of medieval calendars. Even dates could be misrepresented or, as was usually the case with chronological information, left out entirely. To generalise, the primary function of the calendar did not involve informing the users of historical detail, but rather reminding them of due liturgical action; nor did it matter whether it was 21 January or the fifth day before the kalends of October – what mattered was whether it was Sunday, and if so, which Sunday, and whether there were one or several commemorations to consider, and if so, what their hierarchy was.

---

38 Note that the notice on Matthias is given as "*Electio Mathei,*" seemingly referring to Matthew. As the difference between the genitives *Mathei* and *Mathie* is minute, this is likely a scribal error.

39 The same goes for other similar sources. For numerous Roman examples, see Egidi, *Necrologi*. Some exceptions in the S. Pietro books of anniversaries include entries with extensive dating formulas of their own, possibly copied in from other documents. See footnote 13 above.

40 See Kaasik, *The Calendars of S. Pietro and S. Maria Maggiore* in the present volume.

41 Three such entries in ACSP H57 can be found on fols. 44r–v.

42 Kajanto, "Dating in the Latin Inscriptions," 44–45. The practice also undermines the possibility of ever holding jubilee commemorations based on calendar entries. Yet, the concept of celebrating the passing of centuries, for instance, was very much present in Rome from 1300 onwards. See Jasmin Lukkari's article in the present volume.

4 Conclusions – Inertia, Tradition, and the Potential for Change

As it was not necessary to deal with dates at all when using liturgical calendars, it may seem tempting to ascribe the persistence of the relatively cumbersome Roman dating system in medieval calendars to the inertia of convention. It must also be considered, however, that despite being supplanted in most contexts, Roman dating did retain its position in papal letters.[43] It is, therefore, possible that its continued use had to do with maintaining a sense of authority through continuity and tradition. There is a crucial difference between papal documents and liturgical manuscripts, however, in that there seem to have been no centralised structures in place for the production of the latter in medieval Rome. On the contrary, what little evidence remains seems to point to a predominantly grass-roots production, with churches and religious communities relying on their own resources and benefactors to acquire the literature they needed.[44] Therefore, the continued use of Roman dating in calendars must have been largely self-perpetuating.

In this regard, it is noteworthy that Roman dating made a smooth transition to liturgical incunabula, which might have provided a natural opportunity for innovation in favour of a simpler dating system.[45] It seems, then, that some emphasis was put on how calendars were expected to look, either due to the ritual context for which they were mostly produced, the commercial considerations of printers anticipating the attitudes of their clerical clientele, or both.[46] This may very well be the case, as such editions were at times prepared in collaboration with religious houses.[47] Further still, Roman dating was retained in the 16th century, even as printing houses made a break with the visual tradition of manuscript production and developed a new aesthetic of a

43 Thomas Frenz, *Papsturkunden Des Mittelalters Und Der Neuzeit*, 2., aktualisierte Auflage, Historische Grundwissenschaften in Einzeldarstellungen, Bd. 2 (Stuttgart: Franz Steiner Verlag, 2000), 23, 25; Cf. Kajanto, "Dating in the Latin Inscriptions," 47.

44 See Kaasik, *The Calendars of S. Pietro and S. Maria Maggiore* in the present volume.

45 For examples printed in Rome, nowadays available through the Digital Vatican Library, see Inc.IV.10, S. Maria Magg. 105, Stamp.ACSP A41.

46 It has been pointed out, for instance, that at the time of the printing of his missal editions, Ulrich Han, or Udalricus Gallus, one of the earliest printers operating in Rome, worked on a commercial basis, rather than being dependant on curial patronage. See M.D. Feld, *Printing and Humanism in Renaissance Italy: Essays on the Revival of the Pagan Gods*, ed. Cynthia M. Pyle (Roma: Roma nel Rinascimento, 2015), 156–59.

47 In Rome, the Franciscans of S. Maria in Aracoeli were active in this regard. See, for example, the colophon in S. Maria Magg. 105, fol. 291r: "*...hoc opus ... a ftratribus sacri conventus Areceli recte ac fideliter emendatum...*".

printed book, and even as running dates were introduced to calendars alongside their Roman equivalents.[48] This practice of twofold dating is revealing indeed, as it implies that while Roman dates were still something generally expected to be seen on calendar pages, the pressures of practicality had by now mounted high enough to allow for a parallel column of running dates as well.

While factors contributing to the persistence of Roman dates in medieval calendars may have ranged from the aesthetical and the ceremonial to the scribal and the commercial, in practice the use of calendars in their most common context was fairly detached from questions of dating. As the dating apparatuses in calendars had limited practical applications, there was no drive to adopt a more convenient system. In a sense, then, the people producing and using the calendars could afford to be conservative, even at the expense of clarity and at the risk of errors and confusion.[49] The core practical value of the medieval calendar was in coordinating the different cycles of liturgical time. For the many people who used and modified calendars, such as ACSP E14, questions of date and temporal distance seemed irrelevant when compared to the overwhelming significance of the week and the dynamics between the two grand cycles of the liturgical year – the fixed feasts of the *sanctorale* and the ever-shifting *temporale* with Easter at its centre. Indeed, such calendars were not used to coordinate date with event, but rather the week with sequences of commemorations and liturgical services. The effects of these, in turn, would radiate well beyond the sphere of the liturgical, affecting, among other things, the juridical life of medieval Rome in determining when payments were due and when courts could convene.

---

48 See footnote 7 above; see also Peikola, "Instructional Aspects of the Calendar," 99. Yet another factor to consider is that the advent of the incunable coincides with rising interest in classical epigraphy, reflected in the resurgence of Roman dating in Roman epitaphs. See footnote 11 above.

49 This conservative attitude is reflected in the slow pace with which Arabic numerals were introduced into Latin calendars, even though their practical value was well recognised by the 15th century. See Peikola, "Instructional Aspects of the Calendar," 94–96. In fact, the adoption of Arabic numerals in Latin calendars seems to have gone hand in hand with the introduction of running dates. Cf. the examples listed in footnote 7. Calendar ACSP B67 is an oddity in this regard, as the dates on its January page inexplicably alternate between Roman and Arabic numerals. A further curiosity in the calendar is *a. d. XVIII kal. oct* (14 September), where the date numeral is indicated as *X8*.

Bibliography

Sources

Beda Venerabilis. *De Temporum Ratione Liber*. Edited by Ch. W. Jones. Corpus Christianorum – Series Latina, CXXIII B. Turnholti: Brepols, 1977.

Bede. *The Reckoning of Time*. Translated by Faith Wallis. Liverpool: Liverpool University Press, 1999.

Caffari, Stefano. *Memorie di una famiglia della Roma del Quattrocento*. Edited by Alba Ingletto and Stefania Santi. Miscellanea della Società romana di storia patria 54. Roma: Presso la Societa alla Biblioteca Vallicelliana, 2009.

Gherardi, Giacomo. *Il diario romano di Jacopo Gherardi da Volterra dal VII settembre MCCCCLXXIX al XII agosto MCCCCLXXXIV,*. Edited by Enrico Carusi. Rerum Italicarum Scriptores II, 23.3. Città di Castello: Casa editrice S. Lapi, 1904.

Lello Petrone, Paolo di. *La Mesticanza di Paolo di Lello Petrone*. Edited by Francesco Isoldi. Rerum Italicarum Scriptores II, 24.2. Città di Castello: Casa editrice S. Lapi, 1910.

Sodi, Manlio, and Achille Maria Triacca, eds, *Missale Romanum: Editio Princeps (1570)*. Città del Vaticano: Libreria Editrice Vaticana, 1998.

Stornajolo, Cosimo. *Inventarium Codicum Manuscriptorum Latinorum Archivii Basilicae S. Petri in Vaticano Maxima Ex Parte e Recensione Cosimi Stornajolo (Arch. S. Petri, H 99–100) Deprompturn, TOMUS III (F-H)*. Biblioteca apostolica Vaticana: Sala Cons. Mss. 411 (3) (unpublished), 1968.

Literature

Clemens, Raymond, and Timothy Graham. *Introduction to Manuscript Studies*. Ithaca: Cornell University Press, 2007.

Dubois, Jacques, and Jean-Loup Lemaitre. *Sources & méthodes de l'hagiographie médiévale*. Paris: Éditions du Cerf, 1993.

Egidi, Pietro. *Necrologi e libri affini della Provincia romana*. 2 vols. Roma, Forzani e c., tipografi del Senato, 1908.

Feld, M.D. *Printing and Humanism in Renaissance Italy: Essays on the Revival of the Pagan Gods*. Edited by Cynthia M. Pyle. Roma: Roma nel Rinascimento, 2015.

Frenz, Thomas. *Papsturkunden Des Mittelalters Und Der Neuzeit*. 2., aktualisierte Auflage. Historische Grundwissenschaften in Einzeldarstellungen, Bd. 2. Stuttgart: Franz Steiner Verlag, 2000.

Ginzel, F.K. *Handbuch der mathematischen und technischen Chronologie, das Zeitrechnungswesen der Völker*. Vol. III. Leipzig: J.C. Hinrichs, 1914.

Gray, Nicolette. "The Paleography of Latin Inscriptions in the Eighth, Ninth and Tenth Centuries in Italy." *Papers of the British School at Rome* 16 (1948), 38–162.

Harper, John. *The Forms and Orders of Western Liturgy from the Tenth to the Eighteenth Century: A Historical Introduction and Guide for Students and Musicians*. Oxford: Clarendon Press, 1993.

Herbermann, Charles G., Edward A. Pace, Condé B. Pallen, Thomas J. Shahan, and John J. Wynne, eds, *The Catholic Encyclopedia ; an International Work of Reference on the Constitution, Doctrine, Discipline and History of the Catholic Church, Volume I*. New York: Robert Appleton Company, 1907.

Hughes, Andrew. *Medieval Manuscripts for Mass and Office: A Guide to Their Organization and Terminology*. Toronto: University of Toronto Press, 1995.

Kajanto, Iiro. "Dating in the Latin Inscriptions of Medieval and Renaissance Rome." *Arctos: Acta Philologica Fennica* XI (1977), 41–61.

Klauser, Theodor. "Ein Kirchenkalender aus der römischen Titelkirche der heiligen Vier Gekrönten." In *Scientia sacra: theologische Festgabe zugeeignet Seiner Eminenz dem Hochwürdigsten Herrn Karl Joseph Kardinal Schulte, Erzbischof von Köln, zum 25. Jahrestage der Bischofsweihe 19. März 1935*. Edited by Carl Feckes, 11–40. Köln, Düsseldorf: J.P. Bachem, L. Schwann, 1935.

Maddalo, Silvia. "A proposito di un calendario dipinto a Roma, nel Duecento, e di una committenza cardinalizia." In *Immagine e Ideologia. Studi in onore di Arturo Carlo Quintavalle*. Edited by Arturo Calzona, Roberto Campari, and Massimo Mussini, 302–12. Milano: Electa, 2007.

Maddalo, Silvia. "Rappresentare il tempo a Roma nel duecento: I calendari dipinti tra tradizione laica e riproposta cristiana." In *Medioevo: La Chiesa e Il Palazzo. Atti Del Convegno Internazionale Di Studi Parma, 20–24 Settembre 2005*. Edited by Arturo Carlo Quintavalle, 583–97. Electa, 2007.

Magionami, Leonardo. *I manoscritti del Capitolo di San Lorenzo di Perugia*. Roma: Jouvence, 2006.

Muñoz, Antonio. *Il restauro della chiesa e del chiostro dei SS. Quattro Coronati*. Roma: Danesi, 1914.

Paola, Supino Martini. *Roma e l'area grafica romanesca (secoli X–XII)*. Alessandria: Edizioni dell'Orso, 1987.

Peikola, Matti. "Instructional Aspects of the Calendar in Later Medieval England, with Special Reference to The John Rylands University Library MS English 80." In *Instructional Writing in English: Studies in Honour of Risto Hiltunen*. Edited by Matti Peikola, Janne Skaffari, and Sanna-Kaisa Tanskanen, 83–104. Pragmatics & Beyond New Series 189. Amsterdam: John Benjamins, 2009.

Pfaff, Richard W. "Why Do Medieval Psalters Have Calendars?" In *Liturgical Calendars, Saints, and Services in Medieval England*, VI, 1–15. Variorum. Aldershot: Ashgate, 1998.

Pietro dello Schiavo, Antonio di. *Il diario romano di Antonio di Pietro dello Schiavo: dal 19 ottobre 1404 al 25 settembre 1417*. Edited by Francesco Isoldi. Rerum Italicarum Scriptores II, 24.5. Città di Castello: Casa editrice S. Lapi, 1916.

Pontani, Gaspare. *Il Diario Romano Di Gaspare Pontani, Già Riferito al "Notaio Del Nantiporto" [30 Gennaio 1481–25 Luglio 1492]*. Edited by Diomede Toni. Rerum Italicarum Scriptores II, 3.2. Città di Castello: Casa editrice S. Lapi, 1907.

Salzman, M.R. "New Evidence for the Dating of the Calendar at Santa Maria Maggiore in Rome." *Transactions of the American Philological Association* 111 (1981), 215–27.

Spinks, Bryan D. "The Growth of Liturgy and the Church Year." In *The Cambridge History of Christianity, Vol. 2: Constantine to c. 600*, 601–17. Cambridge University Press, 2007.

CHAPTER 7

Calendars in Use: Comparing S. Pietro in Vaticano and S. Maria Maggiore in Rome

Holger Kaasik

Medieval liturgical calendars have been studied from a variety of perspectives over several centuries. Studies range from thorough examinations of individual calendars, aimed at shedding light on the specifics of particularly significant calendar traditions at fixed points in time, to comparisons of groups of calendars based on, say, geographical grounds, in order to chart the level of cultural and cultual exchange between various localities.[1] The approach I apply in the article at hand may be closer to the latter in terms of the quantity of sources, but there are novelties regarding both the methodology and the questions posed. I shall examine a corpus of 40 medieval calendars from a rather more statistical perspective than what has previously been undertaken, focusing less on "exactly what" and more on "what types of things and how much" was written in the calendars. Instead of specific liturgical connections, then, my aim is to examine the very use and function of the calendars in the hands of the communities that produced or possessed them. Specifically, how the use of calendars as repositories of communal memory could differ from community to community, in addition to their general function as coordinative devices for navigating the various overlapping cycles of liturgical time. Specific attention will also be dedicated to examining how this mnemonic function of medieval

1 For the former, see Pierre Jounel's analysis of the 12th-century Vatican calendar tradition, which, although Jounel utilises various types of auxiliary material, is based on a single surviving manuscript calendar, already published as an object of scholarly interest by Cardinal Giuseppe Maria Tomasi in the 17th century: Pierre Jounel, *Le culte des saints dans les basiliques du Latran et du Vatican au douzième siècle*, Collection de l'Ecole Française de Rome 26 (Rome: École Française de Rome, 1977); Giuseppe Maria Tomasi, "Responsoriale et antiphonarium Romanae Ecclesiae de circulo Anni iuxta veterem usum Canonicorum Basilicae Vaticanae Sci. Petri," in *Responsorialia et antiphonaria Romanae Ecclesiae a s. Gregorio Magno disposita* (Romae: Typis Iosephi Vannaccij, 1686), 1–215; For the latter, see for example Elena Petterlini, "Il Santorale Marciano e Il Santorale Della Chiesa Di Verona," in *Da Bisanzio a San Marco. Musica e Liturgia*, ed. Giulio Cattin (Venezia: Fondazione Ugo e Olga Levi, 1997), 300–13; Anna Vildera, "Il Santorale Marciano e Il Santorale Della Chiesa Di Padova," in *Da Bisanzio a San Marco. Musica e Liturgia*, ed. Giulio Cattin (Venezia: Fondazione Ugo e Olga Levi, 1997), 271–300.

 | DOI:10.1163/9789004436251_008

calendars relates to that of other texts that could be used in a similar manner, especially martyrologies. Additionally, the statistical approach, in revealing the extent to which local calendar traditions were internally coherent or diverse, allows for observations on the textual genealogy of calendars used by particular communities.

The analysis is based on my transcripts of 40 calendars, divided into three groups: ten from the basilica of S. Pietro in Vaticano, six from S. Maria Maggiore in Rome, and an outgroup of 24 calendars of diverse backgrounds, mostly from central and northern Italy and mostly from the 13th and 14th centuries.[2] These can be divided further into subgroups based on origin and provenance, as is especially the case with the S. Pietro group, where a distinction exists between calendars with entries pertaining to the basilica written in original hand, and those that only carry such entries as later additions.[3] The aim is, through a series of comparisons between the three groups of calendars, to observe what kinds of functions the calendars served, and, ultimately, how the needs of the communities around these two major Roman basilicas differed regarding their manuscript calendars. A crucial point of comparison is provided by the two surviving medieval martyrologies–books of anniversaries from S. Pietro, whose contents, structure, and function partly overlap with the Basilica's calendars.[4] The focus throughout will be on the most substantial

2 The S. Pietro calendars are: Città del Vaticano, Biblioteca Apostolica Vaticana (henceforth BAV), Archivio del Capitolo di San Pietro (henceforth ACSP), A9, B79, B84, B85, B87, B88, E14, E15, F23, and F27; The S. Maria Maggiore calendars are: BAV, S. Maria Maggiore (henceforth SMM), 50, 91, 97, 99, 119, and 120; The outgroup calendars are: Avignon, Bibliothèque Municipale, 100; BAV, ACSP B63, B64, B65, B66, B67, B68, B69, B71, B72, B81, B82, E1, E4, F28; SMM 41, 48, 106, 112, 113, 114; Stamp.ACSP A41; Vat. Lat. 4406; Roma, Biblioteca Angelica, 1098. Note that the incunable SMM 94 is not included in the analysis, despite showing signs of use in S. Maria Maggiore. This is due to the calendar missing half of January and February. To keep the material as comparable as possible, I have included only whole calendars in the statistics.

3 For more details, see Kaasik, *The Calendars of S. Pietro and S. Maria Maggiore* in the present volume.

4 BAV, ACSP, H56 and H57. These, too, are discussed in more detail in Kaasik, *The Calendars of S. Pietro and S. Maria Maggiore* in the present volume. Further comparative material is derived from the 515 medieval calendars available through the online Calendoscope tool, derived from classic studies by Hermann Grotefend and Bollandist scholars. See, D. Muzerelle, *Calendoscope, Logiciel d'aide à l'identification Des Calendriers Liturgiques Médiévaux* (Institut de recherche et d'histoire des textes, 2013), http://calendoscope.irht.cnrs.fr/; Hermann Grotefend, *Zeitrechnung des deutschen Mittelalters und der Neuzeit. Band 1. Glossar und Tafeln* (Hannover: Hahn'sche Buchhandlung, 1891); Hagiographi Bollandiani, ed., *Catalogus codicum hagiographicorum latinorum antiquiorum saeculo XVI qui asservantur in Bibliotheca nationali Parisiensi, tom. III* (Bruxellis, 1893).

layer of contents in medieval liturgical calendars – feasts of saints and other commemorative entries.[5]

The structure of the study is as follows: after defining the terminology and setting out the limitations of the methodology, I briefly explore the overall contents of the corpus of 40 calendars in total and discuss their implications for the use of calendars as memory aids, concluding with a comparison with martyrologies. Next, I move on to compare the calendars of S. Pietro and S. Maria Maggiore, to highlight the differences between the two Roman basilicas' practices. Here particular attention is dedicated to identifying the layer of locally significant entries, as distinct from feasts that appear more or less consistently throughout the totality of the material. Moving on, I conduct a similar analysis on the layers of modifications visible on the calendar pages, especially concerning calendars not originating at the basilicas but adapted to their use at a later point in time. This, again, is to highlight the extent to which the calendars were used to store locally significant information and the differences that lay between the two communities in their needs and use of calendars. Further on, I turn my attention to the internal cohesion of the calendar groups, especially the more ample S. Pietro material. This serves to dispel any potential illusion that calendars within a particular local tradition form a monolith group – indeed, the focus is on internal divisions, the overlapping existence of various textual traditions, the diversification brought about by the adoption of calendars of diverse origins, and whether this may have been considered problematic at the time.

1 Methodological Notes

It is not difficult to find scholarly assessments of the number of feasts in the Latin liturgical year at various periods in the Middle Ages.[6] The standard

5 For entries of other types, e.g. astrological notices, see Marko Halonen's article in the present volume.

6 See, for example, James A. Brundage, *Medieval Canon Law* (Abingdon: Routledge, 2013), 83; Philippe Rouillard, "Sainteté III. Culte Des Saints," in *Catholicisme. Hier, Aujourd-Hui, Demain*, ed. G. Mathon, G.-H. Baudry, and L'institut catholique de Lille, vol. 13 (Paris: Letouzey et Ané, 1993), 719; Victor Saxer, "Il culto dei martiri romani durante il Medioevo centrale nelle basiliche Lateranense, Vaticana e Liberiana," in *Roma antica nel Medioevo: mito, rappresentazioni, sopravvivenze nella "Respublica Christiana" dei secoli IX–XIII, atti della quattordicesima Settimana internazionale di studio, Mendola, 24–28 agosto 1998* (Vita e Pensiero, 2001), 133–36; Victor Saxer, *Sainte-Marie-Majeure : Une Basilique de Rome Dans l'histoire de La Ville et de Son Église, Ve–XIIIe Siècle*, Collection de l'École Française de Rome 283 (Rome: École française de Rome, 2001), 305–08.

row-by-day structure of the calendars, too, seems to offer possibilities for simple quantitative analyses. Yet, this is not as straightforward a task as one might expect. Ultimately, the challenges boil down to defining the basic unit of analysis – a question of profound impact to understanding what exactly is being looked at, and, consequently, what types of conclusions may be drawn from the analysis.

In this study, I have adopted three fundamental methodological positions: firstly, the conceptual separation of calendar entries from liturgical feasts; secondly, qualifying entries as "commemorative," to encompass entries of not only liturgical, but also of biblical, historical, and, at times, personal character; finally, leaving it up to the calendars themselves to define where an entry ends and another begins, rather than imposing outside knowledge derived from, to name the most common examples, hagiographical scholarship and dictionaries of saints.

The logic behind distinguishing between calendar entries and liturgical feasts can be put simply: the existence of a calendar entry, that is the presence or absence of certain information in a calendar relative to other calendars, is usually an easily verifiable fact, while the liturgical or other annual commemorative action that may have followed from the entry is not. That being said, various mechanisms were used in composing medieval calendars to distinguish between entries of differing import, implying qualitative differences in the actions that were to follow, the most common being rubrication and indicating feast grades. The use of these devices, however, varies greatly from calendar to calendar and inconsistencies exist even within calendars. For instance, some scribes tended not to change ink colour mid-row and consequently rubricated everything on dates where one might expect to find only one of the entries in rubrics. It is also easy to find calendars where rubrication was not employed at all, while in others additional colours, such as blue and gold, were used alongside the usual red to distinguish between various types of significant entries. This implied hierarchy of entries is further complicated by the occasional use of red ink in later additions, although the vast majority of these were written in black. Thus, although elements such as rubrication and feast grades should not be ignored, they become less and less helpful as the corpus of manuscripts becomes more extensive and heterogeneous.

In the end, it is impossible to know what level of commemoration followed from any given calendar entry, if indeed any at all, even in cases where rubrication or indications of feast grades were used. This applies to heavily modified calendars in particular, suggesting considerable distance between the context of their origins and that of their subsequent use – the case for several of the adopted S. Pietro and S. Maria Maggiore calendars.[7] Consequently, in comparing the contents of 40 medieval calendars in their entirety, my focus in the analysis that follows is specifically on the presence and absence of calendar entries and not on the celebration of liturgical feasts.

As the borders between liturgical, historical, biblical, and personal information are vague at times, the calendar entries considered in this study can best be described as commemorative. Consider, for instance, the Ascension of the Prophet Elijah, found in some Umbrian Franciscan calendars on 17 June,[8] or the supposed historical date of the crucifixion that occasionally features in medieval calendars.[9] While the former refers to a biblical event, it also marks the end of the life on earth and the beginning of the life in Heaven of a significant saintly figure – similarly to the Ascension of Mary or, indeed, most liturgical entries commemorating saints. As for the latter, although it is clearly separate from the annual and movable celebration of Easter, it would be difficult to consider it merely an expression of historical curiosity, separate from the commemoration of anniversaries and void of any spiritual meaning. In a similar manner, it is in some cases difficult to establish whether an obit was written down as a prescription for anniversary services or whether it may simply have been of collective or personal sentimental value, especially when found alone in calendars which were not otherwise utilised as obituaries or necrologies.

As for the delimitation between separate calendar entries, it needs to be pointed out that medieval calendars display a considerable degree of variation

7 Calendar SMM 50, for instance, carries a series of original entries pertaining to Genoa and a plethora of Pavian additions. Spatial and temporal distance were not the only factors that could render a calendar's contents inconsequential, as liturgical instructions could be neglected for a range of reasons. In early 15th-century Rome, politically turbulent times and upheavals led to the recurring dereliction of liturgical services, as recorded in the Roman *diario* of Antonio di Pietro dello Schiavo, a Vatican beneficiary.

8 Found in S. Pietro calendars ACSP B84 and F23, as well as the latter's likely relative B88.

9 Such entries have occasionally been confused with the celebration of the Holy Week in modern scholarship, leading to misguided attempts in dating the manuscripts in question. See, for example, the date proposed for Perugia, Biblioteca Capitolare Dominicini, Ms. 5 in Leonardo Magionami, *I manoscritti del Capitolo di San Lorenzo di Perugia* (Roma: Jouvence, 2006), 43. Magionami's conclusion is baffling, considering the calendar has further biblical entries expressed in the exact same indented and rubricated register. See, for example, the 17 April notice "*Ingressio Noe in archam.*" An even more recent case can be found in Anna Welch, *Liturgy, Books and Franciscan Identity in Medieval Umbria* (Brill, 2015), 144. Welch, however, indicates her assessment is derivative of Marcella Degl'Innocenti Gambuti, whose study I have not been able to consult for this article. See Marcella Degl'Innocenti Gambuti, *I codici miniati medievali della Biblioteca comunale e dell'Accademia etrusca di Cortona* (Firenze: S.P.E.S., 1977).

with regard to groups of saints, even some of the more common collectives. This is not surprising, given that many such groups were products of the blending of several hagiographical traditions, resulting in different compositions and variations in the titles and names of the saints from one town, diocese, century, and calendar to the other.[10] It is therefore extremely difficult to know from a modern perspective where one feast ends and the next begins. In previous scholarship, not all have resisted the temptation to reduce such variation inherent to the sources for the convenience of analysis, operating with a supposition of knowledge of the "actual" compositions of such groups.[11] This kind of a disposition to "correct" one's sources is also displayed in some editions and reconstructions of calendars, especially concerning variations in the dates of liturgical feasts – another staple of medieval calendars.[12]

10 As an example, consider the 19 January group of saints: Marius/Maurus, Martha, Audifax, and Abacuch. For the profound hagiographical complexities behind the formation of this group, see Richard W. Pfaff, "The Hagiographical Peculiarity of Martha's Companion(s)," in *Liturgical Calendars, Saints, and Services in Medieval England*, Variorum (Aldershot: Ashgate, 1998), IV, 1–22. In the 40 calendars examined in this study, the composition of the group varies as follows: Marius/Maurus 37 appearances; Martha 38 appearances; Audifax 37 appearances; Abacuch 36 appearances. Martha is the only constant in the group, appearing in each calendar that indicates saints for the date in question. Marius/Maurus is missing in one calendar where all three other saints are present, and Audifax is missing in one case where only Marius/Maurus and Martha are indicated. In addition to being omitted once with Audifax, Abacuch is also missing in one calendar where all three other saints are indicated.

11 For the case of the saints mentioned in footnote 10, see Hagiographi Bollandiani, *Catalogus codicum hagiographicorum*, 611–12; Cf. Muzerelle, *Calendoscope*. As noted by D. Muzerelle in the critical notes to the online Calendoscope tool, the Bollandist scholars, whose calendar collation Calendoscope draws upon, tended to simplify the material. Indeed, their *Catalogus* simply indicates the group "*Marius, Martha, Audifax et Abacum*" is always found as such in the more than 60 *Bibliothèque nationale de France* calendars the saints appear in. Considering the variation observed within a smaller corpus here, such uniformity seems highly unlikely.

12 See, for example, van Dijk's handling of the date of the feast of Bartholomew, going against the bulk of his sources. Stephen Joseph Peter van Dijk, *Sources of the Modern Roman Liturgy : The Ordinals by Haymo of Faversham and Related Documents (1243–1307), v. 2. Texts*, Studia et Documenta Franciscana (Leiden: E.J. Brill, 1963), 372, 382. Cf. page 168 of the present volume. Andrew Hughes, on the other hand, warns against such practices. Andrew Hughes, *Medieval Manuscripts for Mass and Office: A Guide to Their Organization and Terminology* (Toronto: University of Toronto Press, 1995), 280: "It has become increasingly apparent ... that the *only* reliable source of the liturgical date is a manuscript liturgical Kalendar *for the area and years in question*. The date in a Kalendar for Rome may differ from that in a Kalendar for Hereford, and those two dates from the date in every other Kalendar. When manuscripts are not available a printed reference work may be the sole recourse, but the possibility of legitimate variants must be remembered." Italics original.

I have refrained from such interpretation wherever possible, leaving the questions of dates and delimitations of calendar entries up to the calendars themselves. Regarding groups of saints, I have treated those appearing consistently together as single calendar entries, while separating any saint whose presence is not always attested into an entry of their own. Concerning dates, if a feast falls a day earlier in one calendar than in the rest of the material, I have not ventured to equate the two. It is likely, therefore, that some of the variation in my data can be attributed to scribal errors, though such minor deviations in dating do not amount to a significant number of cases in the material. On the other hand, some calendar entries explicitly attribute this kind of variation to diverse liturgical traditions, making it all the more important not to erase any such information by default.[13]

Thus, it is necessary to reiterate the distinction maintained here between calendar entries and liturgical feast, even as there is likely to be a great deal of overlap between the two. As a consequence, the figures presented here are not to be equated with numbers of saints' feasts given in other literature, as my occasional division of groups of saints into separate calendar entries creates a certain number of entries that should not, in all likelihood, be taken to represent liturgical feasts in their own right. As such cases mostly relate to saints, whose names appear relatively frequently throughout the material, this results in a somewhat greater number of entries forming the "common core" of the medieval calendar in my presentation than in other literature. Moreover, since my division of entries is entirely dependent on the calendars consulted, any addition of further material would likely result in even further divisions. Therefore, although I present exact numbers, these should not be taken to be final in any sense. They do, however, serve to indicate general trends: as I have treated all calendars and entries according to the same principles, the figures I present provide insight into the relative consistencies and inconsistencies between calendars in medieval Rome.

13 E.g. the 13 February added entry in calendar SMM 41 "*Valentini in civitate interampn*(?)," which suggests the medieval inhabitants of Terni celebrated their patronal feast a day earlier than what was held to be St Valentine's day in most of Christendom. A similar addition reading "*Bernardini ordinis minorum in Aquila*" in calendar SMM 48 indicates that Bernardino of Siena was celebrated in Aquila one day ahead of his common feast date. Such added notices can also be found in calendar SMM 50, for example the 5 July "*Margarite secundum lombardos*" and the 20 July "*Margarite secundum gallicos.*" See also the Roman date for the feast of St Bartholomew on page 168, fn. 24 of the present volume.

2 The Selection and Character of Contents

A typical exemplar among the 40 calendars considered in this study has somewhere between 200 and 300 entries.[14] The number of distinct entries I have identified across the material amounts to 974, out of which 475, nearly half, appear in just a single calendar. It would seem, then, that such unique entries are a common occurrence, but the high number here could be partly due to the diverse origins of the outgroup calendars. Looking at the other end of the spectrum, only 55 entries appear in each and every calendar in the material, while allowing for absence in four out of 40 calendars sees the figure rise to 135. It is likely that the number of feasts somewhat universal in Rome and parts of central and northern Italy between the 12th and 15th centuries lies somewhere between these two figures. Within the context of the material at hand, these amount to roughly between 6 and 14 % of the distinct entries attested.

The figures give a tentative idea of the plurality we are dealing with when studying medieval calendars, with rare entries being common, and common entries being very much in the minority. It seems, then, that it was an essential characteristic of the calendars to carry the names of lesser-known saints. This tendency should not be confused with the inclusiveness displayed by martyrologies, however, which often contain information on several saints for each day of the year. In comparison, calendars are sparse in content and result from a much stricter process of selection.[15]

These crucial differences between calendars and martyrologies regarding the selection of contents can be sufficiently demonstrated considering material available from S. Pietro. Let us focus on the prominence awarded to the altars and chapels of the basilica in its manuscript calendars and martyrologies, in particular the patronage of Popes Nicholas III (1277–1280) and Boniface VIII (1294–1303). Both pontiffs were major patrons of the Vatican Basilica and both erected chapels there, in which they instituted liturgies and were

---

14 The extremes are ACSP B66 with only 100 entries and Avignon, Bibliothèque Municipale, 100 with a total of 364 entries. The average count of entries per calendar is 264, the median 267.

15 While this is the general rule, there are exceptions, such as Perugia, Biblioteca Capitolare Dominicini, Ms. 31, where the calendar has been filled in its entirety. Many entries in the calendar give geographical information, e.g. "*Rome natalis sancte merentiane virginis*" and "*Epheso natalis sancti Timothei martyris*" on fol. 3r. As such a practice is atypical of calendars, but common in martyrologies, it is perhaps likely the contents of the calendar have been extracted from a martyrology. On such "culling" of calendar entries from martyrologies, see Edward B. Garrison, "Three Manuscripts for Lucchese Canons of S. Frediano in Rome," *Journal of the Warburg and Courtauld Institutes* 38 (1975), 10.

ultimately buried.[16] Both have extensive obituary notices in the basilica's martyrology-based books of anniversaries, detailing their commissions and generosity towards the basilica and summarising the appropriate commemorative services.[17] These far outshine the laconic obits of other popes – that is, of those who have made it into the books in the first place.[18] Yet, out of the two, only the works of Boniface VIII are attested in the Basilica's calendars, and quite prominently at that: not only do the calendars note the chapel founded by the pope, both in conjunction with its dedication and the feast of Boniface IV, whose remains he had translated into the chapel and whose cult he promoted, there is also a mention of the consecration of Boniface VIII as pope, an exceedingly rare type of entry.[19]

One can only speculate on the reasons behind this. Perhaps this is partly due to the period of exceptional flourishing the basilica saw under the papacy of Boniface VIII. The establishment of the Christian jubilee tradition alone brought with it far-ringing prestige, as it posited the Roman basilicas of Peter and Paul at the centre of all of Western Christendom.[20] Furthermore, substantial funds generated by the jubilee were invested in properties and estates for the benefit of the Vatican chapter.[21] It is true that such investments attributed to Boniface VIII outshine those of Nicholas III in the basilica's books of anniversaries, where the jubilee indulgences, too, have been deemed to

16 See Sible de Blaauw, *Cultus et decor: liturgia e architettura nella Roma tardoantica e medievale: Basilica Salvatoris, Sanctae Mariae, Sancti Petri.* 2, Studi e testi 356 (Città del Vaticano: Biblioteca Apostolica Vaticana, 1994), 631–32, 673, 706, 856–57; Jean-Claude Maire Vigueur, *The Forgotten Story: Rome in the Communal Period*, trans. David Fairservice (Roma: Viella, 2016), 36, 299, 301. On papal burials, Blaauw indicates a notable shift in the 13th century in favour of the Vatican Basilica over the Lateran, although many popes were buried outside of Rome due to practical reasons.

17 ACSP H57, fols. 118r, 144v–145r.

18 See, for example, Eugene III (1145–1153) on fol. 96r in H57, Innocent III (1198–1216) on fol. 99v, and Honorius III (1216–1227) on fol. 40r. As an example more contemporary to Nicholas III and Boniface VIII, consider that Nicholas IV (1288–1292) is not mentioned at all, although there are four obits on 4 April, the date of the pontiff's death. See fol. 47v.

19 The distribution of entries relating to Boniface in the Vatican calendars is as follows: 23 January, consecration as pope: B87, added entry; 6 May, dedication of the altar of the chapel of Boniface: A9 and E15, original entries. F23 has the entry as an original on 7 May. B84 and B87, added entries; 4 September, feast of Boniface IV: A9 and F23, original entries. B84, added entry. F27 has the entry added on 5 September; Cf. also footnote 28 below.

20 The very announcement of the Jubilee was laden with symbolism, taking place in S. Pietro on 22 February, the feast of the Chair of St Peter. See Jasmin Lukkari's article in the present volume.

21 Eugenio Dupré Theseider, "Bonifacio VIII, papa," in *Dizionario biografico degli Italiani v. 12*, ed. Alberto Maria Ghisalberti (Roma: Istituto della Enciclopedia italiana, 1970).

merit special mention.[22] Such prominence could have stood out in the minds of contemporaries, especially when compared to the preceding reign of Celestine V – brief, conducted from and under the influence of Naples, and ending up in resignation; and the subsequent pontificate of Benedict XI – another brief reign, followed by the Papacy's relocation to Avignon.[23]

Whatever the causes, the consequences were long-lasting, as the calendars carried the mentions of Boniface VIII through the 14th and well into the latter half of the 15th century.[24] These were also added to several calendars originating elsewhere but later adapted to the S. Pietro tradition.[25] The works of Nicholas III, on the other hand, are nowhere to be found, despite similar circumstances pertaining to patronage and the physical structure of the basilica, as well as the relatively short temporal distance between the two papacies. There is, perhaps, some irony in the contrast, considering it was Nicholas III who had done much to institutionalise commemoration at the basilica, bringing about regulations regarding how anniversaries were to be recorded into the basilica's books of anniversaries and how these entries were to be read.[26] Yet, Nicholas was not the only benefactor excluded from the basilica's 14th-century calendars, as there are further dedications of altars recorded in the books of anniversaries but not in the calendars.[27] This goes to further demonstrate that,

22 ACSP H57, fols. 144v–145r.

23 Blaauw, *Cultus et decor*, 631, 707; Michael J. Walsh, ed., "Boniface VIII," in *A Dictionary of Popes*, 3rd ed. (Oxford University Press, 2015).

24 ACSP A9, E15, and F23 all carry entries pertaining to the Vatican Basilica and Boniface VIII in original hand. All are likely to date from the 14th century and were still used in the mid-15th century, as indicated by the addition of Bernardino of Siena into the calendars.

25 ACSP B84, B87, and F27.

26 Marc Dykmans, "Les Obituaires Romains. Une Définition Suivie d'une Vue d'ensemble," *Studi Medievali, Serie Terza* 19, no. 2 (1978), 624–25. The future Boniface VIII was acting as a notary to Nicholas III at the time.

27 E.g. on 28 February, where "*consecratio altaris sancti Georgii martyris*" is written into the martyrology section in H57, fol. 30r. The lower register, where obits usually appear, reads: "*In isto festo consecrationis dicti altaris expendatur de camera inter canonicos, beneficiatos et clericos manualiter florenus unus.*" Both texts can be found as additions in H56, fol. 30r, written by two different hands. A similar example is 27 September, fol. 136r in H57, with the martyrology section ending: "*Consecratio altarium sancte crucis et sancti Martini sitorum infra basilicam.*" The text is an addition in H56, fol. 136v. The exclusive nature of calendars is also evident from the fact that the obits of neither Nicholas III nor Boniface VIII, extensive as they are in the books of anniversaries, appear in any of the S. Pietro calendars, although, admittedly, these do not register obits in general. As for Nicholas III, not even the 16 May translation of the pope's remains is noted in the calendars, despite being written in original hand into the martyrology sections of both books of anniversaries, fol. 68v in both cases: "*Eodem die sancta translatio corporis sancte memorie domini Nicolai pape tertii in monumento novo.*"

although the basilica's martyrologies–books of anniversaries and calendars are related, the latter are much more exclusive concerning the selection of content.[28]

3 The Extent of Local Traditions

Keeping in mind the calendars' exclusive nature on the one hand, and their potential to carry rare entries on the other, let us look closer at the calendars of S. Pietro and S. Maria Maggiore, in particular the extent to which they exhibit contents distinct from other calendars. The ten S. Pietro calendars have between them a total of 200 entries that do not appear even once in the 24 outgroup calendars. As 123 of these are found in just two calendars not originally from the Basilica, let us narrow our selection down to entries that appear more than once within the S. Pietro group.[29] This brings the figure down to 41. These, in practice, are the entries that bring S. Pietro calendars closer together, while serving to differentiate them from the outgroup calendars. Moving on to include other Roman material, namely the S. Maria Maggiore group in the analysis, we can observe that these 41 entries are in fact particular to S. Pietro calendars to the extent that only one of them makes a single appearance in a S. Maria Maggiore calendar.[30]

Next, let us consider how big of a fraction these distinct S. Pietro entries form out of entries found in the group's calendars in general. It is still a good idea to ignore entries that appear in just one S. Pietro calendar, as a good 80 % of such entries are from the five adopted calendars. Thus, the total number of entries that appear in more than one S. Pietro calendar is 344. It is worth keeping in mind that this figure includes common feasts, such as the Nativity of Christ and the feasts of the apostles – in other words, the between 55 to 135 entries already established to be "universal" within the whole corpus of 40

28 Conversely, there is only one entry pertaining to Boniface VIII that can be found in the S. Pietro calendars but not the books of anniversaries. This is an addition on 23 January in calendar ACSP B87, noting the consecration of Boniface as pope, as well as the indulgences granted to the Vatican Basilica to mark the occasion. This is the only mention of a consecration of a pope I have encountered in medieval calendars. Mentions of indulgences in calendars are relatively rare as well and seem to concentrate in 15th-century material. Cf. ACSP E4, 29 January and 20 March; SMM 119, 20 March; Roma, Biblioteca Angelica, 1098, 20 March and 18 November.

29 ACSP B85 and F27 – both calendars with a high number of additions and somewhat idiosyncratic contents to begin with.

30 St Concordia on 13 August in calendar SMM 50.

calendars included in the study.[31] Consequently, we may observe that the 41 distinct S. Pietro entries do indeed form a substantial part of the calendars' 344 non-unique entries – closer to 12 % to be precise. In practice, then, every eighth or ninth non-unique entry in the S. Pietro calendars brings two or more S. Pietro calendars closer together, while distancing them from all calendars both in the outgroup and the S. Maria Maggiore group.

Indeed, as we are now looking at substantial chunks of the festival substance of the S. Pietro calendars, there is good ground to argue that the calendars represent a distinct tradition that goes beyond the less than a dozen entries that mention its altars or otherwise establish an explicit link with use in or around the basilica.[32] Even more so, as these figures have been reached with a very rigid comparison – by only looking at entries that are completely absent from the outgroup. This restriction does not reflect historical realities very well, however, as the calendar traditions of individual communities were by no means isolate. It is indeed likely that some of the outgroup calendars, especially those from Rome or nearby areas, have their points of contact with the calendars of the Vatican Basilica. For instance, looking at entries appearing more than once in the S. Pietro group and allowing for one appearance between the 24 outgroup calendars increases the figure from 41 to 55, and further to 66 when allowing for two outgroup appearances.

Moving on to the S. Maria Maggiore group, the six calendars have between them 45 entries that do not appear in the outgroup. As with S. Pietro, the majority of these are due to the particularities of individual calendars: this time 34 such entries are from calendar SMM 50, likely foreign to Rome both in its original content and the first substantial layer of additions.[33] However, focusing on entries that appear more than once in the S. Maria Maggiore group but are absent from the outgroup returns just three entries. This is a stark contrast to the 41 entries identified in the S. Pietro group using the same criteria. Allowing for one or two outgroup appearances, the figures are five and 11, still falling a long way short of the comparable figures from S. Pietro, 55 and 66.

This discrepancy between the S. Pietro and S. Maria Maggiore calendars as groups is illustrated in graphs 7.1 and 7.2 below. The figures display the non-unique entries found within the groups, distributed according to their frequency of appearance in the 24 outgroup calendars, divided into four-quarters of six calendars each. Thus, entries in the "Outgroup 1. quarter" appear in one to six outgroup calendars, those in "Outgroup 2. quarter" in seven to 12, and so forth.

31 See *The Selection and Character of Contents* above.

32 On such entries, see Kaasik, *The Calendars of S. Pietro and S. Maria Maggiore* in the present volume.

33 Cf. footnote 7 above.

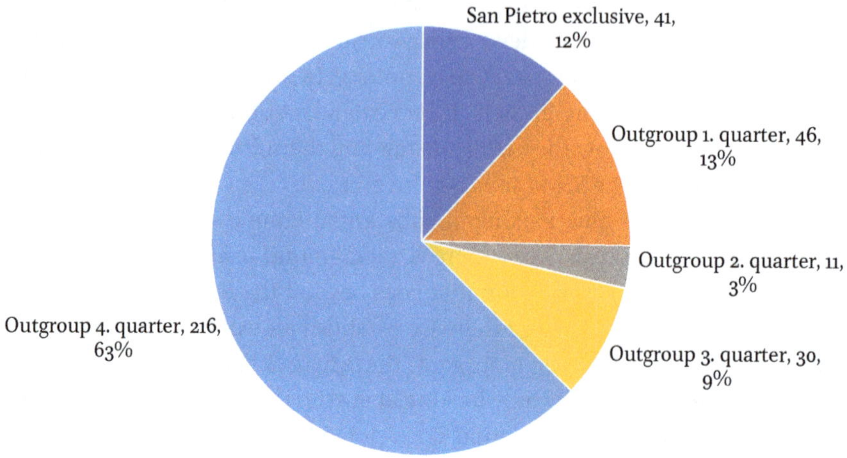

GRAPH 7.1 San Pietro non-unique entries relative to outgroup.

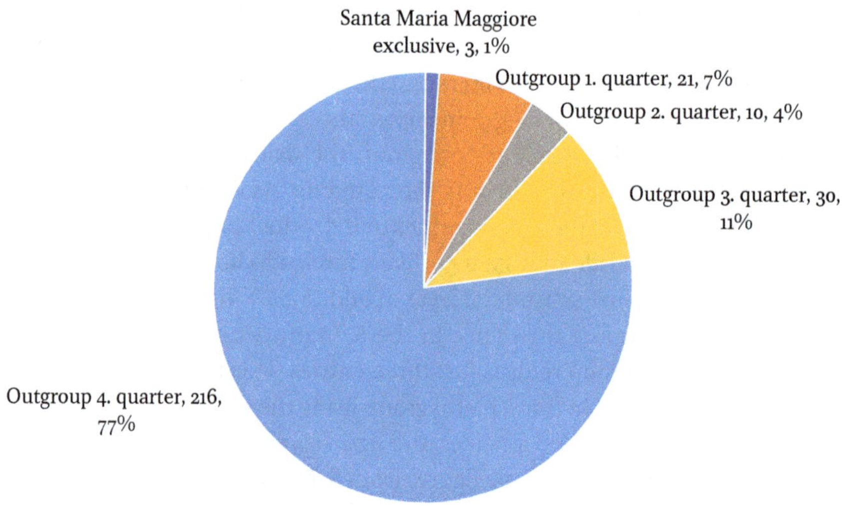

GRAPH 7.2 Santa Maria Maggiore non-unique entries relative to outgroup.

I have once again ignored unique entries, as these are mostly found in calendars originating from outside of Rome and only later adapted to the local traditions, as established above. This also mitigates the effect of instances where a feast appears a date earlier or later in one calendar than in the rest of the material, most of which are either the result of scribal errors or regional and local practices foreign to Rome.

There are two main points to take away from the figures. First, it is clear, albeit perhaps unsurprising, that the core contents of the calendars in both groups consist of relatively common entries. On the other hand, however, there is a profound difference between the two groups in that in S. Pietro, entries that are either very rare in the outgroup calendars or indeed exclusive to the basilica's calendars amount to a quarter of the total entries listed, while in S. Maria Maggiore this figure remains below 10 %. At the same time, the number of entries in the top three-quarters of the outgroup calendars are roughly the same in absolute terms: 216, 30, and 11 for S. Pietro and 216, 30, and ten for S. Maria Maggiore. It is therefore possible to establish that there is no difference between the groups regarding common and moderately common entries. Indeed, the fundamental difference lies with rare entries alone, the contrast being particularly prominent between entries exclusive to the basilicas' traditions. Based on just this comparison, it is impossible to say whether this discrepancy is due to an exceptionally strong calendar tradition at S. Pietro, a particularly weak one at S. Maria Maggiore, or both. What is clear, however, is that there was a marked difference in how calendars were used in these two major Roman basilicas.

4 Layers of Modifications and the Intensity of Use

The contrast between S. Pietro and S. Maria Maggiore calendars is further reflected in the way the calendars were modified. Considering S. Maria Maggiore, the layer of additions made to the calendars is thin.[34] The most consistently added entries pertain to the dedication of the basilica itself and the translation of the relics of Jerome.[35] The former, however essential the feast of the basilica's dedication may have been to its liturgical year, can in no way be considered particular to its calendar tradition, as it is well attested throughout my material and can be found even in calendars outside of Italy.[36] Curiously enough, the

34 With the exception of the incunable SMM 94. The many additions in the calendar remain isolates, however, not only within the S. Maria Maggiore group of calendars, but the material considered in this study as a whole. See Kaasik, *The Calendars of S. Pietro and S. Maria Maggiore* in the present volume.

35 For more on these feasts in the Basilica's liturgical sources, see Saxer, *Sainte-Marie-Majeure*, 320–30; Saxer, "Il culto dei martiri," 153–55.

36 The feast appears in 33 out of the 40 calendars examined in this study, including 20 out of the 24 outgroup calendars. In general, the situation is similar to that of the dedications of other major Roman basilicas: S. Pietro in Vaticano, S. Paolo fuori le mura, and the Lateran.

dedication is missing from the adopted S. Maria Maggiore calendar SMM 97, which exhibits other signs of modification at the Basilica.[37]

The translation of Jerome on 9 May, on the other hand, is much more exclusive to the calendars of S. Maria Maggiore.[38] Added to calendars SMM 50, 91, 97, and 99, the entries in the last two specifically mention the basilica as the destination of the translation. The absence of the feast from the two calendars produced for the basilica towards the end of the 15th century – SMM 119 and 120, dating from 1480 and 1478 respectively – may be explained by considering the addition in calendar 99 at full length:

> Translatio divi Hieronimi de Betheleem ad sanctam Mariam maiorem. Sed nunc celebratur in vigilia Ascensionis de mandato pape Pii II.

As the feast was evidently attached to the *temporale* – the movable celebrations centred around Easter – somewhere during the pontificate of Pius II (1458–1464), there is some logic to it disappearing from calendars, which were largely the domain of feasts with fixed dates. Indeed, although information on movable feasts was occasionally entered into the margins of calendar pages, that this was a markedly rare practice is revealing with regard to the dynamics between the two cycles of liturgical time.[39] A particularly notable example in this regard is the case of the Veil of Veronica. The Veil, carrying on it an imprint of the face of Christ, was displayed annually in Rome, in a grand procession held in January, a tradition established by Innocent III (1198–1216). However, despite being one of the Vatican Basilica's most venerated possessions, often turned to in times of crisis, there is no mention of the procession in the basilica's calendars, as it took place on the Sunday following the octave of the

---

For examples outside of Italy, see 5 August in the Grotefend corpus on Muzerelle, *Calendoscope*.

37 E.g. the 9 May addition "*Translatio sancti Hieronimi doctoris ad sanctam Mariam maiorem.*"

38 The feast does not appear in any of the 515 Calendoscope calendars, for example. Within the context of the material examined here, the feast can be found in only two calendars outside the S. Maria Maggiore group: ACSP B72 and E4, both with ties to the pontificate of Nicholas V. I am aware of a further appearance in calendar London, Brompton Oratory, 12737. The feast's appearance in this calendar is hardly surprising, however, as the manuscript was produced for the Hieronymite monks of Santi Bonifacio ed Alessio in Rome. See N.R. Ker, *Medieval Manuscripts in British Libraries: 1. London* (Oxford: Clarendon Press, 1969), 173–74.

39 For an example of how a celebration without a fixed date could be expressed in calendars, see ACSP F23, fol. 7r, where an indented calendar entry reads: "*Officium pro animabus patrum et matrum fratrum fiat in ultimo die feriali ante adventum.*" It is worth reiterating, however, that this was a potentially cumbersome and an exceedingly rare practice.

Epiphany and thus lacked a fixed date.[40] It is, therefore, not out of place that the translation of Jerome, the feast most distinctive of the S. Maria Maggiore *sanctorale*, should have disappeared from the Basilica's calendars, even as the celebration may have retained its importance.

While it is not possible here to delve deep into the reasons behind this apparent lack of a distinctive festival cycle in S. Maria Maggiore's calendars, three contributing factors can be identified. As the basilica was both the oldest and the most prestigious among the Roman churches dedicated to Mary, it could very well be that its canons were simply too focused on the Marian cult to develop a distinct cycle of saints' feasts. Indeed, many of the calendars likely to have been modified at S. Maria Maggiore carry Marian feasts as added entries. The dedication of the basilica itself can be counted among these as well, as it was usually marked in medieval calendars as *festum nivis*, in reference to its foundational legend of Our Lady of the Snows.

Table 7.1 below shows the distribution of significant Marian entries in the S. Maria Maggiore group, as well as the frequency with which they appear in the 24 outgroup and ten san Pietro calendars for comparison.[41] Two observations can be made from the table at a glance: for one, the feasts are not nearly as exclusive to S. Maria Maggiore as the translation of Jerome's relics, which,

40 On the Veil of Veronica, see Iacopo Stefaneschi, *De centesimo seu iubileo anno: la storia del primo giubileo : 1300*, ed. Claudio Leonardi, Paul Gerhard Schmidt, and Antonio Placanica (Firenze: SISMEL – Edizioni del Galluzzo, 2001), 48–49, endnote 18. For an example of the use of the Veronica in times of need, see Antonio di Pietro dello Schiavo, *Il diario romano di Antonio di Pietro dello Schiavo: dal 19 ottobre 1404 al 25 settembre 1417*, ed. Francesco Isoldi, Rerum Italicarum Scriptores II, 24.5 (Città di Castello: Casa editrice S. Lapi, 1916), 26: " ... *fuit facta processio per totum clericatum, ac etiam per omnibus ordinibus Urbis, in basilica Principis apostolorum de Urbe pro pace, et fuit ostensa Veronica de mandato domini cardinalis de Sancto Angelo et dominorum Urbis conservatorum ... quod Deus mictat pacem per totum mundum, et fiat unio Pape, etcetera: hac etiam propter maximam caristiam panis, qui tunc tempore non inveniebatur per Urbem per nullum denarium* ... ". For further examples, see ibid. 25, 38, 73, 108. For a mention of such practices in the S. Pietro books of anniversaries, see ACSP H57, fol. 65r: " ... *sacrosanctum sudarium quod in basilica nostra ex divino institutione ac dispositione est mirabiliter collocatum, ostenditum ad multam consolationem, peccatorum remissionem cuncto populo christiano* ... ".

41 Note that I have chosen to omit the oldest and most established Marian feasts from the table, as these are universal enough to be attested in all 40 calendars transcribed for this study, regardless of origin and dating. As such, visualising them would only serve to add clutter to the table. The entries are the *purificatio* (2 February), *annunciatio* (25 March), *assumptio* (15 August), and *nativitas* (8 September) of Mary, as well as the octave of the Assumption (22 August), which, although not entirely universal, is only missing in three outgroup calendars in the material. For an overview of Marian feasts in other S. Maria Maggiore liturgical sources, see Saxer, *Sainte-Marie-Majeure*, 310–13, 318–28; Saxer, "Il culto dei martiri," 152–54.

considering the prominence of Mary, is hardly surprising;[42] secondly, despite the numerous additions, none of the Marian feasts are attested in each and every one of the S. Maria Maggiore calendars, once again in contrast with the translation of Jerome, which is only missing in the two calendars produced after the feast was detached from its fixed date in the latter half of the 15th century. Consequently, the translation of Jerome remains the singular most distinctive feature of S. Maria Maggiore's calendars for the preceding centuries.

Another explanation for this lack of a *sanctorale* could be S. Maria Maggiore's role as one of the most important sites for the Roman *stationes*, papal liturgical processions, prompting Victor Saxer to style it the "auxiliary cathedral" of Rome.[43] As most *stationes* were tied to feasts of the *temporale*, they rarely feature on calendar pages.[44] It is doubtful, however, whether these two factors are enough to explain away the contrast between the calendars of S. Maria Maggiore and S. Pietro, as the latter, too, featured among the stational churches of Rome and was the centre for the cult of one of the most important biblical saints. Yet another possible explanation for the Basilica's lack of a distinct cycle of feasts is its high degree of dependence on Lateran liturgical life, also proposed by Saxer. Given the unfortunate lack of calendar sources for the latter, this is a hypothesis no longer possible to verify.[45]

The modifications made to the S. Pietro calendars paint a very different picture. Here the adapted calendars carry a range of added entries particular to the Basilica's tradition. ACSP B84, for example, contains a total of 75 additions, 20 of which are entries exclusive to the S. Pietro group of calendars, shared between B84 and its peers in the group.[46] Moreover, it should be kept in mind

42 Cf. footnote 38 above.

43 Saxer, "Il culto dei martiri," 151–52, 160.

44 Note that the adopted S. Maria Maggiore calendar, SMM 50, is a rare exception here, displaying information on the Christmas *stationes* on its December page – perhaps an indication of an emphasis on the *stationes* at the Basilica.

45 Saxer, "Il culto dei martiri," 160. For more on Lateran calendar sources, see Kaasik, *The Calendars of S. Pietro and S. Maria Maggiore* in the present volume.

46 The additions as found in ACSP B84, with frequencies of appearance across the ten S. Pietro calendars in parentheses: 13 January, *Margarite* (4/10); 19 February, *Gabini* (5/10); 25 March, *Consacratio altaris maioris facta per papam Calistum* (5/10); 5 May, *Translatio corporis Stephani protomartyris* (6/10); 6 May, *Consacratio altaris Bonifatii* (4/10); 6 June, *Vincentii* (5/10); 7 July, *octava Pauli* (4/10); 8 July, *Marcialis* (3/10); 21 August, *Privati* (3/10); 31 August, *Paulini* (3/10); 2 September, *Çennonis et filiorum* (6/10); 3 September, *Ordinatio et electio Gregorii* (5/10); 4 September, *Bonifatii, requiescit in cappella Bonifatii* (3/10); 9 September, *Iordanis* (4/10); 25 September, *Eustachii cum uxore et duobus filiis* (3/10); 28

TABLE 7.1 Non-universal Marian entries in Santa Maria Maggiore calendars.

Date	119	120	50	91	94	97	99	Entries in other groups
July 2	#Visitatio virginis Marie	#Visitatio virginis Marie			#Visitatio sancte Marie	/Festum visitationis\		12/34
July 26	#Sancte Anne matris matris Christi	Sancte Anne matris matris Christi	/Sancte Anne matris domine nostre\		Anne matris virginis			19/34
Aug. 5	#Festum nivis	#Festum Nivis	/Festum nivis\	/Festum nivis virginis Marie\	#Festum nivis		/Festum nivis\	28/34
Sep. 15		Octava sancte Marie	Octava gloriose virginis		Octava sancte Marie			25/34
Nov. 21	/Presentatio beate Marie virginis\	/Presentatio beate Marie virginis\					/Presentatio beate virginis, de mandato reverendissimi cardinalis domini Cesarinis arc{...} est missa canonicalis 1508\	3/34
Dec. 8	#Conceptio gloriose virginis Marie	#Conceptio sancte Marie	/Conceptio sancte Marie\	#Conceptio beate Marie virginis	#Conceptio gloriose virginis marie		#Conceptio beate Marie virginis	29/34

Legend: #entry in rubrics; /added entry\; {text illegible due to damage to manuscript}

that this figure is the result of applying rigorous rules of differentiation between calendar entries – taking into account minor variations in date, erasures, and relocations of feasts would result in higher numbers still, as would allowing for infrequent appearances in the other calendar groups.[47] Enforcing the same criteria on the S. Maria Maggiore group returns only a single added entry, the 19 July feast of St Epafra, added to the calendar pair SMM 119 and 120.[48]

The discrepancy between the figures suggests that the calendars of S. Maria Maggiore and S. Pietro served at least partly differing purposes. While the S. Pietro calendars were used to keep track of a particular cycle of feasts, their counterparts at S. Maria Maggiore served merely as guides to what was more or less the common Roman *sanctorale* at the time. This difference in the intensity of use can be further illustrated by looking at the average count of entries in the groups: the average S. Pietro calendar contains 288 entries – 296 if we discount the relatively sparsely filled calendar B79, the only 12th-century exemplar in my material – while the same figure for the S. Maria Maggiore group is 259. The latter figure is much closer to the average count of entries across the whole material of 40 calendars, sitting at 264.

Furthermore, at S. Pietro, the local tradition was deemed important enough to merit significant intervention to calendars originating elsewhere. S. Maria Maggiore, on the other hand, only had a single feast somewhat characteristic of its calendar tradition, and even this is absent from the two calendars specifically produced for its canons, as it had come to be attached to the *temporale* by the latter half of the 15th century. Indeed, the average additions per calendar amount to 34 for S. Pietro and a meagre eight for S. Maria Maggiore, the average across all 40 calendars being 16.[49] The contrast between the two Basilicas' practices is evident.

September, *Vinceslay* (4/10); 6 November, *Consacratio Marie de Cancellis* (7/10); 13 November, *Iohannis Grisosthomi* (7/10); 4 December, *Iuliane* (4/10); 6 December, *Savini* (2/10).

47 See the section on methodology above.

48 Note that this selection does not feature the translation of Jerome, as it appears in two outgroup calendars. See footnote 38 above.

49 The figures for average count of additions in the S. Pietro and S. Maria Maggiore groups have been calculated discounting the two calendars with a significant number of additions not pertaining to the Roman basilicas – ACSP F27 for the former group and SMM 50 for the latter. Note that all figures on additions include some corrections to the dates of feasts, where an entry has been removed from one row and added to another in the calendar.

5 The Cohesion of S. Pietro Calendars

Let us now consider the cohesion of calendars within a group. One of the few things the S. Pietro and S. Maria Maggiore calendar groups have in common is that both contain a fair amount of material not originally produced for use at the two basilicas. Indeed, five out of the ten S. Pietro calendars identified for this study are likely to originate elsewhere, while the same is true for four out of the six S. Maria Maggiore calendars. As the only two calendars produced directly for S. Maria Maggiore with some certainty are relatively late occurrences, and as the proportion of original to adopted material is more evenly spread in S. Pietro, the following analysis shall focus on the Vatican Basilica's material.

Graph 7.3 shows the spread of entries by frequency of appearance between all ten S. Pietro calendars. Judging by the figure, the group we have thus far analysed as a monolith, would in fact appear to be rather heterogeneous. Indeed, as roughly 40 % of the entries in the calendars are unique occurrences, the group seems only slightly more uniform than the whole selection of 40 calendars considered in this study.[50] This diversity, however, is to a great extent due to the group's adopted calendars, ACSP B85 and F27 in particular, which together account for around 60 % of such unique entries.

Removing adopted calendars from the equation and focusing just on the Basilica's five original calendars paints a different picture altogether, as shown in graph 7.4. Here unique occurrences constitute only 15 % of the entries, while nearly half of the contents of the original S. Pietro calendars can be found in all five calendars and another 40 % in two or more. Comparing the two figures brings into focus the extent of the diversification brought about by the adoption of calendars from various backgrounds. Considering this, it is remarkable how rare it is to find signs of removal of calendar entries. Indeed, the sheer quantity of remaining unique entries indicates this was a rare practice. The typical approach seems to have been one of straightforward stratification, with new information being inserted in addition to and not instead of the original contents. Thus, the coexistence of several calendar traditions on a single calendar page must have been perceived as something natural and relatively unproblematic. As it is in many cases possible to distinguish between these various layers of entries even now, it was likely even more so for the people producing, using, and modifying the calendars.

The S. Pietro group contains one notable exception, however. I have treated calendar ACSP F23 as one of the original S. Pietro calendars based on its several

50 See *The Selection and Character of Contents* above.

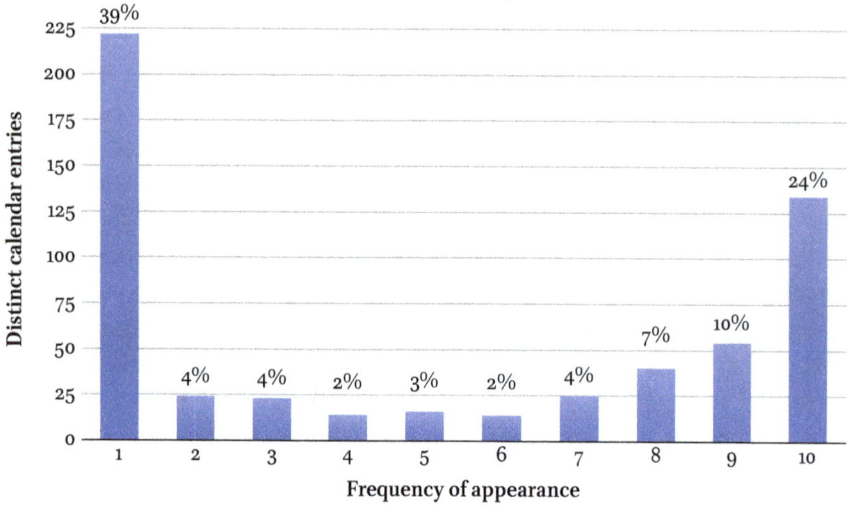

GRAPH 7.3 Number of calendar entries by frequency of appearance in San Pietro calendars.

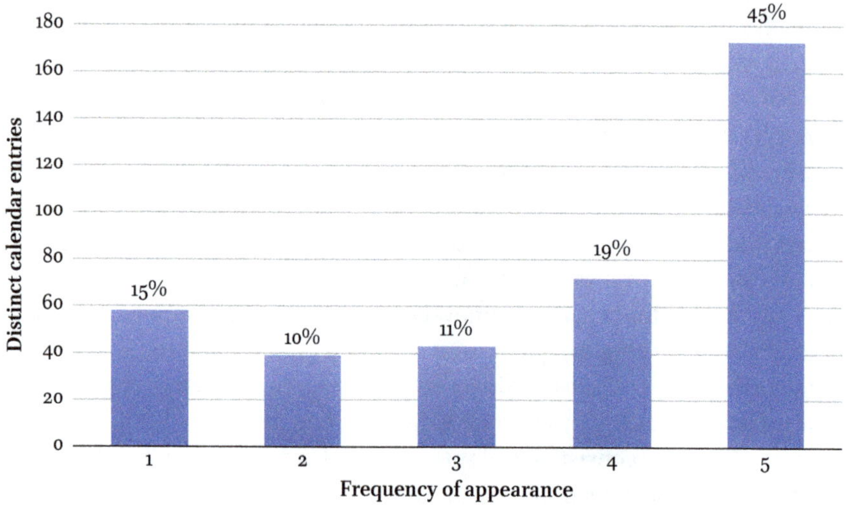

GRAPH 7.4 Number of calendar entries by frequency of appearance in original San Pietro calendars.

original entries mentioning altars of the Vatican Basilica. Yet, the calendar also contains several original entries pertaining to Spoleto, some of which have prompted Pierre Salmon to localise the manuscript in the Umbrian diocese. What is more, the calendar carries several entries of Franciscan character – yet again in original hand.[51]

There were certainly ties between the Vatican Basilica and Spoleto, as the diocese was home to more S. Pietro dependencies than any of its Italian counterparts.[52] It is reasonable to assume that these connections entailed a degree of mobility of people, books, and influence between the Vatican and the religious centres in and around Spoleto, of which calendar F23 may well be a valuable testament. As to exactly how and why the calendar came to carry such diverse elements as original entries, one may only speculate. For one, the manuscript may have been composed for or by someone with affiliations both to Spoleto and S. Pietro, as well as the Franciscans. Another possibility is simply the collation of books of different origins somewhere in the calendar's copying tradition. This would fit in well with what is known of medieval book production and the nature of liturgical manuscripts themselves as compilations of various types of text.[53]

Whatever the mechanisms behind this blending of traditions, its effects were long-lasting. Indeed, many entries that separate F23 from ACSP A9 and E15, both original 14th-century S. Pietro calendars, can be found in the mid- to late 15th-century Vatican calendar ACSP B88.[54] Yet, there are characteristics of F23 absent from B88 as well. Most significant, perhaps, is the lack of any Franciscan elements, while the number of Spoletan saints has diminished too, although there are still some left. Regardless, the many similarities between the calendars, coupled with the original S. Pietro entries in F23 and the Basilica's coat of arms in B88, indicate that such a hybrid calendar tradition did indeed exist at S. Pietro in the 14th and 15th centuries. To generalise, the circumstances that resulted in the creation of F23 in the 14th century ultimately led to the alteration of the Vatican calendar tradition in the 15th-century B88, possibly

51 For details, see Kaasik, *The Calendars of S. Pietro and S. Maria Maggiore* in the present volume.

52 According to Mirko Stocchi, 12 % of the Basilica's Italian dependencies in the medieval and early modern periods were situated in Spoleto. Among these, S. Caterina de Pusterula, a convent of Franciscan nuns *in civitate Spoletana* itself, could be a potential explanation for the many Spoletan and Franciscan references in F23. See Mirko Stocchi, *La Basilica Vaticana e le sue chiese tra Medioevo ed Età Moderna: Volume I, Umbria* (Città del Vaticano: Edizioni Capitolo Vaticano, 2013), 15–16, 33, 123–24.

53 See *Methodology* in Kaasik, *The Calendars of S. Pietro and S. Maria Maggiore* in the present volume.

54 For instance, the Spoletan entries *Ioannis Pinarensis* on 19 March and *Ysaac* on 11 April.

the latest extant medieval S. Pietro calendar. It would therefore seem that the coexistence of different traditions perpetuated by the adoption and modification of calendars of diverse origins, although apparently considered somewhat unproblematic, was not inconsequential.

Taking a broader look at the origins of the adopted S. Pietro calendars, two recurring themes emerge, as there are several calendars that are either central Italian in character or display a degree of Franciscan affiliation. Moreover, in some cases, as indeed with ACSP F23, such elements appear side by side. Calendar B84 can be associated with the Franciscans and Perugia, B87 with the Franciscans, and B85 with central Italy more broadly. In the case of S. Maria Maggiore, too, Franciscan elements appear in calendars SMM 50 and 97. These manifest in a variety of ways, from the noticeable presence, even overrepresentation, of feasts of a particular affiliation on calendar pages, to *incipits* and other forms of emphasis found within the liturgical manuscripts proper.[55] It is difficult to assess what kind of a role such calendars played in the liturgical lives of the Roman communities of canons and whether they were consulted by the chapters at large or remained in personal use by individuals who modified them according to their own preferences. Whatever the case, the frequency of such occurrences indicates that such fluid integration of traditions was indeed commonplace.[56]

Returning to graph 7.4 above, we can observe that a quarter of the entries in the original S. Pietro calendars are attested in only one or two instances. However, as a portion of these can be attributed to the F23–B88 hybrid tradition within the calendar group, let us narrow the view further and assess the consistencies between the group's calendars pairwise.

Graph 7.5 shows the proportion of shared entries out of the total number of distinct entries found between any two S. Pietro calendars. The calendars have been arranged according to their average level of consistency with the other calendars, proceeding from the top left towards the bottom right. The bottom row displays the totals of entries for individual calendars, which helps give a sense of proportion and explain certain trends observable in the table.

Let us start with the extremes: that ACSP B79 is the least consistent with the other original S. Pietro calendars comes as no surprise, as it is the only 12th-century calendar in the material, the others dating from the turn of the 13th

55 For examples of various kinds, see Kaasik, *The Calendars of S. Pietro and S. Maria Maggiore* in the present volume.

56 This observation is further supported by the strong Augustinian character of the late 15th-century S. Maria Maggiore calendars, SMM 119 and 120, in combination with the manuscripts' *incipits*, stating they were produced specifically "*ad usum ... canonicorum ... ecclesie sancte Marie maioris de urbe.*" Cf. Kaasik, *The Calendars of S. Pietro and S. Maria Maggiore* in the present volume.

	A9	B88	F23	E15	B79
A9		77%	78%	79%	60%
B88	77%		83%	73%	57%
F23	78%	83%		73%	54%
E15	79%	73%	73%		62%
B79	60%	57%	54%	62%	
Avg.	73%	73%	72%	72%	58%
Entries	310	279	324	288	217

GRAPH 7.5 Consistency of original San Pietro calendars pairwise.

and 14th centuries onwards. This temporal distance is reflected in the differing volume of content, which alone is enough to explain the calendar's relative lack of consistency with the other exemplars. At the other end of the spectrum, we can observe that the highest level of consistency is exhibited between calendars B88 and F23, supporting the hypothesis of a common origin presented above. Finally, calendars A9 and E15 form another pair, being each other's highest matches. Although the margin is quite small for A9, the distances from E15 to B88 and F23 are already markedly greater. Considering that A9 and E15 display similarities even on the level of wording, the numbers seem to add up all around.[57] This pair can probably be considered to represent the Vatican tradition at its "purest," given that F23 is a blend of Vatican, Spoletan, and Franciscan elements, and that B88 is to an extent derivative of this hybrid trend. The fact that A9 and E15 score higher than average consistency with B79, the oldest known S. Pietro calendar, seems to support this hypothesis.[58]

Thus, even within the small group of five original Vatican calendars, two distinct traditions can be identified. While the A9–E15 line may have represented the older S. Pietro tradition, that traces of the F23 tradition made it into

57 This is especially true for the longer entries mentioning the structure of the Vatican Basilica, which the two calendars share verbatim: 25 March "*Consecratio altaris beati Petri facta a papa Calisto secondo*"; 6 November "*Consecratio altaris sancte Marie de canzellis anno domini M° CC° XVI°*"; 17 November "*Dedicatio altaris Salvatoris de pace.*" While all of these entries appear in some of the other original and adopted Vatican calendars, no similarities of this type can be found between any other pair of calendars.

58 Furthermore, Albert Poncelet has observed that the contents of A9 are derivative of older Vatican Lectionaries, manuscripts ACSP A2 to A6, which places the manuscript within a broader tradition of Vatican liturgical manuscripts. Albert Poncelet, *Catalogus codicum hagiographicorum latinorum bibliothecarum romanarum praeter quam Vaticanae* (Bruxellis, 1909), 29.

the 15th-century B88 demonstrates it was – or perhaps became – more than just an isolated anomaly at S. Pietro. Indeed, the two traditions seem to have coexisted, as indicated by the fact that A9, E15, and F23 all carry the name of Bernardino of Siena as an addition, while B88 has it as an original entry. It seems likely, therefore, that calendars of both traditions were used simultaneously around the mid-15th century.

6 Using Calendars – Memory and Temporal Orientation

Taking another look at graph 7.5 above, we may observe that even where there are higher levels of correlation between some calendars, not even the closest-knit pairs are more than around 80 % consistent in their contents, meaning that at least one in five entries in the calendars appears in one but not the other. This has implications for how the calendars were used, especially with regard to the relationship between dates and feasts. On the one hand, this relative obscurity of a good portion of calendar entries suggests that calendars and related texts were essential memory aids, serving to remind their users of the very existence of certain saints and commemorations. On the other hand, while dating events according to the feasts of saints was common medieval practice, this level of diversity even within a single calendar tradition means that a great number of calendar feasts could hardly ever have been used as temporal markers in this way. Indeed, considering that around half of the 974 entries identified in the 40 calendars transcribed for this study are unique occurrences, most medieval calendar entries seem to have significantly lacked in universality. Thus, it is reasonable to assume that most medieval Romans would not have associated the majority of these with any date, month, or even season.

This is confirmed manifold by looking at how liturgical feasts were used to reference dates in different types of sources available from medieval Rome. Indeed, whether in inscriptions – funerary or otherwise – statutes, contracts, canonisation records, or diaries, the liturgical feasts mentioned in conjunction with or instead of calendar dates are overwhelmingly, if not exclusively, those belonging to the most common and universal core of celebrations: feasts pertaining to biblical figures, such as Christ, Mary, and the Apostles, the Evangelists, the Doctors of the Church, and other widely popular saints, such as Cecilia, Lawrence, and Michael.[59] This seems to hold true even for writers close to

59 For funerary inscriptions, see Iiro Kajanto, "Dating in the Latin Inscriptions of Medieval and Renaissance Rome," *Arctos. Acta Philologica Fennica* XI (1977), 58–59. For non-funerary

basilicas with particular calendar traditions, such as Antonio di Pietro dello Schiavo, a 15th-century beneficiary of S. Pietro.[60] Thus, the repertoire of feasts that could reasonably function as parts of temporal infrastructures was very narrow when compared to the plethora of entries found in calendars. Furthermore, there would have been little room for local particularity, as such feasts would, with very few exceptions, have to have been recognisable by people from different parts of Latin Christendom.[61]

This partial detachment between common temporal frameworks and many liturgical feasts underlines the importance of procuring calendars with correct intervals of feasts for communities invested in observing such commemorations whether or not they had attained general popularity. Following the S. Pietro calendar ACSP A9 day by day, for example, would alert the reader to the

---

inscriptions, see Urpo Kantola's article in the present volume. For Roman statutes, see Camillo Re, ed., *Statuti della città di Roma* (Roma: Tipografia della Pace, 1880), 43–44. For contracts and related examples, see Stefano Caffari, *Memorie di una famiglia della Roma del Quattrocento*, ed. Alba Ingletto and Stefania Santi, Miscellanea della Società romana di storia patria 54 (Roma: Presso la Societa alla Biblioteca Vallicelliana, 2009), 193, 208, 254; ACSP H57 fol. 129v. For *diarios*, see di Pietro dello Schiavo, *Diario Romano*, 7, 13, 15–16, 19–20, 22, 24, 26–29, 34–38, 47, 50, 54, 64, 70, 82, 87. For canonisation processes and more on *diarios*, see Saku Pihko's article in the present volume. Papal bulls and letters, on the other hand, seem not to reference such feasts even when particularly relevant. This is the case with the declaration of the first Christian Jubilee by Boniface VIII, taking place in the Vatican Basilica on the feast of the Chair of St Peter in 1300. The time and place for this event were obviously carefully chosen and laden with symbolism, yet, the bull itself makes no mention of the liturgical feast, instead following usual dating conventions: "*Datum Rome apud sanctum Petrum, VIII kalendas Martii, pontificatus nostri anno sexto.*" See BAV Arch.Cap.S.Pietro.caps.I.fasc.1. For more on the 1300 Jubilee and its proclamation, see Jasmin Lukkari's article in the present volume.

60 On the diarist, see di Pietro dello Schiavo, *Diario Romano*, XI. For examples, cf. footnote 59. The only exception to the rule I have come across are two dates given with reference to the 15 March feast of St Longinus – far from a universal point of reference, yet, particularly well represented in the S. Pietro calendars. See ibid. 102, 108. The feast can be found in original hand in: ACSP A9, B88, E15, F23, F27; SMM 119, 120; Stamp.ACSP A41; the feast has been added to: ACSP B84, B85, B87, F28. It can also be found outside of Italy, but relatively rarely, appearing in 62 out of the 515 Calendoscope calendars. Cf. Muzerelle, *Calendoscope*. Calendar ACSP F27 is true to its idiosyncratic nature here, as it styles Longinus a martyr-bishop. In hagiographical traditions, Longinus is identified with several biblical figures around the Passion of Christ, perhaps most notably so with the Roman soldier who pierced the crucified body of Christ with a lance. See *Bibliotheca Sanctorum*, vol. VIII (Roma: Città nuova editrice, 1966), cols. 89–95. It should also be noted that S. Pietro is known for having among its relics Longinus's supposed spear, the acquisition of which, however, appears to have been a relatively late 15th-century development.

61 One such exception is Bartholomew, whose celebration fell a day earlier according to Roman custom than in most of Latin Christendom. See *Methodology* in Kaasik, *The Calendars of S. Pietro and S. Maria Maggiore* in the present volume.

very existence of the more obscure saints in the basilica's tradition, and the foliation numbers added next to calendar entries in this particular case would further prompt them to look up the appropriate readings in the lectionary part of the manuscript. Still, a surprising degree of variation can be found even with regard to feasts of some local significance: while the dedications of some Vatican chapels and altars, to give an example, were deemed important enough to be recorded in both the original and adopted S. Pietro calendars, one can easily find minor deviations in dating even here.[62] It remains unclear what the use was in marking down the consecration of the main altar of the basilica one day too late, as it seems to have been done in calendar B85, as well as how such an entry might have functioned as a memory aid for someone associated with the Basilica.

The flip-side of the coin is of course the common core of the Christian festival year. These are the feasts present in nearly every calendar, as well as in collective consciousness to the extent of functioning freely as independent temporal markers in their own right. Graph 7.6 shows the relative significance of common and uncommon calendar entries, displaying the frequency of appearance for calendar entries across the full corpus of 40 calendars studied here. As we can see, entries that appear in a maximum of ten calendars, that is in only up to a quarter of the material, constitute an impressive 72 % of all distinct entries found in the 40 calendars. At the other end of the spectrum are the entries that are present in a minimum of 31 calendars, that is in at least three-quarters of the material, amounting to a solid 23 % of distinct entries. Entries appearing in two to three-quarters of the calendars, in other words with mediocre frequency, can be seen in the middle column in graph 7.6. Such entries, neither particularly rare nor very common, amount to a meagre 5 % of the total. It is evident, then, that medieval calendars were predominantly used to carry roughly two types of feasts: the universal and the obscure.

The calendars would have served partly differing functions for these two types of entries: for the more or less universal, the calendars were less likely to serve as memory aids, as such feasts were common enough knowledge to be referenced as independent temporal markers. Here, the primary function of the calendar was to coordinate this cycle of feasts with other layers of time, such as the week and the *temporale*. For the rare feasts, however, the calendar would have additionally served as a repository of communal memory for each day of the year. Conversely, calendars occasionally hint towards collective forgetting, as entries were removed from calendars and calendar traditions. Such

62 For some examples, cf. footnote 19 above.

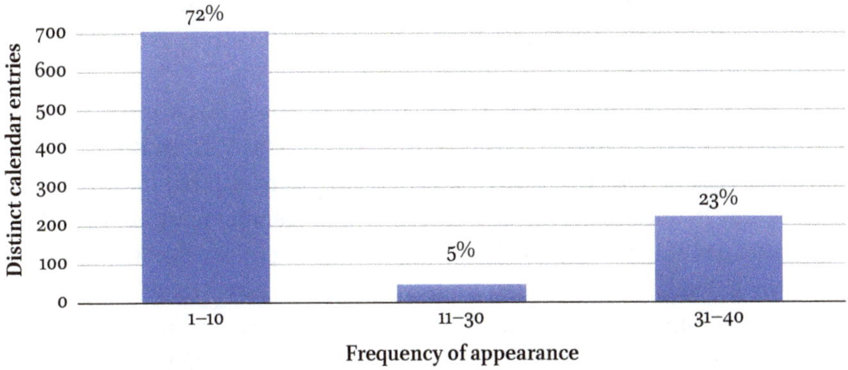

GRAPH 7.6 Number of calendar entries by frequency of appearance in forty calendars.

an observation can tentatively be made by looking at ACSP B88, produced around a century after most of its S. Pietro counterparts. Indeed, the late 15th-century calendar presents a particularly conspicuous example of the phenomenon in its utter silence on any and all construction and consecration at S. Pietro, from the dedication of its main altar to the achievements of Boniface VIII, so characteristic of the Basilica's 14th-century calendars.[63]

This brings us back to the differences between calendars and martyrologies, which could similarly be used as repositories of communal memory. It is worth reiterating, that the selection of contents was much more exclusive in the case of calendars than that of martyrologies, which, in turn, is revealing with regard to local practices and preferences. To give an example, the martyrology-based S. Pietro book of anniversaries, ACSP H57, reports for 2 January:

> Apud Antiochiam passio beati Ysidori episcopi et in Ponto civitate Thomis nativitas trium fratrum, Argei, Narcissi et Marcellini pueri. In

63 There is, however, a 25 May feast of Boniface IV in B88, not given in any of the 14th-century Vatican calendars or the Basilica's martyrologies–books of anniversaries, all of which celebrate the pope in September. The date in B88 is the one eventually adopted in the *Martyrologium Romanum* of 1583. The short martyrology entry, crammed between descriptions of other local feasts, indicates a strong Roman character: "*Romae sancti Bonifacii Papae quarti, qui Pantheon in honorem beatae Mariae ad Martyres dedicauit.*" See *Martyrologium Romanum: ad nouam kalendarii rationem, & ecclesiasticae historiae veritatem restitutum. Gergorii XIII. Pont. Max. Iussu editum* (Romae: ex Typographia Dominici Basae, 1583), 89. This is reinforced by the 25 May feast's complete absence from the 515 Calendoscope calendars. Cf. Muzerelle, *Calendoscope*. Thus, it seems the cult promoted by Boniface VIII remained a part of the Roman tradition, although a change of date had taken place by the mid-15th century.

> Thebayda nativitas sancti Macharii abbatis. Ipso die Thelesphori pape et martyris, sepultus in Vaticano iuxta corpus beati Petri apostoli.[64]

Out of all the saints mentioned in the martyrology, the S. Pietro calendars only name Thelesphorus, clearly of special importance to the Vatican Basilica.[65] On the other hand, unlike the martyrology, the calendars also remind the reader that the day in question is the octave for the 26 December feast of St Stephen. Thus, it is possible to reach a clearer understanding of the functionalities of medieval calendars through first understanding the essential differences between martyrologies and calendars: the former were used as reference books when action was being taken, in this case reading out loud not only the many saints of the day, but also the obits, donations, and dues of the Basilica's benefactors;[66] the calendars, on the other hand, served as instructional texts to help plan and coordinate action yet to take place.[67] Thus, at least ideally, writing the name of Thelesphorus in a martyrology would result in the reading out loud of the saint's name, while writing it in a calendar would result in further liturgical action. Similarly, it would not be necessary to add a purely practical liturgical notice such as the octave of Stephen to the martyrology, simply to be read publicly, but it does belong on the calendar page, which is not only a memory aid, but also a text of instructional and coordinative value.

7 Conclusions

Medieval calendars carry on their pages a great diversity of notices – as we have seen, nearly half of the entries in the 40 calendars examined for this study appear in just one of the calendars. Thus, it may be concluded, that calendars have a tendency to carry names of lesser-known saints. On the other hand, the contents of medieval calendars result from a relatively strict process of selection. Indeed, in san Pietro, although Boniface VIII has a special place in the basilica's 14th-century calendars, other patrons who added to its physical and liturgical fabric failed to find their way onto the calendar pages, while being duly mentioned in the martyrologies–books of anniversaries.

64 ACSP H57, fol. 1v. The manuscript has an unnecessary macron on the *m* in *Thomis*, probably resulting from an error in copying from ACSP H56, fol. 1v.

65 Original entries in A9, B79, and F23; additions in B84 and B85.

66 Dykmans, "Les Obituaires Romains," 623–24.

67 Cf. e.g. John Harper, *The Forms and Orders of Western Liturgy from the Tenth to the Eighteenth Century: A Historical Introduction and Guide for Students and Musicians* (Oxford: Clarendon Press, 1993), 60.

The combination of this function as a store to conserve the memory of the obscure on the one hand, and a certain selectiveness in contents on the other, means that calendars are prone to carry entries that lack in universality but are locally significant. In practice, however, the thickness of the layers that distinguish one group of calendars from others varies greatly according to the needs and conventions of the communities that produced or made use of them. This can clearly be observed by comparing the calendars of S. Pietro and S. Maria Maggiore, both notable Roman basilicas. While manuscript calendars in the latter seem to have served only their most generic purpose – easing coordination between the various overlapping cycles of liturgical time – the S. Pietro material clearly served an additional function of preserving a distinctive local tradition. This much is corroborated by the level of intervention seen by adopted calendars, as the additions in calendars adapted to the Vatican Basilica's use far outweigh the thin layer of modifications made at S. Maria Maggiore.

Yet, despite exhibiting a robust local tradition, the S. Pietro calendars are far from a monolith group. Modification of adopted calendars mostly comprised of inserting the local layer of entries *in addition to* as opposed to *instead of* preexisting strata, that is to say, large-scale removal of entries rarely took place. This brought with it a substantial level of diversification that nonetheless appears to have been considered somewhat unproblematic. On the other hand, it was not inconsequential, as demonstrated by the hybrid Spoletan-Franciscan-Vatican calendar ACSP F23 and the simultaneous existence of two distinct Vatican calendar traditions in the 15th century – the F23–B88 strand carrying remnants of Spoletan influence, and the A9–E15 strand, likely exhibiting the S. Pietro tradition in a form less modified by a single point of external input.

Although a major religious centre with a distinct calendar tradition, even the closest-knit pairs of S. Pietro calendars share no more than around 80 % of their contents. This diversity means that a good part of their contents must have remained somewhat obscure to the general populace surrounding the community of canons. This impression is reinforced by the complete lack of references to such obscure feasts as temporal markers in an array of contemporary sources from epigraphy to *diarios*. Indeed, such references are almost invariably to the more or less universal feasts carried by all calendars, which resonated widely in medieval society, regulating, for instance, payments of debts and dues, market days, processions, and the distribution of bread to the poor, as well as indulgences. Here we return to the dual function of the calendars, as exemplified by S. Pietro and S. Maria Maggiore, where the basic function of calendars was to regulate liturgical time, while also possessing the potential of becoming vessels for communal memory. Delving deeper into the latter opens up possibilities for exploring the culture of remembrance but also of collective forgetting, as hinted by the omission of references to altars and chapels – a staple in 14th-century Vatican calendars – in the 15th-century B88. As for the use of the calendars, while observing their inherent diversity has allowed for various observations, challenges remain especially in tackling the apparent dissonances within particular traditions. One might ask, for instance, how were minor differences in the dates of locally significant feasts in the Vatican Basilica's calendars reconciled in practice.

Bibliography

Sources

Bibliotheca Sanctorum. Vol. VIII. Roma: Città nuova editrice, 1966.

Caffari, Stefano. *Memorie di una famiglia della Roma del Quattrocento*. Edited by Alba Ingletto and Stefania Santi. Miscellanea della Società romana di storia patria 54. Roma: Presso la Societa alla Biblioteca Vallicelliana, 2009.

Hagiographi Bollandiani, ed. *Catalogus codicum hagiographicorum latinorum antiquiorum saeculo XVI qui asservantur in Bibliotheca nationali Parisiensi, tom. III*. Bruxellis, 1893.

Martyrologium Romanum: ad nouam kalendarii rationem, & ecclesiasticae historiae veritatem restitutum. Gergorii XIII. Pont. Max. Iussu editum. Romae: ex Typographia Dominici Basae, 1583.

Pietro dello Schiavo, Antonio di. *Il diario romano di Antonio di Pietro dello Schiavo: dal 19 ottobre 1404 al 25 settembre 1417*. Edited by Francesco Isoldi. Rerum Italicarum Scriptores II, 24.5. Città di Castello: Casa editrice S. Lapi, 1916.

Poncelet, Albert. *Catalogus codicum hagiographicorum latinorum bibliothecarum romanarum praeter quam Vaticanae*. Bruxellis, 1909.

Stefaneschi, Iacopo. *De centesimo seu iubileo anno: la storia del primo giubileo : 1300*. Edited by Claudio Leonardi, Paul Gerhard Schmidt, and Antonio Placanica. Firenze: SISMEL – Ed. del Galluzzo, 2001.

Literature

Blaauw, Sible de. *Cultus et decor: liturgia e architettura nella Roma tardoantica e medievale: Basilica Salvatoris, Sanctae Mariae, Sancti Petri. 2*. Studi e testi 356. Città del Vaticano: Biblioteca Apostolica Vaticana, 1994.

Brundage, James A. *Medieval Canon Law*. Abingdon: Routledge, 2013.

Degl'Innocenti Gambuti, Marcella. *I codici miniati medievali della Biblioteca comunale e dell'Accademia etrusca di Cortona*. Firenze: S.P.E.S., 1977.

Dijk, Stephen Joseph Peter van. *Sources of the Modern Roman Liturgy : The Ordinals by Haymo of Faversham and Related Documents (1243–1307), v. 2. Texts*. Studia et Documenta Franciscana. Leiden: E.J. Brill, 1963.

Dupré Theseider, Eugenio. "Bonifacio VIII, papa." In *Dizionario biografico degli Italiani v. 12*. Edited by Alberto Maria Ghisalberti. Roma: Istituto della Enciclopedia italiana, 1970.

Dykmans, Marc. "Les Obituaires Romains. Une Définition Suivie d'une Vue d'ensemble." *Studi Medievali, Serie Terza* 19, no. 2 (1978), 591–652.

Garrison, Edward B. "Three Manuscripts for Lucchese Canons of S. Frediano in Rome." *Journal of the Warburg and Courtauld Institutes* 38 (1975), 1–52.

Grotefend, Hermann. *Zeitrechnung des deutschen Mittelalters und der Neuzeit. Band 1. Glossar und Tafeln*. Hannover: Hahn'sche Buchhandlung, 1891.

Harper, John. *The Forms and Orders of Western Liturgy from the Tenth to the Eighteenth Century: A Historical Introduction and Guide for Students and Musicians*. Oxford: Clarendon Press, 1993.

Hughes, Andrew. *Medieval Manuscripts for Mass and Office: A Guide to Their Organization and Terminology*. Toronto: University of Toronto Press, 1995.

Jounel, Pierre. *Le culte des saints dans les basiliques du Latran et du Vatican au douzième siècle*. Collection de l'Ecole Française de Rome 26. Rome: École Française de Rome, 1977.

Kajanto, Iiro. "Dating in the Latin Inscriptions of Medieval and Renaissance Rome." *Arctos: Acta Philologica Fennica* XI (1977), 41–61.

Ker, N.R. *Medieval Manuscripts in British Libraries: 1*. London. Oxford: Clarendon Press, 1969.

Magionami, Leonardo. *I manoscritti del Capitolo di San Lorenzo di Perugia*. Roma: Jouvence, 2006.

Maire Vigueur, Jean-Claude. *The Forgotten Story: Rome in the Communal Period*. Translated by David Fairservice. Roma: Viella, 2016.

Muzerelle, D. *Calendoscope, Logiciel d'aide à l'identification Des Calendriers Liturgiques Médiévaux*. Paris: Institut de recherche et d'histoire des textes, 2013. http://calendoscope.irht.cnrs.fr/.

Petterlini, Elena. "Il Santorale Marciano e Il Santorale Della Chiesa Di Verona." In *Da Bisanzio a San Marco. Musica e Liturgia*. Edited by Giulio Cattin, 300–13. Venezia: Fondazione Ugo e Olga Levi, 1997.

Pfaff, Richard W. "The Hagiographical Peculiarity of Martha's Companion(s)." In *Liturgical Calendars, Saints, and Services in Medieval England*, IV, 1–22. Variorum. Aldershot: Ashgate, 1998.

Re, Camillo, ed. *Statuti della città di Roma*. Roma: Tipografia della Pace, 1880.

Rouillard, Philippe. "Sainteté III. Culte des saints." In *Catholicisme. Hier, Aujourd-Hui, Demain*. Edited by G. Mathon, G.-H. Baudry, and L'institut catholique de Lille, 713–26. Paris: Letouzey et Ané, 1993.

Saxer, Victor. "Il culto dei martiri romani durante il Medioevo centrale nelle basiliche Lateranense, Vaticana e Liberiana." In *Roma antica nel Medioevo: mito, rappresentazioni, sopravvivenze nella "Respublica Christiana" dei secoli IX–XIII, atti della quattordicesima Settimana internazionale di studio, Mendola, 24–28 agosto 1998*, 131–61. Milano: Vita e Pensiero, 2001.

Saxer, Victor. *Sainte-Marie-Majeure : Une Basilique de Rome Dans l'histoire de La Ville et de Son Église, Ve–XIIIe Siècle*. Collection de l'École Française de Rome 283. Rome: École française de Rome, 2001.

Stocchi, Mirko. *La Basilica Vaticana e le sue chiese tra Medioevo ed Età Moderna: Volume I, Umbria*. Città del Vaticano: Edizioni Capitolo Vaticano, 2013.

Tomasi, Giuseppe Maria. "Responsoriale et antiphonarium Romanae Ecclesiae de circulo Anni iuxta veterem usum Canonicorum Basilicae Vaticanae Sci. Petri." In *Responsorialia et antiphonaria Romanae Ecclesiae a s. Gregorio Magno disposita*, 1–215. Romae: Typis Iosephi Vannaccij, 1686.

Vildera, Anna. "Il Santorale Marciano e Il Santorale Della Chiesa Di Padova." In *Da Bisanzio a San Marco. Musica e Liturgia*. Edited by Giulio Cattin, 271–300. Venezia: Fondazione Ugo e Olga Levi, 1997.

Walsh, Michael J., ed. "Boniface VIII." In *A Dictionary of Popes*, 3rd ed. Oxford University Press, 2015.

Welch, Anna. *Liturgy, Books and Franciscan Identity in Medieval Umbria*. Leiden and New York: Brill, 2015.

CHAPTER 8

Complex Tools for Complex Time: Solar, Stellar, and Lunar Cycles of Time in Medieval Roman Calendars

Marko Halonen

1 Introduction

The attempt to structure multiple and often contradictory cycles of time is a highly typical aspect of a written culture. It reflects the very basic needs that humans have to bring order to their environment, and to make sense of such a senseless and difficult concept as time. The calendar is the contrivance most often used for this challenging task.

In current English usage, as defined by Oxford English Dictionary, the word calendar has three meanings.[1] It can mean a system for organising a year (such as the "Gregorian calendar"), a table or other contrivance that visually represents days, months, and their events (such as a "racing calendar"), or any kind of list or register. This article focuses solely on the second sense of the word, considering a calendar as first and foremost a contrivance for visualising time.

The term "calendar," then, is used in this article in *sensu lato* for all similar contrivances that are used to visualise time. What matters is the format: the medieval calendar is derived from its ancient Roman origin; the first day of the month was called *Kalendae*, a term which etymologically suggests that it "calls" the other days of the month. Thus, this calendar is a tool for organising the days of the month and their tasks, which can then in turn be related to saints, martyrs, prayers, remembering the dead, or indeed any rituals or customs, depending on the context in which the calendar is used. This means that regardless of whether a contrivance is catalogued as a *kalendarium, martyrologium, horarium, necrologium* or almanac, it can still meet the three abovementioned definitions of "calendar." Each type of calendar has a slightly different focus and is used in a slightly different context, but this does not exclude them from the

1 "A table showing the division of a given year into its months and days, and referring the days of each month to the days of the week; often also including important astronomical data, and indicating ecclesiastical or other festivals, and other events belonging to individual days. Sometimes containing only facts and dates belonging to a particular profession or pursuit, as *Gardener's Calendar, Racing Calendar*, etc. Also a series of tables, giving these facts more fully; an almanac." *Oxford English Dictionary.* Accessed at www.oed.com.

© KONINKLIJKE BRILL NV, LEIDEN, 2021 | DOI:10.1163/9789004436251_009

umbrella term of "calendar." Calendars used within the common liturgical context, often labelled *kalendarium,* are analysed in another article of this volume.

This article examines calendrical contrivances that were presumably used in Rome or in its immediate vicinity between the 10th and 16th centuries, specifically considering calendars as "tools for organising time." The essay investigates the nature of these tools via calendar materials that can be associated with medieval Rome, and analyses which elements relating to these topics are present or absent in this material, going on to evaluate the implications for the general perception of time in medieval Rome. This article, then, takes an innovative approach both by selecting as broad a range of material as possible, and also by supporting its arguments with the digital database of calendars compiled for this research. As this is material that has not previously been studied from a broad perspective, the research questions here are inevitably broad and descriptive: What aspects and cycles of time can be found in the calendars and similar contrivances of medieval Rome? What does this tell us about medieval Roman culture?

This is an important field of enquiry because it will help to reveal chronological developments in the ability to organise time in medieval Rome. Creating and using calendars requires mathematical skills as well as understanding the movements of celestial objects, which are significant aspects in the history of learning. Although calendars were certainly only used by a small minority of the medieval Roman population, the patterns of time in the calendars were reflected in the worldview, habits and everyday life of each member of the community.

Calendars are a valuable source for studying these elements because of their format. Because they are practical tools, they have to be quite compact in size. A calendar made up of 12 pages is much easier to grasp than a calendar of 120 pages. And because calendars are so compact – in medieval times they would typically comprise 12 pages, one for each month – they contain only the most essential information, worthy of using up the precious space. The findings of medieval Roman calendars were most likely significant – if not for all Romans, then at least for those who had the possibility to create and use such a time-organising tool. Calendars contain information which was vital not only for the most obvious religious purposes, but also for all other fields of life. The religious holidays of the calendar also affected trade, as well as many popular events. The calendars also contained astrological information, which was believed to affect personal health as well as the general fortunes of whole societies. In a medieval semi-agrarian society, calendars were also significant for the purposes of farming.

The article focuses on general statistical trends rather than individual calendars and links this information to the big picture of the history of time. Since almost no similar research exists for Rome (or any other place), this article aims to open up a new research approach, which can potentially be expanded into a comparative research into other medieval European cities. After introducing the source material and the general structure of calendars, this article will continue with an analysis of three different cycles of time found in the calendars; solar, stellar, and lunar cycles will each be analysed in separate sections. This will be followed by a section that focuses on the chronological development of the calendars and some of their key features throughout the high and late medieval centuries.

It is important to stress the fact that there was no "universal" medieval calendar, any more than there is a universal modern calendar today. In medieval times, different types of calendars were used in different contexts: ecclesiastical calendars (found in liturgical books, for example), juridical calendars (in law books), and personal calendars (the book of hours). Because of the dominant cultural impact of the Latin Church, the Christian elements, such as the saints, were present in most medieval calendars.[2] But to be precise, even the liturgical calendars are actually a combination of two cycles of time: the temporal and the sanctoral. Only the latter deals with saints, whereas the former follows the main Christian feasts such as Easter. Because of its Biblical background, this feast is bound to the movements of the celestial objects, particularly the moon and the sun. However, as will become evident in this article, calendars often contain many features which seem to lack any clear liturgical or even Christian meaning.

This essay is based on the analysis of 111 versions of medieval calendars that have been identified as produced or used in Rome during the Middle Ages. This selection of material is by no means complete, but it still aims to be the most comprehensive account of medieval Roman calendars to date, and large enough to be considered significant. The most extensive collections of medieval Roman calendars are at the *Biblioteca Apostolica Vaticana* in the Vatican City. Many calendars are also found in the Roman *Biblioteca Angelica*, and individual calendars in various European libraries, although it was not possible to visit each of them for this project (see sources for details).

The selection of the material requires some clarification regarding what is meant by "calendar," "Roman" and "medieval." As already mentioned in the

2 In the Roman material used in this essay, saints are always present, but calendars without saints can also be found, at least in Sweden and Denmark. See Marko Halonen, *Aspects of Medieval Nordic Calendars* (PhD diss., University of Helsinki, 2020).

introduction, this article analyses all the contrivances which represent the division of a year into months and days, and give the relevant data for each day. The most typical of these contrivances is listed below. A *kalendarium* in the medieval context is typically a part of a liturgical book, usually found in the very first pages of a *missale.* Its length is usually 12 pages, and it contains the liturgical, computistical, and astrological information for the year. When the calendar is in the first pages of a book, it also serves as a sort of index or contents page. This customarily only contains one or two feasts per day, so a month will normally fit on a single page. The *martyrologium*, or martyrology, is a calendar with a special focus on martyrs. The *martyrologium* and *kalendarium* are structurally identical, being based on the Julian calendar with the Roman dating system and the same basic computistical features, although the most extensive computistical details are usually found in the *kalendaria.* Martyrologies tend to include so many martyrs, and such long texts, that usually only a few days would fit on the same page, and thus their appearance is usually slightly different. The *horarium*, or book of hours, is a prayer book which typically begins with a calendar, and then moves on to the prayers dedicated to particular saints and the feasts of the days. A *necrologium* is a list of the dead, most frequently the dead members of a religious order or a religious house such as a monastery. An *almanac* is a little booklet which contains detailed calendrical information, often listing many different cycles of time (liturgical, agricultural, astronomical and so on) and their corresponding tables with calculations. Despite these distinctions, all of these categories include contrivances which are typically written using the same underlying structure – a "calendar" in a wider sense.

An important point concerning the material is that the calendars in this essay are analysed as "calendar versions." These 111 "versions" stem from 68 different manuscripts or printed books. A version in this context refers to the many cases where a calendar has later been altered by a second hand. In this article, the altered version is considered as a new calendar version, because it means that this specific calendar was considered as a relevant tool for organising time for at least two different people. Regardless of "what" was changed in the calendar, it can be argued that whoever made the alteration considered the calendar to keep its relevance. Leaving out either the altered or the original version, then, would be very misleading. For example, one calendar which is dated to the 12th century includes additions from both the 13th and the 15th centuries.[3] In this analysis all of these three versions are taken into account as individual

3 Città del Vaticano, Biblioteca Apostolica Vaticana (henceforth BAV), Arch.Cap.S.Pietro.H.56. Hands dated in Pierre Salmon, *Les manuscrits liturgiques latins de la Bibliothèque Vaticane. 4,*

versions rather than (quite randomly) leaving one out. In the source material, a second version is marked with an under stroke and number after the *signum*, for example Arch.Cap.S.Pietro.H.56_2.

All calendars that were found and localised to Rome in academic publications and library catalogues have been taken into account. This has been done regardless of the facts that the localisations are often very problematic, and there has been some disagreement among scholars concerning the question. A closer look at the datings found in catalogues as well as in academic literature raises many questions. Only a small portion of the calendars contain some explicit internal evidence linking them to Rome. It appears that the localisation has usually been based on the liturgical year, in which certain feasts are seen as typically Roman. This approach is problematic on two accounts. Firstly, there is no comprehensive study of the Roman liturgical year in the Middle Ages which covers the entire body of liturgical literature. Therefore, we do not know all the variations of the liturgical year in medieval Rome, which makes it difficult to use the features of the liturgical year for localising calendars. This has recently been demonstrated in a study of a wide range of European calendars.[4] Secondly, we know that the surviving medieval material represents only a small portion of what was produced. The surviving material is much more informative about the later than the earlier Middle Ages.

Similarly, palaeographical and art historical localisations (and datings) can be misleading, as we cannot know if the manuscript was made in Rome, or simply in the Roman style. It is equally difficult to say whether a calendar was used in Rome, or even whether it was made there. This is particularly the case with printed calendars. The question becomes even more complex when we attempt to identify which of the Roman churches is connected to a particular calendar. Sometimes one scholar determines that a calendar is "Roman," and then another will decide that it is not. Both scholars either base their arguments on the methods mentioned above, or do not reveal the method at all.[5] Nevertheless, this problem can be resolved to an extent by analysing the calendars from the perspective of how the calendars functioned as

---

Les livres de lectures de l'office, les livres de l'office du chapitre, les livres d'heures, Studi e Testi 267. (Città del Vaticano: Biblioteca Apostolica Vaticana, 1971).

4 Tuomas Heikkilä and Teemu Roos. "Quantitative methods for the analysis of medieval calendars," *Digital Scholarship in the Humanities* 33 (4) (2018), 766–87.

5 For example, the localisation of BAV, Vat.lat.7722 according to Marc Dykmans, "Les obituaires romains. Une définition suivie d'une vue d'ensemble," Studi medievali 19 (Spoleto: Centro Italiano di Studi sull' alto medioevo, 1978) versus Giacomo Baroffio, "Kalendaria Italica. Inventario" *Aevum. Rassegna di scienze storiche linguistiche e filologiche* Anno LXXVII:2.(Milan: Università Cattolica Del Sacro Cuore, 2003).

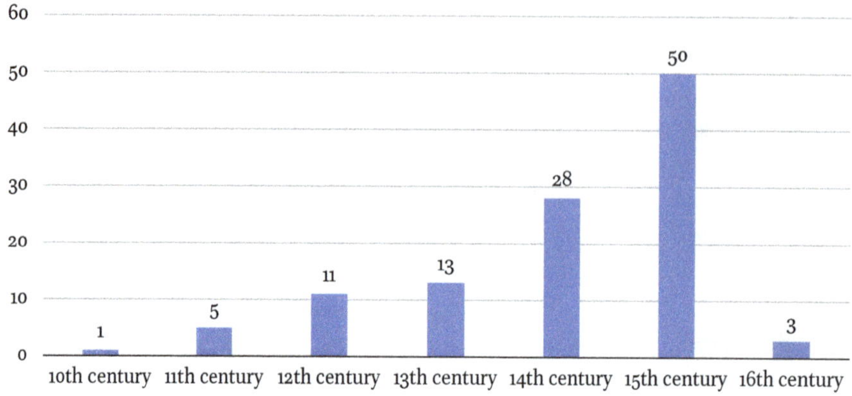

GRAPH 8.1 The numbers of medieval Roman calendars by century analysed for the study.

"tools for organising time" within one city, Rome. From this perspective, the possible problematic localisations of some of the calendars, and the even more problematic localisations to different churches, do not cause the other results to be irrelevant: the big picture still remains valid.

The Middle Ages is a term which even historians understand in surprisingly different ways. In the context of calendars, the division is perhaps slightly simpler, as it is dictated by the source material. Medieval calendars are all structurally based on the ancient Roman Julian calendar, but their content is Christian. The earliest such calendar which is located to Rome is from the 10th century.[6] There is, of course, the "Philocalian calendar" from AD 354 which is a hybrid between the two, but it only survives in much later copies. The deep structure of the calendar remains unchanged until the Gregorian calendar reform of 1582, which can be considered as a true turning point.

In this article, the dating of the calendars is primarily based on the information given in the catalogues and articles dealing with the specific manuscripts. See Appendix 4 for a list of the manuscripts and the sources for information concerning their provenance and dating. In terms of centuries, the calendar versions form the following field.

Unfortunately, little material survives from the early Middle Ages; only one calendar has been dated to the 10th century. The situation is slightly better regarding the centuries of the High Middle Ages, but these three centuries (1000–1200) with their 29 calendars are far distant from the Late Middle Ages, which start c.1300. Eighty-one calendars have been dated to this period, and the 15th century, with 50 calendars, is by far the best documented in this case.

6 BAV, Vat.lat.378.

This naturally means that the conclusions drawn from the material are less reliable in the case of the earlier centuries than the later. With such a small body of material – a handful of calendars – there is a great risk that the surviving material only represents a rather random selection of the reality, which might have been very different. But excluding the entire analysis based on this problem would be a shame, and the four first centuries as a whole contain a reasonable amount of material for determining at least some preliminary general trends, some of which can also be seen in the calendars of later centuries. This will potentially offer a starting point for future research.

It would make no sense to discuss the medieval calendars of Rome without first referring to the calendars of the Roman Republic and the Roman Empire. All western European medieval calendars are fundamentally connected to the ancient Roman calendars, but naturally this is nowhere as evident as in Rome itself. Indeed, late medieval Roman calendars from the 15th century have a very similar structure to the calendars of the imperial period in antiquity. They are even closer to the calendars of Late Antiquity, such as the Philocalian calendar of 354 AD. This similarity is surprising, considering that there is a 600-year gap in the material between the 4th and 10th centuries, when we have the first extant medieval Roman calendar.

Medieval Roman calendars are all based on the "Julian calendar" issued by the Roman *pontifex maximus* and consul Gaius Julius Caesar in 46 BC. At this stage, the old Republican calendar was corrected by adding ten extra days, making the year 365.25 days long (a leap day was added every four years). The lengths of the months were corrected so that there were now 30 days in April, June, September and November, 28 in February, and 31 in all the rest.[7] These "solar" month lengths are, however, merely a convenient way of dividing the year into 12 periods of roughly the same length. In reality, of course, the moon (from which the word month is derived), does not take different times for its rotation around the earth. The moon changes from new to full and back to new in approximately 29.5 days. This is known as the synodic[8] month – the time between two conjunctions of the moon and the sun.[9] Because each lunation takes 29.5 days, for the sake of convenience, medieval calendars considered each "lunar" month's length to be either 29 or 30 days. The varying number of days in the "solar" and "lunar" months seems to have been just as confusing for

7 Dario Sabbatucci, *La religione di Roma antica. Dal calendario festivo all'ordine cosmico* (Milano: Arnoldo Mondadori Editore S.p.A., 1988), 5.

8 From Greek *synodos*, "conjunction."

9 Colum Hourihane, ed., *Time in the Medieval World. Occupations of the months and signs of the zodiac in the index of Christian art* (Princeton: Princeton University in association with Penn State University Press, 2007), 20.

medieval as for modern people; four out of five of the Roman medieval calendars contain a separate mention of how many days the solar and lunar month actually have in that particular month, such as in January: *Januarius h[abe]t dies XXXI et luna XXX.*[10] In the calendars of the 10th and 11th centuries it is actually found in all of them, whereas in the 12th and 16th centuries it is only found in slightly over half of the material.

Other elements that connect the ancient and medieval Roman calendars are language, letters, and numbers. All medieval Roman calendars analysed here were written in the Latin language and alphabet, with only occasional comments in the vernacular. Somewhat surprisingly, the medieval Christian calendars use the same names for months as the ancient Romans did; they are named according to pagan mythology, or (in the case of Julius and Augustus) ancient history: *Ianuarius, Februarius, Martius, Aprilis, Maius, Junius, Julius, Augustus, September, October, November, December*. The only exception is that in almost half of the calendars, the month of May is written in the form *Madius*, which already hints towards the later vernacular *maggio*.[11] This phenomenon, however, remains relatively rare (found in 20–30 % of the calendars) until the 14th century, when is found in over 60 % of the calendars. In the 15th century, it is still found in about half of the calendars, but it disappears completely in the 16th century. This disappearance might be an indication of an attempt to "return" to the original pure Latin forms, which would naturally sit well with the humanistic trends of the Renaissance.

Perhaps surprisingly, in about 10 % of the calendars, the Greek and Hebrew names for the months are also given, and one even has the Egyptian names as well.[12] The Latin names are always also present. Hebrew and Greek are the two main languages of the original Biblical texts, so adding them to the calendar would make it easier to follow the different cycles of time in these three cultures. On the other hand, the Greek names might also be related to the fact that many innovations concerning calendars were made in the classical Greek world, such as the famous 19-year "metonic" cycle which is the best effort of the classical world to synchronise the solar and lunar calendars.[13] These names are also present in the most famous medieval book dealing with the aspects of

10 BAV, Arch.Cap.S.Pietro.A.9, fol. 7r.

11 In 2 % of the calendars, there are also some marginal comments in vernacular.

12 The calendar is divided in two parts: BAV, Vat.lat.4885/Barb.lat.646.

13 Named after the 5th century BC Athenian scholar who came up with the system. Georges Declercq, *Anno Domini. The Origins of the Christian era* (Turnhout: Brepols, 2000), 66; also known as enneakaidekaeternis. Michael York, *The Roman Festival Calendar of Numa Pompilius*, American University Studies, Series XVII, Classical Languages and Literature, vol. 2 (New York: Peter Lang, 1968), 15.

time, Bede's *De temporum ratione.*[14] Adding the Hebro-Greek connection probably gave the calendars more historical and theological prestige as well.

Medieval Roman calendars usually use the Roman numerals. Sometimes, however, the Hindu-Arabic numerals are also used simultaneously for certain purposes in the calendars.[15] The earliest calendars to contain these are from the 13th century, but they remain very rare and few in number until the 16th century when they replace the Roman numbers completely.

The Roman medieval calendars also mainly use the ancient Roman method of counting the days of the month in relation to the marker days of *Kalendae, Nonae,* and *Idus*. Indeed, the calendars always begin with the capital initials KL. In some calendars, the marker days have been written in the ablative case: *Kalendis, Nonis, Idibus.* In the 15th century the ancient habit of calling the day before one of the marked days *pridie* emerges, although it is only found in about one fifth of the material. In the 16th century, however, this habit is found in all three calendars analysed here. Clearly this was a case when the *pridie* was brought back into use according to the fashion of the Renaissance. None of the calendars, interestingly, use the term *postridie* for the day after the marked days, which might have also been adopted in the same fashion. As explained above, the Romans also divided the month into three periods, each followed by either the *Kalendae, Nonae, or Idus.* These groups of seven letters are known as the hebdomadal letters (Lat. *litterae feriales*), and each signifies one day of the week.[16] They are found in more than 90 % of the calendars.

Almost all of the calendars include the *aurei numeri,* a repeating set of Roman numerals from 1 to 19 which are usually on the left-side of a calendar page. They indicate (with the help of a separate table), the days of the new moon on a particular year in a 19-year cycle. These numerals are particularly important for the calculation of Easter, which should be celebrated on the Sunday following the full moon after the spring equinox. But it would be wrong to assume that the only purpose of the *aurei numeri* was the calculation of Easter. If this

14 Beda Venerabilis, *De temporum ratione,* transl. and ed. Faith Wallis (Liverpool: Liverpool University Press, 2004), 11; Charles W. Jones, ed., *Bedae Opera de Temporibus* (Cambridge, MA: The Medieval Academy of America, 1943). The Greek and Hebrew names of the months given in the Roman material vary, and none of them are exactly the same as given by Bede.

15 Found in about a tenth of the calendars. These numbers are popularly called Arabic, as they were introduced into Europe through the Latin translation *Algoritmi de numero Indorum* of the Muslim scholar al-Khwārizmī's. As even the Latin translation suggests, their origin lies in 6th or 7th century India. *Britannica Academic,* s.v. "Al-Khwārizmī," http://academic.eb.com/levels/collegiate/article/4536 (accessed 22 August 2016).

16 Bridget Ann Henisch, *The Medieval Calendar Year* (University Park: Pennsylvania State University Press, 1999), 215.

was the case, why would the scribe bother with the huge task of writing the numbers on every single month, when the only possible months for Easter are March and April? Observing the moon has a Christian background in the Vulgate. The new moon was also important in the Jewish tradition as a special feast day. In the Latin Bible, Vulgate, the term *calendae* is used in a way which is probably causing confusion: does this now refer to the actual day of the new moon, or the first day of the Roman month?[17] This would easily explain why both the golden numbers, and (often palaeographically elaborated) KL initials, are in all of the months. The moon can also be linked to supernatural beliefs, such as causing mental illness, which is referred to in both the Old and New Testaments.[18] Controversially, however, there are also warnings against believing in such effects of the moon: "only an idiot changes like the moon."[19]

2 The Solar Cycle

The first of the analysed cycles is the solar cycle and its related natural phenomena. Perhaps the most universally observed cycle of time in any civilisation, it is based on the rotation of the Earth around the Sun (in c.365 ¼ days), which at the latitude of Rome causes clear changes in both temperature and the amount of daylight during different seasons. This is, of course, universal and by no means limited to Rome. Yet because the entire modern Western world follows a calendar which is based on the Julian and Gregorian calendars (both developed in Rome), it is interesting to determine to what extent medieval calendars work as a bridge between the ancient and the modern. Naturally, a straight continuation from the Roman Empire to the end of the Middle Ages cannot be assumed, as the city of Rome held very different roles depending on the period. But as demonstrated below, there is indeed a high level of continuity from antiquity to the Middle Ages evident in the calendars – to a greater extent than any signs of the Renaissance.

As this essay deals with calendars as tools for organising time, a fundamental question is to determine how and why medieval Roman calendars reflect

---

17 I Regum 20, 18: *"Dixitque ad eum Ionathas: Cras calendae sunt, et requieris."*; I Regum 20, 24: *"Absconditus est ergo David in agro; et venerunt calendae, et sedit rex ad comedendum panem."*

18 Psalmus 121, 5–6: *"Dominus protectio tua super manum dexteram tuam. Per diem sol non uret te, neque luna per noctem."*; Evangelium Secundum Matthaeum 4, 24. *"Et abiit opinion eius in totam Syriam; et obtulerunt ei omnes male habentes, variis languoribus et tormentis comprehensos et qui daemonia habebant et lunaticos et paralyticos et curavit eos."*

19 Ecli 27, 12: *"nam stultus sicut luna mutatur."*

the practical realities of how time is in general perceived by humans. The focus here is on only two major aspects: the seasons that divide the year into units according to the weather, and the amount of daylight, which limited all human activities during the Middle Ages. Both aspects are actually related to the sun. And although arguably it can be said that no celestial objects (including the sun) have true time-measuring capacities, it is nevertheless the sun which regulates all activities on this planet, whether or not we (or medieval Romans) acknowledge it. This chapter, then, aims to provide some insights into how the solar year was perceived in the calendar material. Other celestial objects such as the moon or stars can also be used in measuring time, and indeed these were also observed in medieval Rome, but their significance tends be of a supernatural rather than practical nature. This aspect will be analysed in further chapters.

The seasons have a strong presence in medieval Roman calendars. Most importantly, key information such as *veris initium* ("spring beginning"), is almost always written with red ink, which is usually the colour reserved for important liturgical feasts and important remarks. This implies that the beginning of seasons carried a rather special status. The most commonly mentioned season is spring, which is found in approximately one-third of the calendars. Autumn is marked in approximately one-fifth of the material, and summer and winter in roughly one out of six. This variation could reflect the agricultural cycle; the starting point of the cycle is more important than the middle or end. In addition, the end of the cold winter was probably a highly anticipated occasion, considering the realities of the Roman climate.

This analysis only focuses on spring as the number of mentions regarding the other seasons is not high enough to be considered relevant. Spring is usually marked on 22 February (38 times), and less often on 7 February (five times) or on 24 or 25 February (both twice). All of these dates could indeed be viewed as times when the weather gets considerably warmer in Rome than it is in midwinter (December/January). However, before going further into the analysis, it is important to remember that the beginning of the seasons was also a *topos* of a sort. Many important authors also took an interest in this question, particularly Bede (c.673–735), who recounts that according to Isidore of Seville (c.556–636), the starting dates for the seasons are very late (22 February, 24 May, 23 August, 22 November), whereas the more reliable "Greeks" and "Romans" preferred earlier dates (7 February, 9 May, 7 August, 7 November).[20] This reflects the dates found in the Roman material as well. It is also known that sometimes

20 Beda Venerabilis, *De temporum ratione,* transl. and ed. Faith Wallis (Liverpool: Liverpool University Press, 2004), 100–03. Bede explicitly mentions Pliny the Elder's *Naturalis*

liturgical tendencies affect the entries about the seasons. In some calendars,[21] there are verses linking the end of a season to a certain feast day, such as "end of winter behind the seat of Simon Peter."[22] It is also possible that some of the "Roman" calendars were actually produced elsewhere, or they were copied from exemplars that came from colder parts of Europe – such as France or the German lands – which would also explain the seemingly peculiar starting days of the seasons in terms of the realities of the Roman climate.

But this does not make it irrelevant to study the starting dates of the seasons. Even though a certain bipolarity concerning the starting dates existed, anyone who made a calendar had to choose between the two dates (or possibly a third option), and this decision could have been affected by how relevant it seemed to mention that spring begins at the beginning or end of February, for example. Particularly revealing are those cases in which there are multiple options for the starting date of the season, or when there are other comments regarding natural phenomena. In these cases, it seems that the maker of the calendar was uncertain, perhaps copying from two sources, or perhaps even considering a date given by the source too late or too early, and decided to insert them both. This is the case with the calendar BAV, Arch.Cap.S.Pietro.B.79 dated to the 12th century (and the modified second version to the 13th century).[23] The calendar found in BAV, Arch.Cap.S.Pietro.H.28 also mentions how on 13 February, the Ides of February, birds start to sing, which was a sure sign of spring.[24]

Furthermore, it is difficult to say in which direction the cause-effect relation worked in when it came to "authoritative" starting dates for the seasons. Both the options for "spring" (early or late February) can very likely be linked to the agrarian year, which might have then been "copied" to the Christian year, vaguely visible in liturgical literature as *pars hiemalis*, or by a verse linking the beginning of a season to a certain feast. Bede and all the other authors were probably just as much affected by the agrarian as the Christian year, which brings us back to the relevance of the starting dates of the seasons; they reveal a great deal about the local agrarian year, which in turn is dependent on the local realities of natural phenomena.

The starting dates of spring give the following interesting pattern. The early date for spring is only found up until mid-13th century, when it disappears

---

Historia, but ironically there the dates are different, and less logical: 8 February, 10 May, autumn equinox (i.e. around 24 September in those days), 11 November.

21 BAV, Arch.Cap.S.Pietro B79 and B87.

22 BAV, Arch.Cap.S.Pietro B79, fol. 1r: *"Cedit hyemps ret[ro] cathedrato symo[n]e petro"*

23 BAV, Arch.Cap.S.Pietro.B79.

24 BAV, Arch.Cap.S.Pietro.H.28, fol. 2r: *"Hic aues incipiu[n]t cantare."*

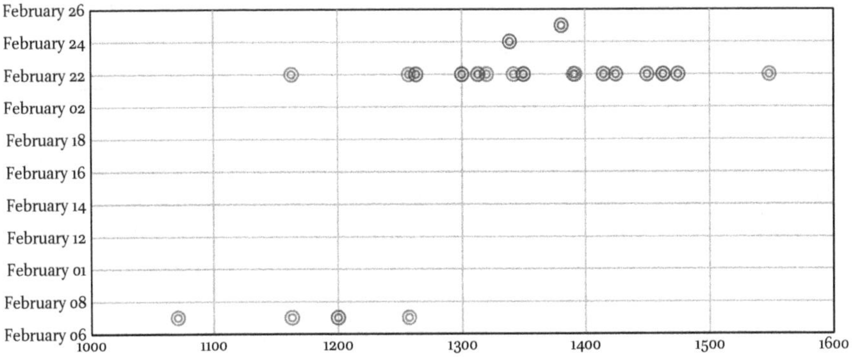

GRAPH 8.2 Start of spring.

completely. The difference is so great that it cannot be explained by the error of the Julian calendar, for instance, as that is only about one day in a century. Therefore, the information implies that spring was considered to arrive sooner in the 11th and 12th centuries than in the 15th century. It is generally believed that the European climate was indeed somewhat warmer in the high Middle ages. A recent multidisciplinary research project has, in fact, identified a particularly cold period of the late Middle Ages known as the Spörer Minimum (SPM) for the years of c.1421–1550.[25]

In a general European context, calendars quite commonly also include illustrations which are related to the typical activities for each month. In scholarship, these are usually referred to as the labour or occupation of the month. In his extensive collection of occupations for the months, Colum Hourihane has demonstrated that the origins of these occupations can quite often be traced to ancient Greece, from where they found their way, via Rome, to medieval European works of art – including calendars.[26] He argues rather convincingly that the occupations the months are both a representation of the agricultural year as well as a general reference to the passing of time, with its religious and astrological connections. Yet Hourihane views the occupations and the zodiac signs as very closely connected: "[It]is impossible to separate the signs from the scenes [of the occupations of the months.]"[27] But in the Roman calendrical context, the occupations are quite rare, being found in only a

25 The SPM defined as in Chantal Camenisch et al., "The 1430s: a cold period of extraordinary internal climate variability during the early Spörer Minimum with social and economic impacts in north-western and central Europe," *Climate of the Past* 12 (2016), 2107–26.

26 Hourihane, *Time in the Medieval World*, liv–lv.

27 Hourihane, *Time in the Medieval World*, lii.

TABLE 8.1 A comparison of illustration motifs of the months between a general pattern given by Maiello (1994), in the calendar found in BAV, Chigi VI.174 (*c.* 1300) and BAV, Vat.lat.14701 (*c.* 1450).

Month	Maiello	Chigi VI.174	Vat.lat.14701
January	A feast	Janus	A feast
February	A fireplace	A fireplace	A fireplace
March	Cutting the vineyard	Cutting the vineyard/trees	Sowing?
April	Seigneurial activities	Man with flowers	Cutting trees
May	Seigneurial activities	Man with an eagle	Man with an eagle
June	Sheep shearing	Sowing or digging a ditch	Reaping
July	Working on the fields	Harvesting grain	Sowing a field
August	Reaping grain	Threshing grain	Threshing grain
September	Sowing a field	Collecting grapes	Collecting grapes
October	Pressing wine	Pressing wine	Pressing wine
November	Collecting acorns	Falling a tree	Man with dogs
December	Killing a pig	Killing a pig	Killing a an ox

couple of calendars, whereas the zodiac signs (which will be discussed further later) are very common.

According to Francesco Maiello, approximately half of the calendars in the books of hours made between 1485 and 1510 have the same illustration motifs of the seasons.[28] Maiello argues that more realistic images of the seasons only appeared in the 16th century, and earlier they were merely symbolic.[29]

In the Roman material, there are just two cases where such images of the labours for the month are present. The first of these is from c.1300 (BAV, Chig.VI.174), and the second from the mid-15th century (BAV, Vat.lat.14701). Their motifs are compared above (since table 8.1 is above on the same page) with the general pattern given by Maiello.

28 Francesco Maiello, *Storia del calendario. La misurazione del tempo, 1450–1800* (Torino: G. Einaudi, 1996), 31.

29 Maiello, *Storia del Calendario*, 32.

The Roman material quite closely follows the general picture presented by Maiello, but at least in these two cases it can be argued that the motifs may actually reflect reality, to a certain extent. All the activities presented here seem to fit quite well into the agrarian season, and there are no major contradictions. It is hard to estimate the validity of some of the activities (such as hawking), but at least they are not considered impossible. Interestingly, neither of the calendars have any Christian images of the seasons, although the calendars where these images are found represent the liturgical year.

Medieval Roman calendars also pay a surprising amount of attention to the amount of daylight in each month. Almost half of the calendars mention at the beginning of each month how many hours the night and the day have (out of 24 hours). This is most commonly found in the calendars of the 14th century (in 68 %) and it is most rare in the 10th and 11th centuries (less than one fifth).

The amount of daylight usually follows the following pattern: January, eight hours; February, ten; March, 12; April, 14; May, 16; June, 18; July, 16; August, 14; September, 12; October, ten, November, eight; December, six. It is quite interesting that this pattern does not, however, really represent a Roman reality. A day's length in Rome is about 15 hours on summer solstice. Even if the roughly half an hour of civil twilight on both ends is counted, 16 hours of daylight is the maximum for Rome. The latitude of Paris is much closer (16 hours of daylight, 17.5 with twilight), and that of Dublin even closer (17 hours of daylight, 18.8 with twilight). As it is already quite hard to write during the civil twilight, the actual day length is perhaps more likely to be defined by sunrise and sunset. If this is the case, it seems likely that the day length pattern was actually created in Ireland or northern England, both areas which were very active in computistical matters during the early Middle Ages. This pattern, as with so many other aspects during the medieval period, was then copied again and again, without too much attention being paid to the actual lengths of the day. Perhaps this discrepancy was finally noticed in Rome as well, because starting from the 14th century, the amounts of daylight are sometimes closer to the truth than in the general pattern presented above, although their number is small, being found only in 12 calendar versions.[30]

Noting the amount of daylight probably had some sort of a practical purpose; it is a good reminder for anyone who intends to read or write, in a period with limited availability of artificial light. Nevertheless, in the context of the

30 Found in BAV, Chig.C.VI.174, Inc.IV.10, S.Maria.Magg.91, S.Maria.Magg.97 (2 versions), S.Maria.Magg.99, Stamp.Arch.Cap.S.Pietro.A.41; Roma, Biblioteca Angelica, Mss.24 (2 versions), Z.14.10/1, Nürnberg, Germanisches Nationalmuseum, Inc.40.4113, München, Bayerische Staatsbibliothek, Ink.M.471.

liturgical calendar as well as a book of hours, (*horarium*), the amount of daylight helped to organise the daily prayers, which also followed the rhythm of the sun.

Another feature related to the amount of daylight are the equinoxes and solstices, which are also commonly found in medieval Roman calendars. It is clear that the spring equinox is present in so many calendars (approximately 40 %) because it is needed to calculate the correct day for Easter Sunday. But this fails to explain why other days connected to the same phenomenon are also present, namely the autumn equinox (26 %) as well as the summer (18 %) and winter (25 %) solstices.

The presence of the equinoxes and solstices should be seen in a double context: classical and Biblical. The former implies that medieval Roman calendars could and should be viewed as a form of continuity from the classical Greco-Roman world. The Biblical context is at least as significant. The Jewish world, which naturally was well connected to Greek as well as Egyptian and Babylonian cultural spheres, was heavily centred around the celestial objects, in particular the sun and the moon, which are also present in the Bible. Their movements were seen as a demonstration of the power of God. An interest in these aspects was not, therefore, merely a continuity from the Hellenistic world, but it was also probably constantly maintained by the passages of the Vulgate, as it was God who controlled light and darkness.[31] This idea of God as the creator of everything was also constantly present in the liturgy of the church; God can be discovered in everything through the senses, and the liturgy connects the visible and the invisible.[32] The medieval Roman calendars were primarily used for the liturgical purposes of following the liturgical year, and therefore their connection to these various liturgical aspects is essential.

Most of the medieval Roman calendars have a separate entry for the spring *aequinox* (equinox), meaning the day when the night and the day are equal in

31 E.g., Psalmus 18, 2–7: *"Caeli enarrant gloriam Dei, et opera manuum eius adnuntiat firmamentum. Dies diei eructat verbum, et nox nocti indicat scientiam. Non sunt loquelae neque sermones, quorum non audiantur voces eorum. In omnem terram exivit sonus eorum, et in fines orbis terrae verba eorum. In sole posuit tabernaculum suum; et ipse tanquam sponsus procedens de thalamo suo, exsultavit ut gigans ad currendam viam. A summo caelo egressio eius, et occursus eius usque ad summum eius nec est qui se abscondat a calore eius."*; Job 38, 19–20: *"In qua via lux habitet, et tenebranum quis locus sit; ut ducas unumquodque ad terminos suos?"*

32 Eric Palazzo, "Les cinq sens dans la littérature liturgique du haut Moyen Âge," in Carolyn Marino Malone, ed., *Consuetudines et regulae: Sources for monastic life in the Middle Ages and the early modern period* (Turnhout: Brepols, 2014), 245.

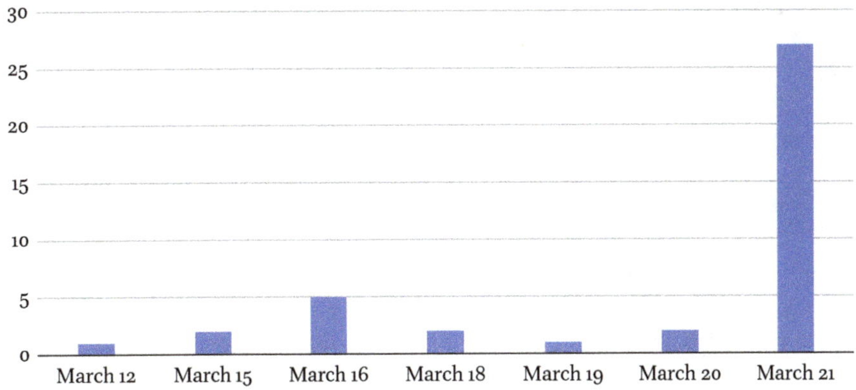

GRAPH 8.3 Various dates given for the spring equinox in the Roman calendars.

length. What makes this day intriguing is that it is not always marked on the same day; rather, different options are found in the calendars, as seen below.

The most common option is 21 March, which is the date decided at the Council of Nicaea in 325 AD.[33] The reason for alternative competing times for the equinox is a clear indication of the capability of medieval Romans for making critical observations, even in the midst of the "dark" Middle Ages. Because of the miscalculated length of the solar year in the Julian calendar, the calendar of 365 and a quarter days includes a discrepancy of 0.0078 days (11 minutes and 14 seconds) compared to reality. In 128 years, a whole extra day appears which causes the calendar to lag behind.[34] During the Middle Ages, the real equinox moved century by century: it moved to 14 March in 1100, to 13 March in 1250, and 11 March in 1500.[35] Apparently this was also noted by the scribes of some of the medieval Roman calendars because they also marked the spring equinox on different days. This is probably due to observations made or heard by the producers of the calendars. Indeed, the dates of the equinox follow a very clear pattern where the date falls earlier and earlier as the centuries progress, as seen below.

33 Faith Wallis, "Introduction," in Faith Wallis, transl. and ed., *Beda Venerabilis, The Reckoning of Time* (Liverpool: Liverpool University Press, 2004), xxxix; Jones, *Bedae Opera de Temporibus*, 22.

34 Declercq, *Anno Domini*, 56. This is why we still speak of the Russian "October revolution." It took place in October according to the Julian calendar, but according to the (more accurate) Gregorian calendar, it actually occurred in November.

35 Sam Owen Jansson, "Dagjämningar och solstånd," *Kulturhistoriskt Lexicon för Nordisk Medeltid från vikingatid till reformationstid*, 11 (Helsinki: Ab. Örnförlaget, 1957), 610.

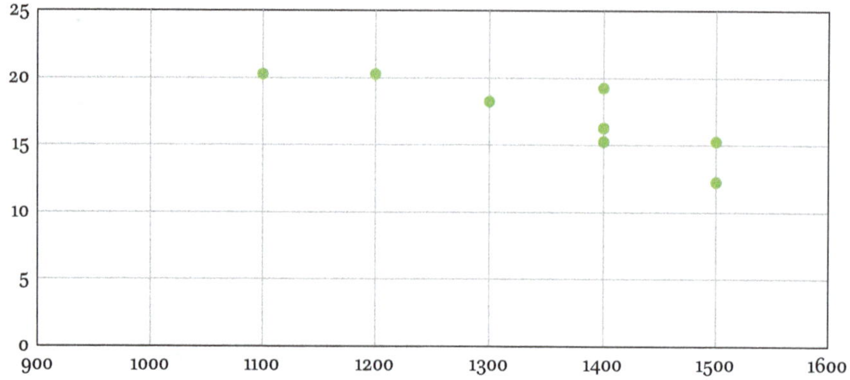

GRAPH 8.4 Date of the spring equinox (Y) by centuries (X).

This demonstrates that attention was paid in Rome to the true equinox in addition to the ecumenical Nicaea decision of 21 March. However, there is no clear pattern where these correct observations were made evident, as the calendars that include a different date for spring equinox are from many different churches.

A further problem of the lunisolar year to which the medieval Roman calendars paid attention is the leap year. Inconveniently, the solar year's length is not an exact number of days, but about 365.24219 days (365 days, five hours, 48 minutes and 46 seconds).[36] But in the calculations for the Julian calendar, the great Greek-Egyptian scholar Sosigenes (1st century BC) made a slight miscalculation. He thought that the solar year was exactly 365.25 days long, and therefore the extra six hours (a quarter of a day) could be compensated by adding a leap year every four years.[37] This was done by doubling 24 February, or using the Roman calendar style, making a *bis sextilis*, or "double sixth," meaning the sixth day before the *kalendae* of March.[38] It is not clear why this day was chosen, but apparently this was still important in medieval Rome as well, as 73 % of the calendars contain a separate reminder of the location of this day.[39]

Medieval Roman calendars were inevitably obliged to follow the restrictions of the solar year. However, there was considerable variation among the

36 Robert Hannah, *Time in Antiquity* (London and New York: Routledge, 2009), 12.

37 This system did not start off well, because leap years were accidentally added every fourth year, until Caesar's heir Augustus ordered a correction. Hannah, *Time in Antiquity*, 12.

38 Known as *dies bissextilis* "double sixth day," meaning six days before the first day of March (see the chapter on months below).

39 This is marked with a (usually red) entry *locus bissextilis* on the day.

calendars, particularly when it comes to the seasons. It is not wise to draw sensational conclusions on the topic, but it can be tentatively suggested that the Roman calendrical evidence seems to point towards the idea that the information concerning both the seasons and the amount of daylight did not come from Rome. In the case of the daylight, it seems clear that Roman calendars are repeating a tradition that is most likely coming from the British Isles or France. This would make sense historically. The case might be similar for the seasons, but in this case the possibility of the calendrical tradition reflecting climate change cannot be ruled out. Still, calendars also tend to cause a certain amount of chaos in relation to time, because they cause people to expect that the clear pattern of the calendars indeed reflects reality. When reality is different to the calendar (for example when spring is late or the summer is hotter than before), people tend to create new explanations, or, as with the current climate change, also rely on denial.

3 The Stellar Cycle

Medieval Roman calendars also contain a significant amount of information concerning the stars. Over half of them mention the signs of the zodiac. This is usually done with a separate (typically red) entry indicating the sun's position within a particular sign, such as *"sol in scorpione."* They remain a popular feature throughout the period, with only a slight drop in the 12th and 13th centuries, when they are found in slightly less than half of the calendars. Only two early calendars (from the 11th and 12th centuries) use the system of *litterae signorum*, letters marked next to each day of the month, which with the assistance of a separate table can be used to determine the zodiac sign.

The zodiac signs are at least theoretically connected to the actual natural phenomenon of certain stars creating groups when perceived from a certain latitude on the Earth. In this case they are nevertheless seen as "superstitious" beliefs because they are not actually based on real observations, since the times of the sun's supposed entry into the sign were already decided by the ancient Babylonians, and by the Middle Ages these times had changed due to the gradual changes in the Earth's orbit.[40]

Astronomical beliefs have a long tradition in Rome, stemming from both a classical and a biblical background. Bright stars and comets were often believed

40 Robert Hannah, *Greek and Roman Calendars: Constructions of Time in the Classical World* (London: Duckworth, 2005), 12; E.G. Richards, *Mapping Time: The Calendar and Its History* (Oxford: Oxford University Press, 1998), 25.

to mark the importance of person, such as those seen during the funeral of Caesar.[41] Augustus seems to have believed in catasterism, the belief that deserving humans join the heavens as stars in constellations.[42] Augustus's horoscope sign of Capricorn can be found in many portrayals, from coins to gemstones.[43] Even his birthmarks were said to represent the constellation of Ursa Major.[44] In Trimalchio's feast there is a platter with the 12 zodiacal signs.[45]

The same importance of the stars can also be found the Vulgate concerning the birth of Jesus and the appearance of a star as its herald.[46] And in the end, when the day of Judgement comes, the sun, the moon and the stars will fade away.[47] But on the other hand, there are also clear warnings "against" observing (let alone worshipping) celestial objects in the Bible.[48]

Colum Hourihane has collected an impressive index of hundreds of representations of both the signs of the zodiac and the labours of the month in medieval art. He argues that both of these were seen by medieval people as part of God's work, and the people in turn worked under the heavenly signs, reinforcing their part in the grand picture and bridging the gap between the mundane and the eternal.[49] He also states that "[t]ogether they formed a complete calendar, and it is impossible to separate the signs from the scenes."[50]

If this is the case in medieval Europe in general, as Hourihane implies, then Rome forms an exception; the zodiac signs are found in over half of the calendars, whereas the labours of the months are almost non-existent (found only

41 Ovid Met. 15.843.

42 Paul Rehak, *Imperium and Cosmos: Augustus and the Northern Campus Martius* (Madison: University of Wisconsin Press, 2007), 63, 70. Aeneis 9.641, 8.680. 1.286.

43 Such as the *aurei* issued, the breastplate of Primaporta, Actium Victory Cameo, Gemma Augustea. Rehak, *Imperium and Cosmos*, 72–73.

44 Suetonius Aug 80.1.; Aeneis 6.791–97.

45 Petronius, Satyricon 35.

46 Evangelium Secundum Matthaeum 2, 2: "*Ubi est qui natus est rex Iudaeorum? Vidimus enim stellam eius in oriente et venimus adorare eum.*"

47 Evangelium Secundum Matthaeum 24,29–30: "*Statim autem post tribulationem dierum illorum sol obscurabitur; et luna non dabit lumen suum, et stellae cadent de caelo, et virtutes caelorum commovebuntur; et tunc parebit signum Filii hominis in caelo, et tunc plangent omnes tribus terrae et videbunt Filium hominis venientem in nubibus caeli cum virtute multa et maiestate.*"

48 Deuteronomium 4,19: "*- - ne forte, elevatis oculis ad caelum, videas solem et lunam et omnia astra caeli et errore deceptus adores ea.*"; Iob 31, 26–28: "*si vidi solem cum fulgeret et lunam incedentem clare et laetatum est in abscondito cor meum, et osculatus sum manum meam ore meo: quae est iniquitas maxima, et nagatio contra Deum altissimum.*"; Isaias 3, 18: "*In die illa auferet Dominus ornamentum calceamentorum et lunulas.*"

49 Hourihane, *Time in the Medieval World*, lii,lv.

50 Hourihane, *Time in the Medieval World*, lii.

in two, as discussed above). Although all cycles of time are always connected, in the case of the signs and the labours it is hard to see as strong a link as Hourihane suggests. This strong link can be seen, however, between the liturgical year and the stellar year, as the passages from the Vulgate also suggest. The labours of the month lack this level of connection. Although there are also references to these in the Vulgate, their meaning is inevitably different as the seasons are utterly different in Palestine than in central Europe, whereas the difference in the visibility of the zodiac is actually less noticeable.

In some calendars there are also other types of star-related aspects: the so-called *dies aegyptiaci* as well as *dies caniculares.* The *dies egyptiatici,* "Egyptian days," are usually marked in the calendars with a D-sign, or with the text "*dies aegyptiaci*" or "*dies mali,*" "bad days." Such an entry is marked in two-fifths of the calendars. These bad days also have a long tradition in Rome. They can already be found in the Philocalian calendar from 354 AD, and before then they had a long tradition in antiquity.[51] According to an ancient Egyptian habit, during these days it was forbidden to practice blood-letting, as the days were considered ill-omened.[52]

Even more common is to include a rhymed verse at the beginning of each month that indicates these bad days, such as "*Prima dies iani timor est et septima uani.*"[53] These verses are found in almost half of the calendars. Their historical origin is not clear; Francesco Maiello argues that C.W. Jones was wrong to attribute these verses to Bede, as they are instead an invention of the 11th century.[54] Some calendars also mention a separate reminder concerning blood-letting, such as "*A*[*nte*] *xiiii k*[*alendas*]*l aug*[*ustas*] *usq*[*ue*] *ad no*[*nas*] *septemb*[*ri*]*s n*[*e*] *minuas sanguem.*" All of these are, however, connected to the same phenomenon.

The Egyptian days are most common in the calendars of the 14th century, being found in over two-thirds of the materials, but in all centuries they can be found in at least a quarter of the calendars. The same pattern also applies to the verses, although they are not present in 11th century.

According to Don C. Skemer, every Egyptian day was also paired with a specifically ominous hour in medieval beliefs.[55] Although this phenomenon

51 Michele Renee Salzman, *On Roman Time: The Codex-calendar of 354 and the rhythms of Urban life in Late Antiquity* (Berkeley, Los Angeles and Oxford: University of California Press, 1990), 13.

52 Baroffio, "Kalendaria Italica," 450.

53 BAV, Arch.Cap.S.Pietro.A.9, fol. 7r.

54 Maiello, *Storia del Calendario*, 32.

55 Don C. Skemer, "'Armis gunfe': remembering Egyptian days," *Traditio* 65 (2010), 75.

is present in the medieval Roman calendars as well, it seems just as common to have the Egyptian days only, without the hour. Out of the 43 Roman calendar versions which contain the Egyptian days, there are only 20 calendars which also mention a particular hour. The pairing of Egyptian days and hours is not present in the calendars before the 13th century. They are very common in the 14th century, when they can be found in half of the calendars. In the 15th century they become rare and disappear completely in the following century.

In approximately a quarter of the calendars, the "dog days" or *dies caniculares* are also marked. This refers to the period when Sirius (the dog star) is visible. The *dies caniculares* are most popular in the 15th century, where they can be found in a third of the calendars, and least popular in the 12th and 13th centuries where they are only found in less than 10 % of the calendars. This period is also often the hottest time of the summer, July–17 August.[56] In Roman calendars, however, there is a certain amount of variation within this period, as there are five different options for the start of the period, namely the 8, 12, 14, 15 and 16 July. This probably means that either there were actual observations regarding the visibility of Sirius, or, perhaps less likely, these refer to the actual beginning of the hottest period, which in Rome is probably closer to 16 July than 8 July. It is most likely that like the zodiac, this was a tradition that was followed with superstitious implications, and following a Ptolemaic tradition that was also used in medieval medicine.

Roman calendars from the Middle Ages demonstrate a great extent of continuity from antiquity, which lasts throughout the medieval period. In the case of the "supernatural" beliefs, this is particularly well perceptible. Medieval calendars kept the same time format as the ancient world, simply adding to it a Christian meaning (which, on the other hand, already had a presence in the ancient world as well.)

There is a similar situation in relation to the supernatural beliefs of the effects of the celestial objects; the original idea of correctly predicting the visibility of a particular zodiac becomes an illusion, as the zodiac signs are not actually visible – even as they are not modern horoscope calendars. The traditional times of visibility are inherited from the ancient (Babylonian) past.[57]

---

56 Sometimes the starting point is also 14 July or 19 July, Sam Owen Jansson, "Hundagar" *Kulturhistoriskt Lexicon för Nordisk Medeltid från vikingatid till reformationstid*, Band VII (Helsinki: Ab. Örnförlaget, 1962), 81–82.

57 Hannah, *Greek and Roman Calendars*, 12; Richards, *Mapping Time*, 25.

4 The Lunar Cycle

Like the sun, the moon also had a practical function for illumination, but naturally at night. The same goes for stars which can naturally be used for orienteering. Many important Christian authors such as Bede and Ambrose (c.337–397) also mention that the moon affects the plants, humidity, and the sea in many ways.[58] Thus, it is no surprise that the Moon is also present in the medieval Roman calendars.

The *aurei numeri,* the golden numbers, are very commonly used in medieval Roman calendars, and found in about four-fifths of all the calendars. They indicate the days on which the new moon falls during a 19-year cycle. This cycle is needed because solar years ("normal years") and lunar years are not equally long, and it takes roughly 19 years before the solar and lunar cycles match again. The use of the *aurei numeri* is particularly popular in the calendars dated to the 14th century, and can be found in almost all of them.

Already in the Middle Ages, many started to notice that the golden numbers do not actually reflect the truth. This is why in some late medieval calendars there are two columns of golden numbers: the originals (*aurei numeri pro primatione*), and the corrected ones (*aurei numeri pro incensione*).[59] In Rome these are quite rare. Out of the calendars analysed here, they are only found in one: BAV, Vat. lat. 7658.[60] Perhaps the inaccuracy of the lunar cycle was not observed in Rome, or more likely, it was not added to the calendars but dealt with in a different way.

In only three calendars of the material is there a set of numbers known by the slightly contradictory term *aureus numerus niger*.[61] This is a number which directly marks the Easter Sunday, following the same 19-year cycle, in a very similar fashion to modern calendars. This development now finally gets around the laborious task of repeating the *aurei numeri* throughout the year, but as explained above, their significance was most likely connected more to the other supernatural beliefs than to the seemingly poorly mathematical *computus*, the attempt to calculate the correct day for Easter (which required knowledge of the spring equinox and the full moon, as well as the seventh day of the week, Sunday).

58 Bede, *De temporibus ratione*, 28; Ambrose, *Hexameron* 8:31.

59 Sam Owen Jansson, "Incensio lunae," *Kulturhistoriskt Lexicon för Nordisk Medeltid från vikingatid till reformationstid,* Band VII (Helsinki: Ab. Örnförlaget, 1962), 370.

60 BAV, Vat.lat.7658.

61 BAV, Barb. lat. 609; BA, MSS 24; BA, Z 14 10/1.

Some calendars (13 %) also include symbols which are used to quickly identify the phase of a moon on a particular day.[62] One could, as we have seen, of course simply use the *aurei numeri* for this purpose by counting forwards from the new moon. There were, however, "inactive or stupid people" (as Bede puts it[63]), who instead of counting forwards from the golden number, used separate lunar letters for each day so that they could immediately see the age of the moon on a certain day. As discussed above, the synodic month (period between two full moons) is c.29.5 days long. A common way of working around the half day was to use 59 signs called *litterae lunares* (two synodic months) with letters, and dots either before or after them: A-V (*litterae nudae*), A.-V. (*litterae postpunctatae*), and .A-.T (*litterae prepunctatae*).[64][65] Some calendars used only the first set of these letters, excluding "A," which made it 19 letters. According to Nils Lithberg, this system was adopted in Merovingian France, but it was only common in the Early Middle Ages (up to c.1000 AD), and after Bede's time, these lunar letters gradually disappeared as useless (perhaps scared by Bede's harsh judgement!)[66] In the Roman context, however, they were used throughout the Middle Ages, but never became a very common system.

Some calendars also contain other computistical features. One of these is the *claves*, "keys," which are days marked in the calendar from which a certain amount of days depending on the year can be counted in order to discover the correct day for Pentecost, for example. The *claves* first appear in the calendars of the 12th century as a rather rare system but starting from the 14th century

62 Raymond Clemens and Timothy Graham, *Introduction to Manuscript Studies* (Ithaca and London: Cornell University Press, 2007), 200.

63 *Bedae opera de temporibus*, 224: *"Quod si adeo quisque deses vel hebes est ut absque omni labore computandi lunae cursum scire voluerit - -"*. Translation by the author.

64 O. Pedersen, "The Ecclesiastical Calendar and the Life of the Church," in G.C. Coyne et al., eds, *Gregorian Reform of the Calendar: Proceedings of the Vatican Conference to Commemorate its 400th Anniversary 1582–1982* (Città del Vaticano: Pontificia Academia Scientiarum & Specola Vaticana, 1983), 65. Nils Lithberg, *Computus med särskild hänsyn till Runstaven och den Borgerliga Kalendern,* Nordiska Museets Handlingar 29 (Stockholm: Nordiska Museet, 1953), 67 uses the terms *subnotatae* and *supernotatae.* Note again that there was usually no letter J in the medieval alphabet, making the number of symbols mentioned above 59.

65 Sam Owen Jansson, "Lunarbokstäver," *Kulturhistoriskt Lexicon för Nordisk Medeltid från vikingatid till reformationstid,* Band XI (Helsinki: Ab. Örnförlaget, 1966), 6.

66 Sam Owen Jansson, "Computus ecclesiasticus," *Kulturhistoriskt Lexicon för Nordisk Medeltid från vikingatid till reformationstid,* Band II (Helsinki: Ab. Örnförlaget, 1957), 589; Sam Owen Jansson, "Kalendarium I," *Kulturhistoriskt Lexicon för Nordisk Medeltid från vikingatid till reformationstid,* Band VIII (Helsinki: Ab. Örnförlaget, 1963), 90; Clemens and Graham, *Introduction to Manuscript Studies*, 200.

they are found in about half of the calendars. Another very similar computistical system is to mark the first and last possible dates for Easter and related feasts. This is done by marking a day with the words *primus/prima/primum* or *ultimus/-a/-um*, followed by the feast in question. These are common in each century, and in every other calendar on average.

5 Chronological Trends

As in the beginning of this essay, it should again be emphasised that dating the calendrical contrivances is often very problematic, and there is a high risk of circular reasoning when using palaeographical, liturgical, or stylistic methods, which have all been commonly used in the past. Yet with such a large number of calendars employed in this study, it can rather confidently be argued that certain chronological trends are clearly visible in the material. This is demonstrated in two ways: firstly, by a comparison of centuries and how common certain aspects were during each of them, and secondly by a comparison of specific aspects and how they developed during the centuries.

Starting with the general developments, the figure below offers an overview of how common certain aspects of the calendars were during the different centuries.

The number of calendar versions from the 10th and 11th centuries is only six, and thus they are analysed here together with the 12th century. This gives a total of 17 calendar versions which gives a somewhat better general picture. The 16th century is excluded from this graph, as only a handful of the analysed calendars are dated then.

As the graph shows, some features are only rarely found in this material compared to the average. These include the number of months, weeks, and days, equinoxes and solstices in general, and hours with the Egyptian days as well as the *claves*. The marking of summer is surprisingly popular in these early centuries; it is mentioned almost as often as in all the other centuries combined. The Egyptian days are also a very common feature in these early centuries.

Many aspects which were missing from the earliest calendars appear in the 13th century, but there are also cases where there is decline. Compared to the next centuries, however, the 13th is not particularly rich in features. Only two elements were most popular in the calendars of the 13th century: autumn equinox and winter. Even in these cases, the difference compared to the other centuries is not great. The dog days are least common in the 13th century, but here again the absolute number is close to the other centuries.

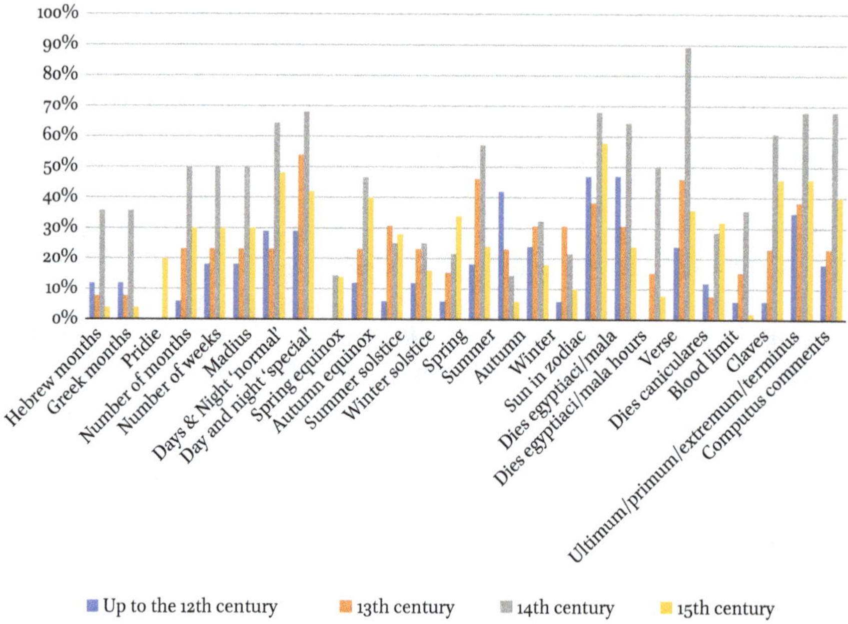

GRAPH 8.5 Main chronological developments and differences between the selected time periods (up to the 15th century).

The calendrical contrivances of the 14th century are different to the rest of the data in many respects. The use of the Greek and Hebrew names for the months declines rapidly when moving on to the following centuries. The same applies to the number of months, weeks, and days, as well as the lengths of days and nights. The sun in zodiac, Egyptian days, the zodiac verse, the limitations on blood-letting, *claves, ultimum,* and other *computus* comments are more common in the 14th century than in any other. Similarly, practically none of the features is rarer in the 14th century than in the preceding or following centuries. This represents a situation in which the elements present in the calendars had become a norm, which only somewhat changed in the next century. From the perspective of the different aspects of the calendars, the 14th century is the richest.

No features emerge as particularly popular in the 15th century, but some do decline, which perhaps points towards a humanistic influence and a tendency to remove unnecessary aspects which are no longer in use. These include the use of Hebrew and Greek month names, mentions of summer and autumn, Egyptian days, and the blood-letting limitations. All of these are aspects that are not really essential for the main purpose of most calendars as liturgical tools.

The general pattern seems to be that the calendars of the 11th, 12th and 13th centuries were somewhat simpler than the calendars of the 14th century, which seem to contain a particularly wide variety of features. In the 15th century some of these aspects seem to have lost their meaning and thus are no longer included in the calendars.

A closer look at both the aspects and centuries reveal clear sub-trends, which can be outlined in three main options of trends:

1) Decreasing trend
   a. *Littarae nudae/punctatae*
2) Fluctuating decreasing trend
   a. Local events
   b. *Madius* for *Maius*
   c. *Dies egyptiatici*
3) Increasing trend
   a. *Locus bissextilis*
4) Fluctuating increasing trend
   a. Local amount of daylight
   b. *Claves*
   c. Red letter days
   d. Verse for bad days
   e. *Dies caniculares*
   f. *Ultimum/primum*
5) Fluctuating trends
   a. Hebrew and Greek month names
   b. Number of months in a year
   c. Number of weeks, days in a year
   d. Day and night standard lengths

Only one aspect shows a very clear decreasing chronological trend. This is the case with the *litterae nudae/punctatae*, which is a set of letters on each row used to determine the phase of the moon. The number of calendars which contain *littarae nudae/punctatae* is quite modest, but the trend is clear. The reason for the gradual disappearance of these letters is probably linked to the increase of other computistical devices, such as the use of the *claves* and *ultimum/primum* -systems (increasing aspects are discussed further below).

Some features here have a fluctuating decreasing trend, meaning that generally the aspect is declining, but it is also showing temporary increase amid this decline. This applies to local historical events and the use of *Madius* instead of *Maius* for May, as well as the Egyptian days, *Dies Egyptiatici*. What makes the implications of this particularly interesting is not only that these

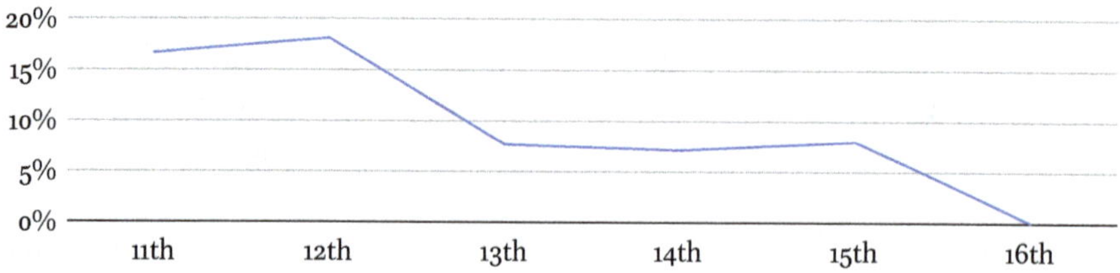

GRAPH 8.6 The percentage of calendars containing *litterae nudae / punctatae.*

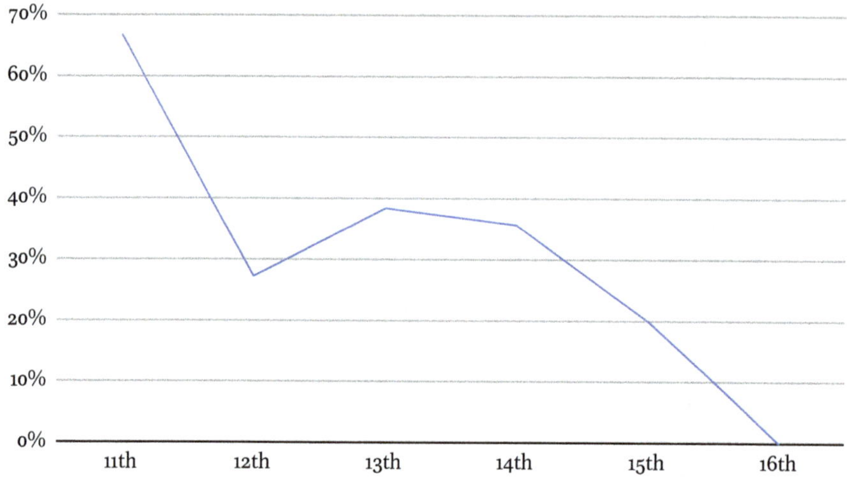

GRAPH 8.7 The percentage of calendars containing local historical events (n=39).

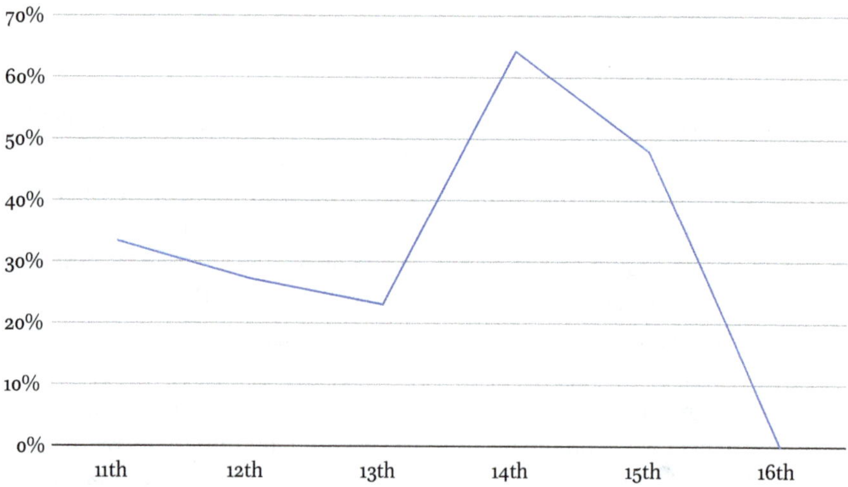

GRAPH 8.8 The percentage of calendars where May is written as *Madius* instead of *Maius* (n=50).

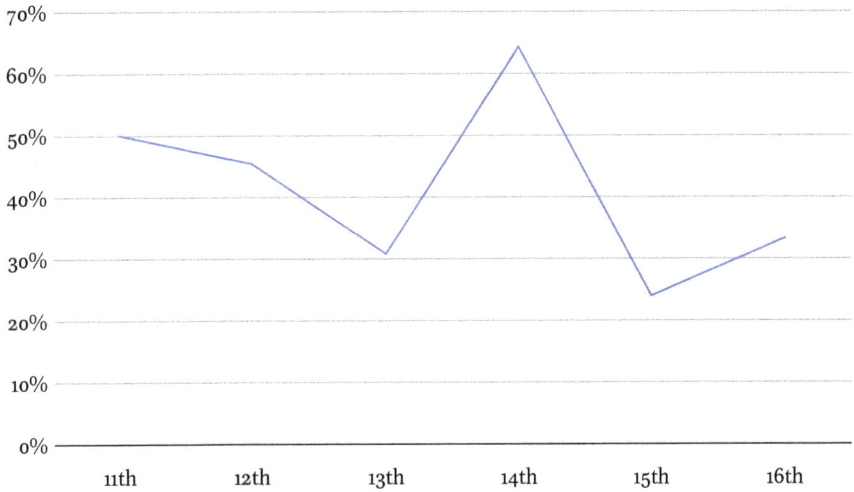

GRAPH 8.9 The percentage of calendars containing the Egyptian days (n=43).

three aspects share a trend which looks similar to a graph, but also that the fluctuations and their timings are very similar.

All three elements first experience a decline from the 11th to the 13th century, then growth until the 14th century, followed by another decline – although in the case of the Egyptian days there is a slight increase again in the 16th century. Calendars of the 13th and particularly the 14th centuries seem to have been richer in terms of these aspects than the preceding or following calendars. This is supported further with four aspects that fluctuate in a similar way, although without the initial high percentage, as shown below.

From this it can generally be supposed that all of these three aspects were perhaps considered to be additional information which had not perhaps yet been acquired in the earlier centuries, and was considered to be useless in the later ones, leaving only a short period when these aspects flourished.

Moving on to the aspects that demonstrate chronological increase, there is one aspect which demonstrates a very clear pattern of growth: the mention of the location of the leap day, *locus bissextilis.*

This is a somewhat surprising trend, as it could be assumed that its position would become better known in time. The explanation probably lies in the increasing interest in *computus* and other mathematical aspects of the calendars in later centuries. This also applies to the six aspects which increase over time, although experiences temporary declines as well, as shown below.

The *claves* system, as well as the *ultimus/primus* system, probably replaced the use of *litterae nudae/punctatae*, as was discussed at the beginning of this

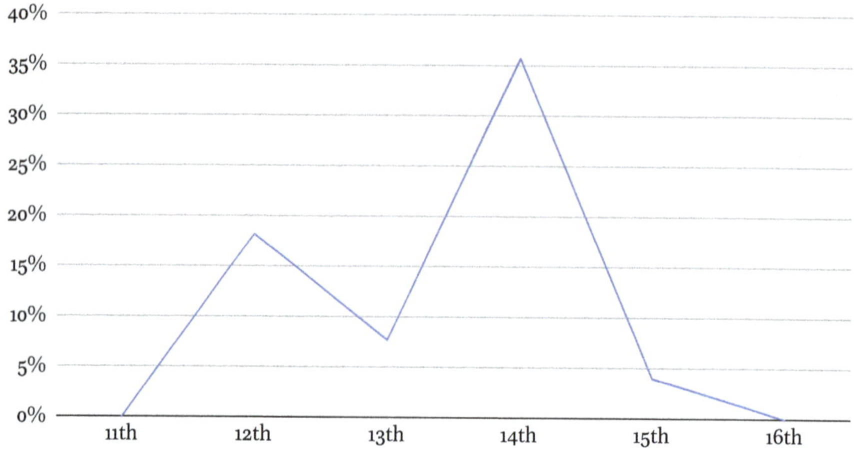

GRAPH 8.10 Calendars containing Hebrew and Greek month names (n=15).

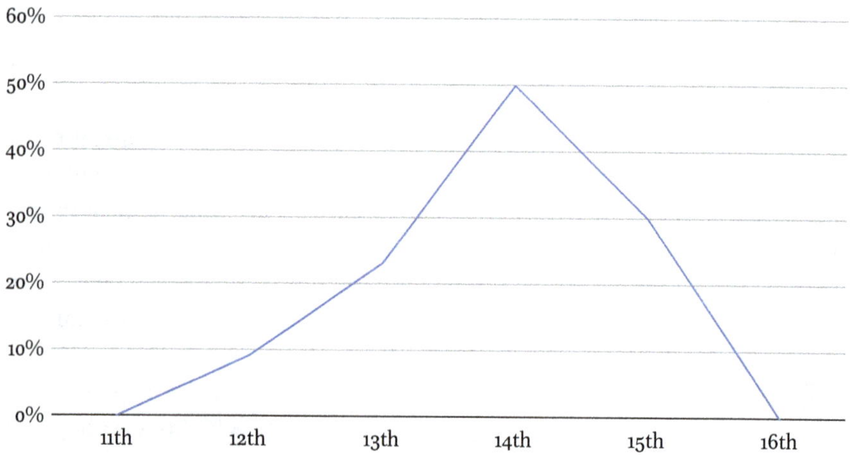

GRAPH 8.11 Calendars containing a mention of the number of months in a year (n=33).

chapter. The local amount of daylight is also a clear indicator of an interest in correcting errors in the calendars, and making calendars a tool that would actually help to organise time, rather than just repeat a pattern adopted from earlier times. Many of the red-letter days, which increase in number over time, were also about computistical aspects. However, the increasing – though fluctuating – number calendars with information regarding the *dies caniculares* and the verses for bad days seems at first to be slightly contradictory, as these elements seem rather superstitious. But as explained earlier in this essay,

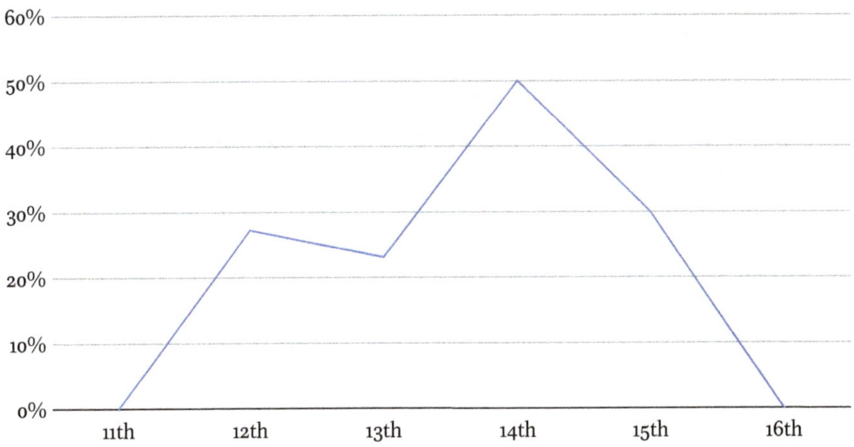

GRAPH 8.12 Calendars containing a mention of the number of weeks and days in a year (n=35).

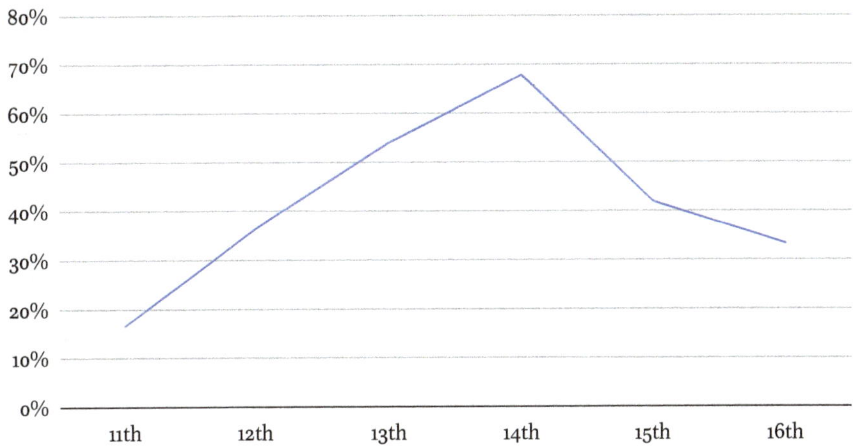

GRAPH 8.13 Calendars containing the standard amount of daylight (n=53).

their meaning was considered on the other hand to be mathematical (to calculate the movements of the celestial objects), and on the other hand Christian, as these objects could be used to follow the will of God. Thus, as new developments were made in *computus*, they were reflected in the calendars as additional features. Ultimately, this development lead to the Gregorian calendar reform, when with the help of calendars and computistical and astrological skills it became apparent that the tools which were supposed to be used to control and organise time were no longer functioning as they should.

6 Conclusions

Medieval Roman calendrical contrivances are tools which were made, copied, and used by people who had an interest in organising and understanding the difficult concept of time, with its many overlapping and sometimes

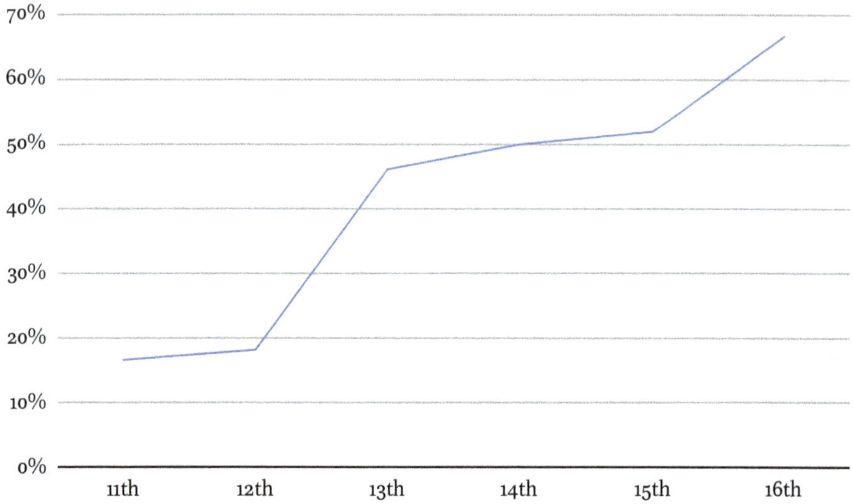

GRAPH 8.14 Calendars containing the *locus bissextilis* (n=51).

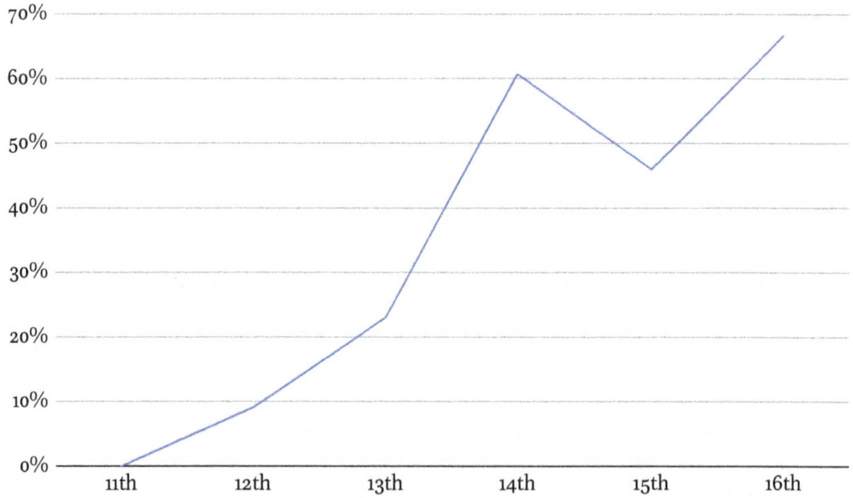

GRAPH 8.15 Calendars containing the *claves* system (n=46).

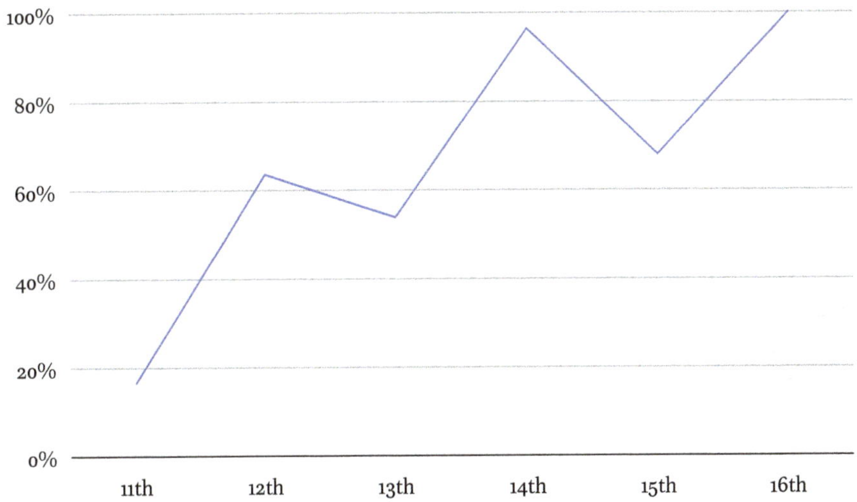

GRAPH 8.16 Calendars containing red-letter days (n=79).

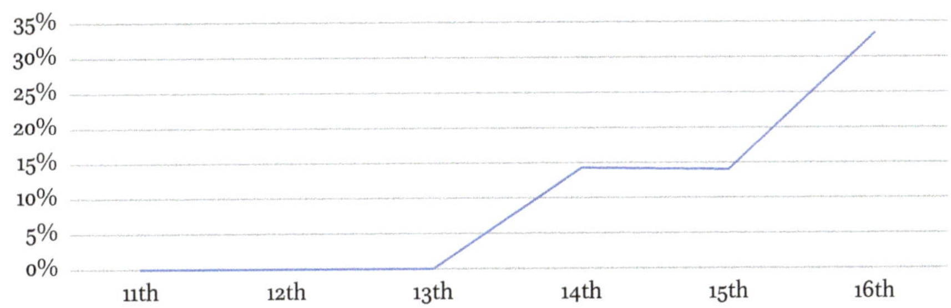

GRAPH 8.17 Calendars containing a local amount of daylight (n=12).

contradicting cycles. The calendars reflect a tradition inherited from antiquity and adopted into Judeo-Christian culture, as well as a genuine interest in the natural phenomena which form the framework of the human experience of time. They should be seen as a rich source group of medieval history which was more varied than has earlier been observed. Although the primarily function of most of the calendars was probably liturgical or at least connected to the liturgical year, the calendars still offer a rich source of information for the study of various mathematical, astrological, and environmental issues.

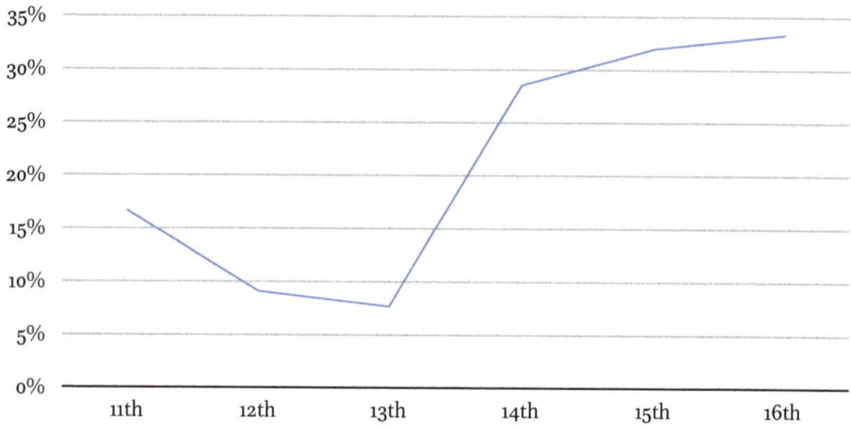

GRAPH 8.18 Calendars containing the *dies caniculares* (n=28).

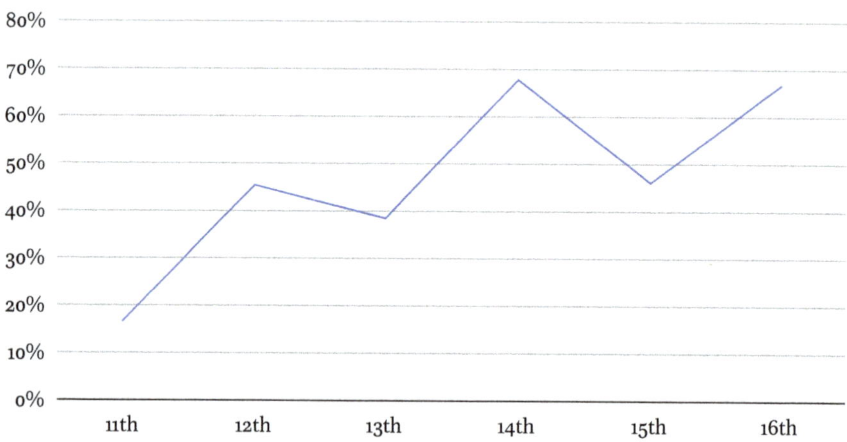

GRAPH 8.19 Calendars containing the *ultimus/primus* -system for *computus* (n=55).

In terms of their basic structure, they have all inherited the Roman Julian calendar's inner logic in terms of their layout, language, and numerals. These features remain rather static throughout the period discussed here, from the 10th to the 16th centuries, which demonstrates great continuity from antiquity through the Middle Ages, rather than a great decline followed by a re-birth.

In the aspects relating to natural phenomena connected to the solar cycle, Roman calendars show both tendencies of innovation and conformism. The starting dates of the seasons seem to follow a pattern which reflects the general European climate change of the period, but this is not visible in all of the calendars. Clearly, some were simply following a tradition when marking the

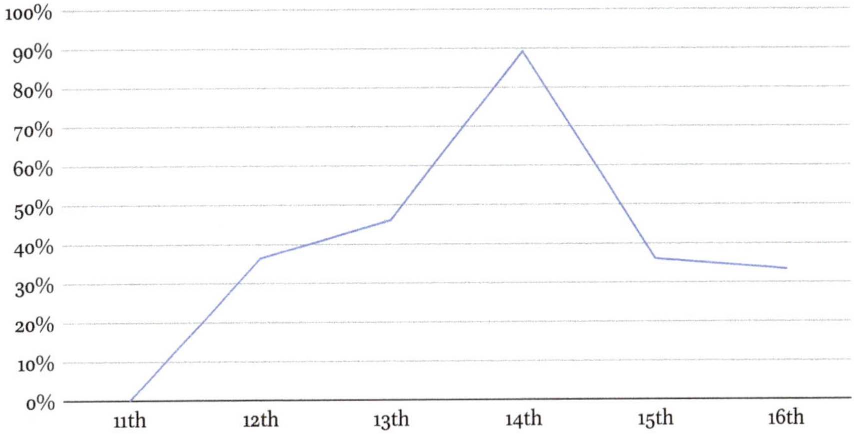

GRAPH 8.20 Calendars containing the verse for bad days (n=54).

dates of the seasons, and this is also reflected in the few images of the "labours of the months" present in the calendars. The dates of equinoxes can also be considered to be simply copied without attention to the reality, but there are also clear indications of constant calculations and according updates.

Roman calendars also contain a large number of aspects which this essay has related to the stellar and lunar cycles of time, but of course they are just as much connected to another belief in the supernatural: the Christian faith. The zodiac signs and Egyptian days are both elements that are quite common in medieval Roman calendars: both are found in approximately half of them, and both were most popular in the 14th century. The dog days, *dies caniculares,* are also present in many of the calendars, but their popularity is lower and follows a different pattern, as they become more popular at the end of the Middle Ages. Some aspects, such as the *litterae nudae/puncatae*, the Egyptian days, *Madius* for May, and local events were more frequently present in the High Middle Ages than later. Some aspects, such as Hebrew and Greek month names, the number of months, weeks, and days in a year, and the amount of standard amount of daylight only emerge in the 13th and 14th centuries, when they seem to briefly flourish in many of the calendars, only to disappear. And on the contrary, computistical aspects such as the *claves*-system and the *ultimum/primum* -system, as well as the local amounts of daylight, demonstrate an increased interest in studying, understanding, and controlling the complexity of the celestial objects and their movements. As these movements form the basis of our concept of time, understanding them essentially aims at understanding time itself. In medieval Christian culture, with where the temporal life was filled with an anticipation of the eternal life in Heaven, calendars were also a way of giving temporal existence a purpose.

Bibliography

Sources

Manuscripts

London, British Library (BL)
Add MS 30034
Add MS 30034_2
Inc. 4° 4113
München, Bayerische Staatsbibliothek
BSB-Ink M-459
BSB-Ink M-471
Nürnberg, Germanisches Nationalmuseum
Inc.40.4113
Paris, Bibliothèque Nationale de France (BNF)
Latin 826
Latin 755
Roma, Biblioteca Angelica (BA)
MSS. 24
MSS. 24_2
MSS. 1098
MSS. 1909
MSS. 1909_2
Rari. I.2.7
Z.14.10/1
Città del Vaticano, Biblioteca Apostolica Vaticana (BAV)
Arch.Cap.S.Pietro.A9
Stamp.Arch.Cap.S.Pietro.A.41
Arch.Cap.S.Pietro.B.65
Arch.Cap.S.Pietro.B.67
Arch.Cap.S.Pietro.B.69
Arch.Cap.S.Pietro.B.71
Arch.Cap.S.Pietro.B.72
Arch.Cap.S.Pietro.B.79
Arch.Cap.S.Pietro.B.81
Arch.Cap.S.Pietro.B.84

Arch.Cap.S.Pietro.B85
Arch.Cap.S.Pietro.B.87
Arch.Cap.S.Pietro.B.88
Arch.Cap.S.Pietro.D.156
Arch.Cap.S.Pietro.E4
Arch.Cap.S.Pietro. E6
Arch.Cap.S.Pietro. E7
Arch.Cap.S.Pietro.E.14
Arch.Cap.S.Pietro.E.15
Arch.Cap.S.Pietro.F27
Arch.Cap.S.Pietro.H.28
Arch.Cap.S.Pietro.H.56
Arch.Cap.S.Pietro.H.57
Arch.Cap.S.Pietro.H.58
Barb.lat.421
Barb.lat.563
Barb.lat.598 (XIII7)
Barb.lat.609
Barb.lat.646/Vat.lat.4885
Chig.C.V.138
Chig.C.VI.174
Inc.IV.10
Ott.lat.3
Ott.lat.38
Ott.lat. 356
S. Maria.Magg. 91
S. Maria.Magg. 97
S. Maria.Magg. 99
S.Maria.Magg.105 (INC)
S.Maria.Magg.106
S.Maria.Magg.119
S.Maria.Magg.120
Vat.lat.378.
Vat.lat. 4406
Vat.lat. 4885 (/ Barb.lat.646)
Vat.lat. 6827 i
Vat.lat. 6827 ii
Vat.lat.7209
Vat.lat. 7658
Vat.lat.7722

Vat.lat.7792
Vat.lat.9490
Vat.lat.12986
Vat.lat.12988
Vat.lat.12988
Vat.lat.14701

Printed Sources

Beda Venerabilis. *Bedae Opera de Temporibus*. Edited by Charles W. Jones. Cambridge, MA: The Medieval Academy of America, 1943.

Beda Venerabilis. *The Reckoning of Time. Translated, with introduction, notes and commentary by Faith Wallis.* Liverpool: Liverpool University Press, 2004.

Biblia Sacra Vulgatae Editionis. Sixti V Pontificis Maximi iussu recognita et Clementis VIII auctoritate edita. Milano: Edizioni San Paolo, 2013.

Ovidius. *Metamorphoses.* Ed. Jeffrey Henderson. Loeb Classical Library 42. Cambridge, MA: Harvard University Press, 1956.

Petronius. *Satyricon*. Ed. Jeffrey Henderson. Loeb Classical Library 15. Cambridge, MA: Harvard University Press, 1969.

Suetonius. *Vitae.* Ed. Jeffrey Henderson. Loeb Classical Library 31. Cambridge, MA: Harvard University Press, 1964.

Vergilius. *Aeneis.* Ed. Jeffrey Henderson. Loeb Classical Library 64. Cambridge, MA: Harvard University Press, 1966.

Literature

Baroffio, Giacomo. "Kalendaria Italica. Inventario." In *Aevum. Rassegna di scienze storiche linguistiche e filologiche. Anno LXXVII.* 2. Milan: Università Cattolica Del Sacro Cuore, 2003.

Bickermann, E.J. *Chronology of the Ancient World.* London: Thames and Hudson, 1968.

Britannica Academic. http://academic.eb.com.

Camenisch, Chantal et al. "The 1430s: a cold period of extraordinary internal climate variability during the early Spörer Minimum with social and economic impacts in north-western and central Europe." *Climate of the Past* 12 (2016), 2107–126.

Clemens, Raymond and Timothy Graham. *Introduction to Manuscript Studies.* Ithaca and London: Cornell University Press, 2007.

Declercq, Georges. *Anno Domini. The Origins of the Christian era.* Turnhout: Brepols, 2000.

Dykmans, Marc. "Les obituaires romains. Une définition suivie d'une vue d'ensemble." *Studi medievali* 19. Spoleto: Centro Italiano di Studi sull' alto medioevo, 1978.

Halonen, Marko. *Aspects of Medieval Nordic Calendars*. PhD diss., Helsinki: University of Helsinki, 2020.

Hannah, Robert. *Time in Antiquity.* London and New York: Routledge, 2009.

Hannah, Robert. *Greek and Roman Calendars: Constructions of Time in the Classical World.* London: Duckworth, 2005.

Heikkilä, Tuomas and Teemu Roos. "Quantitative methods for the analysis of medieval calendars." *Digital Scholarship in the Humanities* vol. 33 (4) (2018), 766–87.

Henisch, Bridget Ann. *The Medieval Calendar Year.* University Park: Pennsylvania State University Press, 1999.

Hourihane, Colum, ed. *Time in the Medieval World. Occupations of the months and signs of the zodiac in the index of Christian art.* Princeton: Princeton University in association with Penn State University Press, 2007.

Jansson, Sam Owen. "Computus ecclesiasticus." *Kulturhistoriskt Lexicon för Nordisk Medeltid från vikingatid till reformationstid.* Band II, 589. Helsinki: Ab. Örnförlaget, 1957.

Jansson, Sam Owen. "Dagjämningar och solstånd." *Kulturhistoriskt Lexicon för Nordisk Medeltid från vikingatid till reformationstid.* Band II, 610. Helsinki: Ab. Örnförlaget, 1957.

Jansson, Sam Owen. "Hundagar." *Kulturhistoriskt Lexicon för Nordisk Medeltid från vikingatid till reformationstid.* Band VII, 81–82. Helsinki: Ab. Örnförlaget, 1962.

Jansson, Sam Owen. "Incensio lunae." *Kulturhistoriskt Lexicon för Nordisk Medeltid från vikingatid till reformationstid.* Band VII, 370. Helsinki: Ab. Örnförlaget, 1962.

Jansson, Sam Owen. "Kalendarium I." *Kulturhistoriskt Lexicon för Nordisk Medeltid från vikingatid till reformationstid.* Band VIII, 90. Helsinki: Ab. Örnförlaget, 1963.

Jansson, Sam Owen. "Lunarbokstäver." *Kulturhistoriskt Lexicon för Nordisk Medeltid från vikingatid till reformationstid.* Band XI, 6. Helsinki: Ab. Örnförlaget, 1966.

Linton, C.M. *From Eudoxus to Einstein. A History of Mathematical Astronomy.* Cambridge: Cambridge University Press, 2004.

Lithberg, Nils. *Computus med särskild hänsyn till Runstaven och den Borgerliga Kalendern.* Nordiska Museets Handlingar 29. Stockholm: Nordiska Museet, 1953.

Maiello, Francesco. *Storia del calendario. La misurazione del tempo, 1450–1800.* Torino: G. Einaudi, 1996.

Palazzo, Eric. "Les cinq sens dans la littérature liturgique du haut Moyen Âge." In *Consuetudines et regulae: Sources for monastic life in the Middle Ages and the early modern period.* Ed. Carolyn Marino Malone, 271–90. Turnhout: Brepols 2014.

Pedersen, O. "The Ecclesiastical Calendar and the Life of the Church." In Coyne, G.V. et al., eds, *Gregorian Reform of the Calendar. Proceedings of the Vatican Conference to Commemorate its 400th Anniversary 1582–1982.* Vatican City: Pontificia Academia Scientiarum & Specola Vaticana, 1983.

Rehak, Paul. *Imperium and Cosmos. Augustus and the Northern Campus Martius.* Madison: University of Wisconsin Press, 2007.

Richards, E.G. *Mapping Time. The Calendar and Its History.* Oxford: Oxford University Press, 1998.

Sabbatucci, Dario. *La religione di Roma antica. Dal calendario festivo all'ordine cosmico.* Milano: Arnoldo Mondadori Editore S.p.A., 1988.

Salmon, Pierre. *Les manuscrits liturgiques latins de la Bibliothèque Vaticane. 4, Les livres de lectures de l'office, les livres de l'office du chapitre, les livres d'heures.* Studi e Testi 267. Città del Vaticano: Biblioteca Apostolica Vaticana, 1971.

Salzman, Michele Renee. *On Roman Time. The Codex-Calendar of 354 and the Rhythms of Urban life in Late Antiquity.* Berkeley, Los Angeles and Oxford: University of California Press, 1990.

Skemer, Don C. "'Armis gunfe': remembering Egyptian days." *Traditio* 65 (2010), 75–106.

Steel, Duncan. *Marking Time. The Epic Quest to Invent the Perfect Calendar.* New York: John Wiley & Sons Inc., 2000.

Wallis, Faith. "Introduction." In *Beda Venerabilis, The Reckoning of Time. Translated, with introduction, notes and commentary by Faith Wallis.* Liverpool: Liverpool University Press, 2004.

York, Michael. *The Roman Festival Calendar of Numa Pompilius.* American University Studies. Series XVII. Classical Languages and Literature, vol. 2. New York: Peter Lang, 1968.

APPENDIX 1

Non-Funerary Epigraphs in Vincenzo Forcella's *Iscrizioni delle chiese ed altri edifici di Roma* up to 1527 AD: A Checklist

For bibliographical information, please refer to the bibliography of Urpo Kantola's article.

This appendix serves a twofold purpose: firstly, together with Appendix 2, to act as an index of inscriptions used; secondly, in order to facilitate future studies on these, and to provide a checklist for non-funerary inscriptions in Forcella's corpus, with indications of erroneous and otherwise omissible material.[1] Inscriptions excluded from the data used in my article are marked with a cross †. Inscriptions with either a double publication in Forcella, or with a better publication elsewhere (in Appendix 2) are marked with an arrow → and reference to the used publications, which contain more detailed information. Places and dates without further references are as indicated by Forcella.[2] The lemmata are in the order given by Forcella. This is basically a topographic one, but because of various additions and corrections, it is advisable to consult Appendix 3 for a topographic summary of the material. For epitaphs mentioned in the discussion, see the end of Appendix 2.

Monuments honouring the deceased are mostly excluded, even in cases where the monument is not necessarily the actual grave monument,[3] or where the inscription includes a minor non-funerary element, such as a signature of the artisan who made the monument. Some exceptions have been made: First, inscriptions recording actions directed in a testament, without a direct funerary reference, are included;[4] therefore I have included one case where this is implemented in a basically funerary inscription.[5] Second, some groups, such as the dedicatory columns in the Lateran for the soul of a relative, belong to a non-funerary context, and feature a style clearly distinct from

1 In these times of modern electronic devices, a scholar interested in such an endeavour should rather contact the author for data in a more convenient format.

2 Except for volume 1, in which Forcella has not provided the information.

3 This division is not always immediately evident from the inscription, but as the language used on the both sides seems to be similar, it is preferable to discuss them together.

4 At least the following: 2:1494, 3:345, 3:348, 3:350, 3:354, 3:545; 4:2, 6:57, 6:1061, 7:19, 7:327, 8:345, 8:363, 8:786, 8:1001, 8:1203, 8:1206, 8:1211, 8:1230–1234, 9:244, 9:712, 12:457.

5 12:1494. Others may exist, but they have escaped my attention when gathering the material, focusing on non-funerary features.

 | DOI:10.1163/9789004436251_010

epitaphs.[6] This applies also to some inscriptions related to sepulchres of saints, or combining funerary and other contents.[7]

The publications referred to in the article are listed in its bibliography and will be referred to with the following shorthands, the parts in parenthesis having been omitted: Blennow (*Latin Consecrative Inscriptions*); Carunchio (*La Villa Lante al Gianicolo*); Claussen, *Magistri*; D'Achille (*Iscrizioni votive e sepolcrali*); Favreau (*Épigraphie médiévale*); Filippi (*S. Paolo fuori le Mura*); Galletti (*Inscriptiones Romanae*); Giovenale (*S. Maria in Cosmedin*); Gray (*Paleography of Latin Inscriptions*); Guillou (*Recueil*); Hülsen (*Le chiese*); Kajanto, *Dating*, *Classical and Christian*, and *Papal Epigraphy*; Kendall (*Allegory of the Church*); KSRM (Claussen, Mondini, Senekovic); Marini and Mai (*Inscriptiones Christianae*); Silvagni = MEC 1. The minor publications are listed in the following note.[8]

Forcella	Place	Date	Other publications	Photos	Referred to on p.
1:1	Capitolium	after 1237 (Silvagni)		MEC 1 tab. 28 no. 2	136, 152
1:2	→ 6:12; double publication as indicated by Forcella in vol. 6. (Non-indicated triple publication 13:2.)				
1:3	– (lost)	1296			130, 151
1:4	– (lost)	1300			92, 117, 132
1:5[a]	–	1300			92, 132
1:6	– (lost)	13th c.			
1:7–10[b]	–	14th c.			
1:11	– (lost)	1425[c]			122, 149
1:12	– (lost)	1425[d]			122, 149

6 8:25–28, and Claussen, KSRM 2, 174 note 731; cf. 9:622.

7 4:623, 5:816, 5:821, 6:53, 7:237, 7:398, 9:622–623, 12:575.

8 Chastel = André Chastel, *The Sack of Rome* (Princeton: Princeton University Press, 1983). – Cecchelli and Persico = Carlo Cecchelli and Edoardo Persico, *SS. Marcellino e Pietro: la chiesa e la catacomba* (Roma: Libreria Cremonese, 1938). – De Rossi, BCAM 1872 = Giovanni Battista De Rossi, "Diploma pontificio inciso in marmo," *Bullettino della Commissione Archeologica Comunale di Roma* 1872, 54–58. – De Rossi, ICUR[1] II = Giovanni Battista De Rossi, *Inscriptiones Christianae Urbis Romae* vol. II[1] (Roma: Guggiani, 1888). – Diehl = Ernst Diehl, *Inscriptiones Christianae Veteres Latinae* (Berolini: Apud Weidmannos, 1925). – Kajanto, Nyberg, and Steinby = Iiro Kajanto, Ulla Nyberg, and Margareta Steinby, "Le iscrizioni," in *L'area sacra di Largo Argentina. 1, Topografia e storia*, ed. Filippo Coarelli, Iiro Kajanto, Ulla Nyberg, and Margareta Steinby (Roma: Musei e monumenti comunali di Roma, 1981). – Petrella = Enrico Donato Petrella, *Le carte lapidarie di Roma* (Città di Castello: Casa tipografico-editrice S. Lapi, 1912).

Forcella	Place	Date	Other publications	Photos	Referred to on p.
1:13	–	1427			122, 149
1:14	– (lost)	1468			122, 149–150
1:15	– (lost)	1468			125, 149–150
1:16	→ Kajanto, *Papal Epigraphy*, 85				
1:17	– (lost)	1472			122, 149
1:18	– (lost)	1475			149
1:19	–	1480			122, 150
1:20	–	1481			122, 149–150
1:21	– (lost)	1483			117, 122, 150
1:22	– (lost)	1486			
1:23	– (lost)	1489			122
1:24	– (lost)	1491			120, 122
1:25	– (lost)	1494			122, 150
1:26	– (lost)	1494			
1:27	– (lost)	1496			
1:28	– (lost)	1498			121–122, 149–150
1:29	– (lost)	1492–1503			122
1:30	– (lost)	1502	Cf. Kajanto, *Papal Epigraphy*, 95 note 16.		122, 149
1:31	– (lost)	1506			122
1:32	– (lost)	1507			122, 150
1:33	–	1508			122, 149
1:34	–	1508			122
1:35	– (lost)	1510			122
1:36	–	1510			122
1:37	–	1514	Cf. Kajanto, *Papal Epigraphy*, 89.		122, 149
1:38	–	1514–1515? (prosopography; 16th c. Forcella)			
1:39	– (lost)	1515			
1:40(1)	→ Kajanto, *Papal Epigraphy*, 98–99				
†1:40(2)	–	1818			

Forcella	Place	Date	Other publications	Photos	Referred to on p.
1:41	–	1521	Cf. Kajanto, *Papal Epigraphy*, 98 note 17.		122, 141
1:42	– (lost)	1522			122
1:48(1)	–	1510			122
†1:48(2), (3)	–	1543			
1:53	–	16th c.			111
1:60	–	16th c.			142–143
1:67	– (lost)	16th c.			
1:415	S. Maria in Aracoeli	13th c.			
1:453	–	1348			
1:456	–	1372			
1:473	–	1160–1190 (Claussen; 14th c. Forcella)	= Claussen, *Magistri*, 60–63	*Magistri*, figs 65–67	
1:474	–	14th c. (Forcella)	Cf. Claussen, *Magistri*, 62.	*Magistri*, fig. 34 (only L. 2)	
1:497	– (lost)	1441			
1:1097	S. Maria ad Martyres = Pantheon (extant?)	1270	= Claussen, *Magistri*, 170		
1:1098	–	1270		Kajanto, *Classical and Christian*, fig. 5	120, 122, 126, 153
1:1104	– (lost)	14th c.			
1:1114	– (lost)	1484–1492			
1:1121	→ Kajanto, *Papal Epigraphy*, 98				
1:1132	→ Blennow 215–217(1–2)				
1:1175	S. Maria del Popolo	1263			145
1:1179	→ 13:1042; double publication not indicated by Forcella.				

Forcella	Place	Date	Other publications	Photos	Referred to on p.
1:1196	–	1472	Cf. Petrella 54 no. 37.		120, 141, 149, 151
1:1197	–	1472	Cf. Petrella 53 no. 35.		120, 125, 141, 149, 151
1:1201	– (lost)	1477	[e]		
1:1209	–	c.1474–1513[f]			
1:1224	–	1489			
1:1226	→ 1:1201[g]				
1:1237	–	1497			
1:1254	– (lost)	1503			
1:1320	– (lost)	16th c.			
1:1567	S. Maria sopra Minerva (lost)	after 1380 (prosopography; 14th c. Forcella)			
1:1629	–	1493	Cf. Petrella 54 no. 41.		120, 141, 149
1:1641	–	1499			
1:1690	– (lost)	16th c.			
1:1691	– (lost)	16th c.			
1:2077	Capitolium (lost)	1471			
1:2078(1)	S. Maria in Aracoeli	1278			
†1:2078(2)	–	1652			
2:5	S. Maria Nuova	1370–1378 or later (prosopography; 14th c. Forcella)			135
2:10	–	13th c. 1/2 (Claussen; 14th c. Forcella)	= Claussen, *Magistri*, 147–148; *KSRM* 1, 478		
2:56	→ Blennow 33–35 no. 4				

Forcella	Place	Date	Other publications	Photos	Referred to on p.
2:59	S. Cecilia in Trastevere	1293 (Claussen; 1294 Forcella)	= Claussen, *KSRM* 1, 262–263[h]		
†2:62	–	14th c.	Too fragmentary.		
2:63	–	14th c.			
2:82	–	1480			141
2:139	→ Blennow 145–154 no. 32				
2:140	→ Blennow 224–225				
2:141	→ Blennow 171–178 no. 37				
2:144	S. Adriano al Foro Romano	14th c.			
2:145	– (lost)	1495			
2:188	S. Biagio al Campidoglio (lost)	1384			
2:191	– (lost)	1420			120
2:193	→ Blennow 116–119 no. 25				
2:205	→ Blennow 228–229				
2:258	S. Gregorio al Monte Celio (lost)	11th c.	Cf. Senekovic, *KSRM* 3, 189 note 20.		
2:289	– (lost)	1513			
2:486	→ Blennow 76–84 no. 16				
2:487	S. Crisogono	1129	Cf. Favreau 253.	*MEC* 1 tab. 24 no. 5	
2:581	S. Simeone Profeta (lost)	1506			141
2:645	SS. XII Apostoli	14th c.			125, 141, 149, 151
2:646	–	early 13th c. (Claussen; 14th c. Forcella)	= Claussen, *Magistri*, 112–115; *KSRM* 1, 116	*Magistri*, fig. 121–122; *KSRM* 1, 116 fig. 74	

Forcella	Place	Date	Other publications	Photos	Referred to on p.
2:662	–	c.1474 (Claussen; 15th c. Forcella)	= Claussen, *KSRM* 1, 114	*KSRM* 1, 115–116 figs 73–74	125
2:664	– (lost)	1471–1484			
2:1035(1)	S. Maria in Trastevere	11th c.			
†2:1035(2)	–	1870			
2:1037	–	12th c.			
2:1054	–	1484			
2:1055	–	1484			
2:1056	S. Maria sopra Minerva	c.1460–1462[i] (15th c. Forcella)			
2:1057	→ 9:740; double publication not indicated by Forcella.				
2:1072	S. Maria in Trastevere (lost?)	16th c.			
2:1080	–	16th c.			118, 120, 141
2:1177	→ Blennow 127–134 no. 28				
2:1178	S. Eustachio (lost?)	12th c.			
2:1179	– (lost)	12th c.			
2:1180	– (lost)	12th c.	= Claussen, *KSRM* 1, 460		
2:1186	–	1336			142, 146, 148
2:1204	– (lost)	1503 or shortly after (1503 Forcella)			
2:1230	– (lost)	16th c.			
2:1333	→ Blennow 112–116 no. 24				
2:1334	S. Maria in Aquiro (lost)	1295			142
2:1484	→ 8:538; double publication not indicated by Forcella.				

Forcella	Place	Date	Other publications	Photos	Referred to on p.
2:1485	S. Stefano del Trullo (lost)	1479			
2:1490	S. Prassede (lost)	1073–1085 (prosopography; 11th c. Forcella)			
2:1491	– (lost)	1073–1085 (prosopography; 11th c. Forcella)			
2:1492	– (lost)	1124–1130			
2:1494	–	1259			143
2:1495	–	1286	Cf. Kajanto, *Dating*, 59 note 54.	*MEC* 1 tab. 28 no. 6	143, 146
2:1591	S. Crisogono (lost)	16th c.			
3:494	S. Giacomo degli Spagnoli	1450			
3:507	– (lost)	1486 (Petrella; 1480 Forcella)	Cf. Petrella 54 no. 40.		120, 141, 149
3:538	– (lost)	1503			143, 151
3:545	– (lost)	1506(?)			
3:721	S. Salvatore in Pensilis, 1285	1285			121, 141, 145, 148, 150, 153
3:722	–	1457			
3:1071	S. Maria dell'Anima (lost)	1514			
4:2	SS. Silvestro e Martino ai Monti (lost)	1259[j]	Cf. Petrella 49 no. 25.		141, 146
4:10	– (lost)	1479			

Forcella	Place	Date	Other publications	Photos	Referred to on p.
4:89	S. Silvestro al Quirinale (lost)	1518			120, 145, 149, 151
4:146	SS. Cosma e Damiano al Foro Romano	1153	= Claussen, *Magistri*, 17; *KSRM* 1, 370	*KSRM* 1, 371 fig. 299	109
4:224	S. Angelo in Pescheria (extant?)	1365			
†4:260	S. Nicola in Carcere	late[k] (Forcella 1127)			
4:261	→ Blennow 88–92 no. 18				
4:263	– (lost)	1277			
4:280	–	1515			120
†4:622	S. Girolamo della Carità	18th c. (cf. 4:621; Forcella 1477)			
4:623	–	1497[l]			
4:714	→ Blennow 96–100 no. 20				
4:742	→ Blennow 68–72 no. 14				
4:743	→ Blennow 65–67 no. 13				
4:744	S. Maria in Cosmedin	1123 (Giovenale; 12th c. Forcella)	= Giovenale 63 no. 7	Giovenale tab. 21 a	
4:746	–	1296–1300 (Claussen; 1295–1304 Giovenale; 12th c. Forcella)	= Claussen, *Magistri*, 213; Giovenale 64 no. 15	Giovenale tab. 17; *Magistri*, fig. 275	

Forcella	Place	Date	Other publications	Photos	Referred to on p.
4:747	–	1286 or earlier (Claussen; 1295–1304 Giovenale; 13th c. Forcella)	= Claussen, *Magistri*, 165–167; Giovenale 64 no. 14	Giovenale tab. 18 d	
4:748	–	1230	= Giovenale 64 no. 12		153
4:749	–	1284? (1286 Forcella; 1289 Giovenale)	= Giovenale 64 no. 13	Giovenale tab. 57 a	153
4:818	S. Marco (lost)	c.1153–1154 (Claussen; 1153 Forcella; 1154 De Rossi)	= Claussen, *Magistri*, 18		
4:821	–	1466	Cf. Petrella 53 no. 33.		114, 120, 141, 149
4:1114	S. Maria in Publicolis	15th c. 2/2 (prosopography; 15th c. Forcella)			150
4:1144	S. Francesco delle Stimate	1298			120, 145, 148
4:1244	S. Clemente	13th c.			
4:1245	–	1299	= Claussen, *KSRM* 1, 345	*KSRM* 1, 346 fig. 275	117
4:1265	S. Paolo alla Regola	1096		*MEC* 1 tab. 21 no. 5	122
4:1286	→ Claussen, *KSRM* 1, 135–136				
4:1287	→ Claussen, *KSRM* 1, 139–140 with notes 30–31				
4:1311[bis p. 545][m]	(→ 11:704; double publication not indicated by Forcella) → Blennow 48–53 no. 8.				
4:1312	→ 11:705; double publication not indicated by Forcella.				
4:1314	S. Pantaleo (lost)	1418			150

Forcella	Place	Date	Other publications	Photos	Referred to on p.
4:1316	– (lost)	1488			
5:223	S. Agostino (lost)	16th c.			137, 149
5:239	– (lost)	16th c.			
5:341	→ Blennow 43–48 no. 7				
5:342	→ Mondini, *KSRM* 3, 308 no. III				
5:343	→ Blennow 92–96 no. 19				
5:344	→ Blennow 134–139 no. 29				
5:345	S. Lorenzo in Lucina (lost)	1427			117, 121, 149, 151–152
5:347	–	1462			
5:433	S. Lorenzo ai Monti (lost)	1259			
5:434	– (lost)	13th c.			
5:435	– (lost)	13th c.			
5:436	– (lost)	1290			120
5:437	– (lost)	1318			
5:686	S. Pietro in Montorio (lost)	1500			
5:696	– (lost)	1523			
5:816	S. Onofrio	1449			
5:821	– (lost)	1501			
5:1015	S. Maria in Campitelli (lost)	1290–1300 (Claussen; 1290 Forcella)	= Claussen, *Magistri*, 214–216	*Magistri*, fig. 276 (drawing)	
5:1120	S. Lorenzo in Panisperna	1383	= Mondini, *KSRM* 3, 531 note 20	*KSRM* 3, 530 fig. 476	142
5:1123	– (extant?)	1508	= D'Achille 88 no. 17		
5:1296	S. Maria della Pace	1504			
5:1353	S. Maria in Monticelli	1143 (Silvagni; 1198 Forcella)	Cf. Blennow 5–6 note 5.	*MEC* 1 tab. 25 no. 2	117
5:1354	– (extant?)	1366			

Forcella	Place	Date	Other publications	Photos	Referred to on p.
6 p. 4	Vatican (lost)	1227–1241			
6 p. 5	– (lost)	1198–1216			
6:4	–	late(?)[n]			
6:5	–	1102 or later[o] (1102 Forcella)	Cf. Petrella 47 no. 23.	*MEC* 1 tab. 23 no. 1	141
6:6	– (lost)	1130–1143			
6:7	→ Kajanto, *Papal Epigraphy*, 60, (1)				
†6:8	–	late(?)[p]			
†6:9	–	late(?)[q]			
6:12[r]	–	1278	= Kajanto, *Papal Epigraphy*, 60, (2)		125, 152
6:14	– (lost)	1288			153
6:17	–	1300	Cf. Petrella 53 no. 30.		92, 132, 141, 149, 151
6:21	–	1327			119–120, 152
6:22	–	1341			
6:24	– (lost)	1344			148
6:25	– (copy, later)	1344			
6:26	– (lost)	1353			117, 145, 152
6:46	–	1431–1447	Cf. Kajanto, *Papal Epigraphy*, 62.		125
6:47	→ Kajanto, *Papal Epigraphy*, 60, (3)				
6:48	–	1438			
†6:50	–	late(?)[s]			
6:51	–	1447–1455			
6:53	–	1447–1455			125, 149
6:56	– (extant?)	1450			
6:57	–	1451			121, 125
6:58	– (lost)	1456			152
6:63	–	1460			

Forcella	Place	Date	Other publications	Photos	Referred to on p.
6:67[bis 40][t]	– (lost)	mid 15th c. (Ratti; 15th c. Forcella)			113
†6:69	–	late(?)[u]			
6:75	– (lost)	1472			125, 136
6:76	– (extant?)	1472			137
6:86[v]	SS. Cosma e Damiano dei Barbieri	1479			151
6:88	Vatican (lost)	15th c.			
6:90	–	1484			
6:92	–	1486			
6:97	– (lost)	1484–1492			
6:103	–	1492	= Kajanto, *Papal Epigraphy*, 88		152
6:105	– (lost)	1494			
6:106	–	1495			152
6:115	–	1492–1503			152
6:118	– (extant?)	c. mid 15th c. – 1506 (prosopography in Forcella's commentary; 16th c. Forcella)			
6:122	→ Kajanto, *Papal Epigraphy*, 95–96				
6:124	– (lost)	16th c.			125
6:125	– (lost)	16th c.			
6:126	– (lost)	16th c.			
6:129	–	1503–1513			
6:130	–	1503–1513			152
6:131	–	1503–1513			152
6:132	– (lost)	1508			

Forcella	Place	Date	Other publications	Photos	Referred to on p.
6:135	–	1511			152
6:136	–	1514			
6:137	–	1517			
6:138	–	16th c.			
6:139–144 & 149–157[w]	–	1513–1521 (15th/16th c. Forcella)			151
6:145	–	1513–1521 (16th c. Forcella)			
6:146	–	1513–1521 (16th c. Forcella)			
6:147	–	1513–1521 (16th c. Forcella)			
6:149–157	→ 6:139–144 & 149–157				
6:165	–	1524			
6:166	– (lost)	1525			
6:168	–	1526			
6:191	– (lost)	early 12th c. (Claussen; late 11th – early 12th c. De Rossi; 16th c. Forcella)	= Claussen, *Magistri*, 10		
6:967	→ Blennow 105–10 no. 22				
6:1060	S. Giacomo alla Lungara (lost)	1158 or earlier (Claussen; 12th c. Forcella)	= Claussen, *KSRM* 3, 12		135
6:1061	– (lost)	1277			146–147
6:1067	– (lost)	14th c.			
6:1072	S. Balbina	1489	= Claussen, *KSRM* 1, 121 note 3		

Forcella	Place	Date	Other publications	Photos	Referred to on p.
6:1093	S. Maria Traspontina	1495			123
6:1173	S. Spirito in Sassia	1471			120–121
6:1174	Ospedale di S. Spirito	1471–1484			
6:1175	–	1471–1484			
6:1183	S. Spirito in Sassia (lost)	1507			
6:1214	– (lost)	16th c.			
6:1647	Vatican	1500			
7:19	S. Giovanni dei Fiorentini	16th c.			
7:121	S. Giovanni Decollato (extant?)	1470–1499 ("14.." Forcella)			
7:171	SS. Salvatore in Lauro	1470			
7:237	SS. Venanzio ed Ansovino (extant?)	1400			122
7:245	–	16th c.			138, 146
7:327	S. Giovanni dei Genovesi	1481			
7:335	S. Tommaso di Cantorbery	1442			
7:395	S. Tommaso in Formis	13th c.	= Claussen, *Magistri*, 91–94	*Magistri*, figg. 108–109	
7:396	–	13th c.	= Claussen, *Magistri*, 92	*Magistri*, figg. 108–109	
7:397	– (lost)	13th c.			
7:398	– (lost)	1213[x]			141–142
7:590	→ Blennow 221–224				
7:591	→ Blennow 178–182 no. 38				
7:592	→ Blennow 199–202 no. 43				

Forcella	Place	Date	Other publications	Photos	Referred to on p.
7:622	S. Sabina	16th c.			
7:623	–	16th c.			
7:624	–	16th c.			118, 120, 151
†7:735	SS. Bonifacio ed Alessio	1456–1458?[y] (15th c. Forcella)			
7:788	S. Barbara dei Librai	12th c. (Silvagni; 11th c. Forcella)	Cf. Petrella 46 no. 21.	*MEC* 1 tab. 41 no. 2	
7:838	→ Blennow 194–198 no. 42				
7:860	S. Lucia del Gonfalone	1511			120, 134, 142, 146, 149
7:982	S. Stefano del Cacco (lost)	1162	= Claussen, *Magistri*, 58–59		
7:1091	→ Blennow 100–104 no. 21				
8:3	S. Giovanni in Laterano	11th c.			
8:4[z]	–	1072	Cf. Blennow 6 note 6.	*MEC* 1 tab. 20 no. 2	117
8:5	–	1114		*MEC* 1 tab. 22 no. 5	128
8:6	– (lost)	1177			
8:7	→ Claussen, *KSRM* 2, 318				
8:12	– (lost)	1287	Cf. Claussen, *Magistri*, 217 note 1243.		
8:13	–	c.1297 (Claussen; 13th c. Forcella)	= Claussen, *Magistri*, 216. – Cf. *KSRM* 2, 205.	*Magistri*, fig. 277; *KSRM* 2, 207 fig. 112 & 209 fig. 114	
8:14	→ Claussen, *KSRM* 2, 341				
8:15	→ Claussen, *KSRM* 2, 343–345				
8:16	→ Claussen, *KSRM* 2, 342–343				

Forcella	Place	Date	Other publications	Photos	Referred to on p.
8:17	– (lost)	1297	Cf. Claussen, *Magistri*, 216 note 1231.		145
8:18	– (lost)	1297			135, 151
8:25	–	1361	= D'Achille 74–75 no. 2		144
8:26	– (extant?)	1363			
8:27	– (lost)	1364	= Claussen, *KSRM* 2, 174 note 731, 3; D'Achille 74–75 no. 1		144
8:28	–	1365	= D'Achille 76–77 no. 3	D'Achille fig. 12	144
8:29	– (lost)	1365	= Claussen, *KSRM* 2, 174 note 731, 1		144
8:30(1)	–	1369	= Claussen, *KSRM* 2, 191–192 note 842(1)		119, 121
8:30(2)	–	1369	= Claussen, *KSRM* 2, 191–192 note 842(2)		119, 121
8:31	–	1372	Cf. Petrella 53 no. 32.		119, 139, 141, 151
8:42	–	1475	Cf. Petrella 54 no. 39.		120, 125, 134, 141, 146, 151
8:44	–	1492			
8:45	– (lost)	1492			
8:54	– (lost)	1525	= Claussen, *KSRM* 2, 65 note 220		
8:301	SS. Salvatore alla Scala Santa	1198–1216			
8:302	–	1198–1216			
8:303	–	1277–1280			

Forcella	Place	Date	Other publications	Photos	Referred to on p.
8:304	–	1277–1280 (Claussen; 13th c. Forcella)	= Claussen, *Magistri*, 208–210	*Magistri*, figs 270–272	
8:305	–	13th c.			
8:338	Ospedale del SS. Salvatore al Laterano	1348			119–120, 127, 130
8:339	–	1348			137
†8:341	–	(A later vernacular copy of 8:340.)			
†8:342	–	1397			119–120
8:343	– (lost)	1404			
†8:344	–	(A later vernacular copy of 8:343.)			
8:345	– (lost)	after 1439 (Kajanto; 1439 Forcella)			
†8:346	–	(A later vernacular copy of 8:345.)			
8:347(1)	–	1462 (Forcella's commentary; 15th c. Forcella)			137
8:347(2)	–	17th c., before 1657[aa] (Forcella 15th c.)			
8:348	– (lost)	1479			

Forcella	Place	Date	Other publications	Photos	Referred to on p.
8:349	– (lost)	1473–1483 (prosopography; 15th c. Forcella)			
8:350	– (lost)	1484			
8:352	– (lost)	1494			120, 141, 152
†8:353	–	(A later vernacular copy of 8:352.)			
8:354	– (lost)	1500			118, 132
†8:355	–	(A later vernacular copy of 8:354.)			
8:357	–	1518			146
8:358	– (lost)	1520			
†8:359	–	(A later vernacular copy of 8:358.)			
8:360	– (lost)	1523			
8:361	– (lost)	1525			132
†8:362(1)	–	(A later vernacular copy of 8:360.)			
†8:362(2)	–	(A later vernacular copy of 8:361.)			
†8:363	–	1528			138, 146
8:485	S. Maria in Domnica	1500			
8:486	–	1500			

Forcella	Place	Date	Other publications	Photos	Referred to on p.
8:498	S. Croce in Gerusalemme (lost)	15th–16th c.?[ab]			128, 151
8:499	→ Claussen, *KSRM* 1, 437				
8:500	– (lost)	1475			134
8:501	→ 13:1328; double publication not indicated by Forcella.				
8:502	– (lost)	1495			
8:504(1)	–	1521–1523			
8:504(2)	–	1521–1523			125, 141, 151
8:504(3)	–	1521–1523			127, 148–149
8:505	–	1520			127, 152
8:538	→ Kajanto, *Papal epigraphy*, 61, (1)				
8:539	S. Stefano Rotondo	1454			
8:717(1)	SS. Quattro Coronati	1111[ac]		*MEC* 1 tab. 22 no. 1; *KSRM* 1, 407 fig. 326	
8:717(2)	–	1111		*MEC* 1 tab. 22 no. 1	116, 141, 145
8:718	→ Blennow 165–70 no. 36				
8:736	SS. Quirico e Giulitta	1475	Cf. Kajanto, *Papal epigraphy*, 81.		
8:737	– (lost)	1475			
8:738	– (lost)	1475			133
8:779	→ Blennow 161–164 no. 35				
8:780	Ospedale della Consolazione (lost)	1472	Cf. Petrella 54 no. 36.		114, 120, 141, 149
8:783	S. Maria della Consolazione	1521			

Forcella	Place	Date	Other publications	Photos	Referred to on p.
8:785	Ospedale della Consolazione	1521	Cf. Petrella 54 no. 42.		120, 125, 141, 151
8:1001	S. Maria Maddalena (lost)	c.1517–1550ad (Forcella 16th c.)			
8:1156	→ Blennow 123–127 no. 27				
8:1192	S. Giovanni in Laterano	c.1230–1240? (Claussen; 11th c. Forcella)	= Claussen, *KSRM* 2, 262–265. – Cf. *Magistri*, 132 note 732.	*Magistri*, figs 146 & 150; *KSRM* 2, 263–264 figs 157–158	
8:1193	–	1195			
8:1203	S. Andrea della Valle (extant?)	1473			
8:1206	S. Maria della Consolazione	1517			146
8:1211	–	16th c.			
8:1230	–	16th c.			138
8:1231	–	16th c.			146
8:1232	–	16th c.			146
8:1233	–	16th c.			146
8:1234	–	16th c.			
9:149	S. Silvestro in Capite	1119		*MEC* 1 tab. 23 no. 5	116
9:244	Ospedale di S. Giacomo	1338			115, 119
9:245	– (extant?)	1435			137
9:531	SS. Salvatore al Ponte Senatorio (lost)	1475			

Forcella	Place	Date	Other publications	Photos	Referred to on p.
9:622	S. Brigida a Piazza Farnese	1391 (Forcella), or later[ae]			121
9:623	–	1391 (Forcella), or later			117
9:624	–	1391 (Forcella), or later			
9:636	S. Ambrogio della Massima (lost)	1411			
9:701	SS. Sergio e Bacco (lost)	16th c.			138
9:712	S. Giovanni della Malva (lost)	1512			
9:713	– (lost)	1518			
9:739	SS. Silvestro e Dorotea	1500			133–134, 153
9:740[af]	– (lost)	1500			134, 151, 153
9:806	→ Blennow 15–19 no. 1				
9:818[ag]	S. Biagio in Cantosecuto = della Pagnotta	1072	= Claussen, *KSRM* 1, 178, 1. – Cf. Blennow 6 note 6.	*MEC* 1 tab. 20 no. 1	117
9:843	S. Biagio della Fossa (lost)	1525			
9:969	S. Giovanni della Pigna (lost)	12th c.			
9:975	– (lost)	1501			
9:981	S. Giovanni in Ayno	16th c.			

Forcella	Place	Date	Other publications	Photos	Referred to on p.
9:999	S. Urbano a Campo Carleo	1263			131
9:1007	→ Blennow 59–62 no. 11				
10:1	S. Giovanni e Paolo (lost)	12th c.	= Mondini, *KSRM* 3, 76 with note 46	*KSRM* 3, 73 fig. 52	
10:2	–	1154–1159 (Kendall; 1157 Forcella; 12th c. Silvagni; mid 12th c. Petrucci)	= Kendall 267–268 no. 133; Mondini, *KSRM* 3, 90–91	*MEC* 1 tab. 41 no. 6; Petrucci tab. 8; *KSRM* 3, 89 figs 65–66	
10:3	– (lost)	1199–1216 (Mondini; 1157 Forcella)	= Claussen, *Magistri*, 95–96; Mondini, *KSRM* 3, 113–114	*KSRM* 3, 116 fig. 95	
10:4	→ Blennow 110–112 no. 23				
10:5	→ Blennow 182–173 no. 39				
10:7	– (lost)	1522			
10:72	S. Anastasia (lost)	1210			
10:152	S. Benedetto in Piscinula (lost)	1412			
10:360	S. Martinello al Monte di Pietà (lost)	1500			
10:362	– (lost)	1521			152
10:364	S. Nicola dei Prefetti (lost)	1250	= Claussen, *Magistri*, 173–174		117
10:448	S. Teodoro	1451			

Forcella	Place	Date	Other publications	Photos	Referred to on p.
10:540 + Guillou no. 117ah	SS. Cosma e Damiano in Trastevere	1065 (Guillou; 1066 Silvagni; 1069 Forcella)	= Claussen, *KSRM* 1, 348. – Cf. Blennow 5–6 note 6.	*MEC* 1 tab. 19 no. 5; *KSRM* 1, 348 fig. 276	117, 142
10:542	– (lost)	1475			134, 141–142, 149
10:543	–	1475			
10:564	S. Agata alla Suburra (lost)	1024			
10:565	→ Blennow 203–204: not an inscription.				
10:568	– (extant?)	1504			
10:569	– (lost)	1517			
10:570	– (lost)	1523–1534			135, 149
10:640	→ Blennow 154–157 no. 33				
10:727	→ Blennow 204–210				
10:728	S. Matteo in Merulana (lost)	1524–1530 (prosopography; 1517 Forcella)			
10:867	S. Maria in Campo Marzo (lost)	1505			121, 127, 152
10:868	–	1520			
10:870	– (lost)	1521			
10:901	S. Sisto (lost)	1222			148
10:902	→ 12:177; double publication not indicated by Forcella.				
10:904	–	14th c.			
10:905	–	1478			
11:1	S. Maria Maggiore	1150		*MEC* 1 tab. 25 no. 4	
11:4	– (extant?)	1259			151
11:5	– (lost)	1289			
11:7	–	1295			
11:8	–	1295			

Forcella	Place	Date	Other publications	Photos	Referred to on p.
11:18	–	1325	Cf. Petrella 53 no. 31.		146
11:44	–	1464	Cf. Petrella 53 no. 34.		114, 120, 141, 149, 151
11:228	S. Bibiana	13th c.	Cf. Claussen, *KSRM* 1, 179 note 4.		148–149, 150, 152
11:229	→ Blennow 220–221				
11:230	– (lost)	13th c.	= Claussen, *KSRM* 1, 180 note 8		
11:246	→ Claussen, *KSRM* 1, 83				
11:248	S. Antonio Abbate (lost)	1481			
11:262	S. Pudenziana	1073–1085 (1073 Forcella)	Cf. Blennow 5–6 note 5.	*MEC* 1 tab. 21 no. 3	117, 129, 139, 141
11:285	SS. Vito e Modesto	1477			
11:297	→ Blennow 120–123 no. 26				
11:298	S. Giovanni(?)	1509			
11:311	S. Prisca (lost)	1456			
11:312	–	1455–1458 (prosopography; 15th c. Forcella)			135, 152
11:313	– (lost)	1513–1521			135, 149, 152
11:321	SS. Salvatore in Primicerio	1113	Cf. Blennow 5–6 note 5.	*MEC* 1 tab. 23 no. 4	117, 142
11:356	S. Vitale	1475			
11:357	– (lost)	1512			
11:374	→ Senekovic, *KSRM* 3, 548 no. 1				

Forcella	Place	Date	Other publications	Photos	Referred to on p.
11:375	S. Lucia della Tinta	1290 (Senekovic; 1279 Petrella; 1278 Forcella)	Cf. Petrella 52 no. 29; Senekovic, *KSRM* 3, 544 with note 10.		141, 145, 148, 152
11:449	S. Eligio dei Ferrari	1494			
11:542	→ Blennow 184–179 no. 40				
11:543	S. Agnese fuori le Mura	1479			
11:544	– (lost)	1387–1388[ai] (1487 Forcella)			119–120
11:545	–	1490			
11:572	S. Pancrazio (lost)	1249			
†11:573	–	1249	Too fragmentary.		
11:596	→ Claussen, *KSRM* 3, 27–29				
11:597	→ Claussen, *KSRM* 3, 16 note 16				
11:609	→ Blennow 226–228				
11:611	SS. Marcellino e Pietro (lost)	16th c.			
11:704[aj]	→ Blennow 48–53 no. 8				
11:705[ak]	S. Maria del Buon Consiglio	1201			
11:723	S. Lorenzolo (lost)	1517	= D'Achille 72–73 note 12		
11:766	→ Blennow 35–39 no. 5				
11:767	→ Blennow 53–54 no. 9				
12:2	S. Paolo fuori le Mura	1030		*MEC* 1 tab. 18 no. 3	

Forcella	Place	Date	Other publications	Photos	Referred to on p.
12:4	–	1063		*MEC* 1 tab. 20 no. 3; Filippi tab. 96	
12:5	–	1070		*MEC* 1 tab. 34 no. 6; Filippi tab. 133	
12:6	–	1070	Cf. Filippi 41.		137
12:11	–	1226	Cf. Filippi 42.		
12:12	–	1235	Cf. Filippi 42.		
12:13	– (lost)	1329	Cf. Filippi 69.		120
12:17	– (lost)	1494	Cf. Filippi 69.		
12:141	S. Maria Porta Paradisi	1522			
12:176	S. Sabba	1205 (Claussen; 1204 Forcella)	= Claussen, *Magistri*, 75–76	*MEC* 1 tab. 28 no. 1; *Magistri*, fig. 90	
12:177[al]	–	1320			117
12:178	–	1463			
12:186	S. Sebastiano fuori le Mura	1466			
12:245	S. Andrea a Ponte Molle	1462			141, 149
†12 p. 311	S. Maria del Sole	late 17th c. or later?[am]			
12:442	–	1475			126, 133
†12:444	SS. Vincenzo ed Anastasio alle tre Fontane	18th c.[an]			108, 152
12:445	–	1221 (Silvagni; 1244 Forcella)	Cf. Blennow 5–6 note 5.	*MEC* 1 tab. 35 no. 4	141–142
12:456	SS. Salvatore a Porta Paolo (lost)	1450			

Forcella	Place	Date	Other publications	Photos	Referred to on p.
12:457	–	14th c. (Forcella; 1218 Petrella)	Cf. Petrella 50 no. 26.		146–147
†12:505	S. Sebastiano al Palatino	17th c.?[ao]			
†12:506	–	17th c.?[ap]			
12:539	→ Blennow 72–76 no. 15				
12 p. 508	→ Mondini, *KSRM* 3, 348–349				
12:565	S. Lorenzo fuori le Mura	1148	= Claussen, *Magistri*, 16; Mondini, *KSRM* 3, 411	*MEC* 1 tab. 26 no. 1; *KSRM* 3, 411 fig. 358	
12:570	– (extant?)	14th c.			
12:575	– (lost)	1492	= Mondini, *KSRM* 3, 351 note 135		
13:1	S. Maria in Domnica	1157		*MEC* 1 tab. 25 no. 5	
13:2	→ 6:12 (and 1:2); triple publication not indicated by Forcella.				
13:3	Porta S. Sebastiano	1327			152
13:4	Walls	1447–1455 (prosopography; Forcella 15th c.)			
13:5	Porta S. Lorenzo	1453			
13:6	Walls	1458			
13:7	–	1461–1474 (prosopography; Forcella 15th c.)			152
13:8	–	1461–1474 (prosopography; 15th c. Forcella)			152

Forcella	Place	Date	Other publications	Photos	Referred to on p.
13:9	Porta Settimana (lost)	1492	= Kajanto, *Papal Epigraphy*, 88–89		
13:10	Walls	1492			
13:11	– (lost)	1512	= Kajanto, *Papal Epigraphy*, 94[aq]		
13:12	–	1512	= Kajanto, *Papal Epigraphy*, 94 note 13 (1)[ar]		
13:13	–	1512	= Kajanto, *Papal Epigraphy*, 94 note 13 (2)[as]		
13:14	–	1512	= Kajanto, *Papal Epigraphy*, 94 note 13, (3)		
13:89	Ponte Cestio	1193		*MEC* 1 tab. 40 no. 1	137
13:90	Ponte S. Angelo	1447–1455 (prosopography; Forcella 15th c.)			
13:91	Ponte Milvio	1458			
13:92	→ Kajanto, *Papal Epigraphy*, 82 no. 2				
13:93	→ Kajanto, *Papal Epigraphy*, 82 no. 3				
13:110	→ Kajanto, *Papal Epigraphy*, 83 with note 63				
13:111	Campo de' Fiori	1483	= Kajanto, *Papal Epigraphy*, 83, (2)	Kajanto fig. 11	
13:112	→ Kajanto, *Papal Epigraphy*, 86–87				
13:113	→ Kajanto, *Papal Epigraphy*, 95, (1)				
13:114	→ Kajanto, *Papal Epigraphy*, 93				
13:121	Via Celsa	16th c.			

Forcella	Place	Date	Other publications	Photos	Referred to on p.
13:141	→ Kajanto, *Papal Epigraphy*, 60–61				
13:142	Piazza di S. Maria in Trastevere (lost)	1492–1503 (prosopography; Forcella 15th c.)	Cf. Kajanto, *Papal Epigraphy*, 95.		
13:143	– (lost)	1509			
13:144	S. Giovanni dei Fiorentini (lost)	16th c. (1509 Forcella)			
13:145	– (lost)	16th c.			
13:247	Castel Sant'Angelo	1493	Cf. Kajanto, *Papal Epigraphy*, 88 note 5.		
13:248	–	1492–1503 (prosopography; Forcella 15th c.)	Cf. Kajanto, *Papal Epigraphy*, 88 note 5.		152
13:249	–	1495	Cf. Kajanto, *Papal Epigraphy*, 88 note 5.		152
13:250[at]	–	1495	Cf. Kajanto, *Papal Epigraphy*, 88 note 5.		
13:251	–	1495	Cf. Kajanto, *Papal Epigraphy*, 88 note 5.		
13:325	Arch of Constantine (lost)	1107			
13:326	Collegio Capracina	1451			
13:327	Palazzo Nardini = del Governo Vecchio	1475			

Forcella	Place	Date	Other publications	Photos	Referred to on p.
13:328	–	1480[au] (Forcella 1475)		Wikimedia Commons[av]	
13:330	Via della Lungarina	1495			
13:331	Vatican	1497			
13:345	Piazza della Suburra	16th c.			
13:422	S. Maria Traspontina (lost)	1230			
13:423	– (lost)	1277			122
13:424(1)	Via del Banco di S. Spirito	1277	[aw]		122–123
†13:424(2)	–	1640	[ax]		
13:425	S. Maria sopra Minerva (lost)	1379			
13:426	–	1422			144
13:427	– (lost)	1476			123, 152
13:428	S. Giacomo degli Spagnoli (lost)	1495			141, 152
13:429	→ 13:250; double publication not indicated by Forcella.				
13:430	Via del Paradiso	1495			123
13:431	Ospedale di S. Spirito (lost)	1495			125, 144
13:432	S. Eustachio	1495			141
13:433	S. Maria sopra Minerva	1495			141

Forcella	Place	Date	Other publications	Photos	Referred to on p.
13:434	Palazzo Caetani (lost)	1495			123
13:435	Via della Sapienza 41 (extant?)	1495			
13:436	Palazzo Caetani (lost)	1514			117, 123, 140, 142, 150
13:520	S. Maria in Campo Santo	1501			
13:586	SS. Cosma e Damiano dei Barbieri (lost)	1344			120
13:587	→ 6:86; double publication not indicated by Forcella.				
13:588	S. Lorenzo in Damaso (lost)	1365			153
13:616	– (lost)	1535–1589 (prosopography; Forcella 16th c.)			
13:786	→ Blennow 19–26 no. 2				
†13:809	S. Girolamo della Carità	18th c. (cf. 4:621; Forcella 1087)			
†13:810	–	18th c. (cf. 4:621; Forcella 1216)			
†13:811	–	18th c. (cf. 4:621; Forcella 1247)			
†13:812	–	18th c. (cf. 4:621; Forcella 1313)			

Forcella	Place	Date	Other publications	Photos	Referred to on p.
†13:813	–	18th c. (cf. 4:621; Forcella 1340)			
†13:814	–	18th c. (cf. 4:621; Forcella 1413)			
†13:815	–	18th c. (cf. 4:621; Forcella 1468)			
13:853	– (lost)	16th c.			149
13:903	S. Maria sopra Minerva (lost)	1468			
13:905	– (lost)	15th c.			
13:1029	S. Clemente (lost)	mid 15th c. (prosopography; 15th c. Forcella)			112
13:1042[ay]	S. Maria del Popolo	1477			
13:1138	S. Maria in Publicolis (lost)	16th c.			
13:1142	S. Pietro in Montorio (extant?)	1500			120, 134, 142
13:1210	SS. Bonifacio ed Alessio	13th c.		*Magistri*, fig. 86; *KSRM* 1, 221 fig. 156	110
13:1279	S. Maria del Popolo (lost)	1500			
13:1328[az]	S. Margherita	1476			
13:1335	Torre dei Conti	13th c. (Silvagni; 1280 Forcella)		*MEC* 1 tab. 40 no. 5	

Forcella	Place	Date	Other publications	Photos	Referred to on p.
13:1336	Via del Pellegrino / Piazza della Cancelleria	1497			
13:1339	Piazza della Bocca della Verità	12th c. (Forcella; late 11th c. Silvagni)		*MEC* 1 tab. 40 no. 2	
13:1340	–	12th c. (Forcella; late 11th c. Silvagni)		*MEC* 1 tab. 40 no. 3	
13:1341	–	12th c.			

a In the Capitoline Museums, courtyard of Palazzo dei Conservatori, between the fragments of Constantine I's colossal statue. Forcella's edition is based on earlier scholars, and the reading is defective: L. 1 *n(ost)ri* and *p(a)p(e)* have been omitted, many abbreviations have not been indicated, and the text has been divided incorrectly into four lines instead of two. See my complete reading (based on examination 13 March 2018) in Jasmin Lukkari's article p. 92 note 140. Note that in L. 1 "*alme urb(is)*" V has been written over E.

b Measures. Counted as one inscription here.

c The transcription lacks abbreviations (and would thus give the impression of a later date than 1425), but this probably derives from the manuscript consulted by Forcella; cf. the surviving 1:13.

d Cf. 1:11.

e The support being unknown, this may or may not be an epitaph (or another commemoration for a deceased person). Cf. 1:1226.

f According to Forcella's commentary, by Pinturicchio; Forcella undated 15th century.

g Compare to 1:1201; obviously the same inscription.

h Lost to Forcella, but found anew in 1981; cf. Claussen.

i Transported to S. Maria in Trastevere in the 18th century; date and origin: Francesco Caglioti, "Altari eucaristici scolpiti dal Primo Rinascimento: qualche caso maggiore" in Jörg Stabenow, ed., *Lo spazio e il culto: relazioni tra edificio ecclesiale e uso liturgico dal XV al XVI secolo* (Venezia: Marsilio, 2006), 54 with note 2.

j Compared to two surviving testamentary donations by the same cardinal Petrus Capoccinus, 2:1494 and 11:597, this lost one raises some doubt about the recorded year date: in this, the formula is simply *actum est anno Domini MCCLIX*, whereas the others have a more extensive formula including the papal year. In any case, the abbreviations (or the lack thereof) in the transcription probably do not correspond with the original inscription.

k According to Forcella, the inscription is "[*d*]*ipinta a lettere moderne sopra la porta principale nell'interno della chiesa.*" Based on his transcription, every stylistic aspect seems to point to a post-medieval date. Therefore the inscription ought to be either a heavily modified version of an earlier inscription, or a later commemoration of a centuries-old restoration.

l In contrast to 4:621–622 and 13:809–815, Forcella does not indicate here that the letters were from the 18th century; moreover, the content does not contradict the possibility that this inscription was contemporary.

m Erroneous double numbering: dissimilar to 4:1311 on p. 540, an 1870 inscription in S. Bartolomeo all'Isola.

n Historical depiction which seems to date back to at least the Renaissance, according to the layout as presented by Forcella. His date refers to the depicted event.

o The inscription seems to be from the 12th century, but in the year dating LL. 1–3 "*anno ab i*(*n*) *carnatio*(*n*)*e D*(*omi*)*ni n*(*ost*)*ri* | (*Ies*)*u* (*Christ*)*i MCII,* (...), *indic*(*tione*) *X, temp*(*or*)*e d*(*omi*)*ni G*(*re*)*g*(*orii*) *VII* | *p*(*a*)*p*(*ae*)*,*" the AD year and indiction are compatible with each other, but Gregory VII had already deceased in 1085.

p Later historical depiction; see 6:4. A later date is strongly suggested by the title *pont. max.*

q Cf. 6:4.

r = 6:12 as indicated by Forcella; however, the triple publication 13:2 is not indicated.

s Cf. 6:4.

t Erroneous double numbering: dissimilar to 6:67 on p. 39, a 1470 epitaph.

u Cf. 6:4.

v = 13:587; double publication not indicated by Forcella.

w Historical depictions which Forcella leaves undated, but the reign of Leo X (1513–1521) is suggested by the placing of the lemmata. Counted as one inscription (historical narration with a contextual temporal reference).

x I wonder if the inscription is from a later date, but as it is lost, this is difficult to assess.

y According to Michaël George, *Le chapitre cathédral de Verdun* (*fin XIIe – debut XVIe siècle*)*: étude d'une communaute eccléstiastique séculière*, (diss. Université de Lorraine, 2016) <https://tel.archives-ouvertes.fr/tel-01752195v2/file/THESE-M.GEORGE_Volume-2.pdf>, 131–132 no. 248a, the inscription is a funerary one. As it is a pavement slab, this may hold true, but still I am not entirely convinced of this, since the contents are atypical for an epitaph, at least in the Roman context. Nevertheless, I have excluded this inscription from my discussion.

z Nearly identical with 9:818 (cf. commentary below).

aa According to Forcella, the previous inscription was relocated in the time of Francesco Gualdi († 1657).

ab Forcella, referring to Lorenz Schrader (*Monumentorum Italiae, Quae hoc nostro sæculo & à Christianis posita sunt* (Helmaestadii: Typis Iacobi Lucij Transyluani, 1592), vol. 2 128), tells the inscription was formerly "*in un marmo incastrato nella parete destra presso l'ingresso della cappella sotterranea di S. Elena,*" but "[*s*]*crive il* [*Raimondo*] *Besozzi* (*La storia della Basilica di Santa Croce in Gerusalemme* [*1750*] *ecc. 61*) *che fu tolta via perchè riconosciuta falsa la storia in questa narrata.*" Because of Schrader's testimony, the inscription must predate 1592, but, in the Forcellan transcription, the style again seems to be of a Renaissance date (e.g. the lack of medieval abbreviations, the spelling *foeminarum*). Thus the date perhaps lies somewhere between the 1430's and 1592, presuming the transcription is accurate enough; otherwise the inscription could date back to an even earlier date.

ac This is either a remake of an earlier inscription from the time of Leo IV (847–855), or was made to look like one.

ad Date based on prosopographic considerations: Juan de Sepúlveda (apparently dissimilar to the famous Juan Ginés de Sepúlveda) became the titular bishop of Corinth in 1517 (Raymond Janin, *Dictionnaire d'histoire et de géographie ecclésiastiques* 13 (Paris: Librairie Letouzey et Ané, 1956), col. 879), and, accordingly, the inscription may post-date this with only few decades.

ae The date 1391 (canonisation of St Bridget) is mentioned in the text, but only excludes an earlier date for the inscription itself. However, even if the inscription has been implemented later, it must have not occurred much later, because it belongs to the medieval style (Forcella: "*scritta a caratteri gotiche*"). The same applies to the two following Bridget inscriptions, 9:623 and 624, which do not mention the year 1391 (9:623 has a poetic expression for 1373, the year when the saint passed away).

af = 2:1057; double publication not indicated by Forcella.

ag Nearly identical with 8:4. The reason for the duplication is enigmatic; a possible scenario is presented by Claussen, *KSRM* 2, 324–325 (cf. *KSRM* 1, 178 note 8)

ah Guillou's reading for L. 14.

ai Forcella's date 1487 obviously contradicts a date in Urban VI's reign, given in the inscription. Either Forcella's typography or his sources are erroneous: the date on L. 5 "MCCCCLXXXVII" has one superfluous C; note that on L. 6 there is also "*tempobe*" (sic). The year 1387 should be a 10th indiction year: thus, either the indiction or the number of the year may have a further error of one year.

aj = 4:1311[bis 545], not indicated by Forcella.

ak = 4:1312; double publication not indicated by Forcella.

al = 10:902; double publication not indicated by Forcella.

am Forcella states that "*[n]el XV secolo poi sotto il pontificato di Sisto IV, ossia nel giubileo del 1475, fu risarcito e dedicato alla Madre di Dio sotto l'invocazione del Sole,*" which would suggest this inscription was from the period, but according to Hülsen (p. 484) "*quello [nome] di S. Maria del Sole non occore prima della fine del sec. XVII.*" Forcella's statement is also contradictory to 12:442, recording the Sixtine restoration of *aedes beati Stephani protomartyris.*

an On the epistyle of the front porticus. Forcella gives the date 1140, found in the inscription, but mentions that the letters are from the previous century. The inscription is a modern production in every other respect, too.

ao Forcella's imprecise 11th century date must be considerably too early: both the content and the style of the inscription in general strongly imply a date in the Renaissance period at the very least. As the church was rebuilt in the 17th century (Forcella vol. 12 p. 413) and as there are no other inscriptions predating 1675 with certainty, a 17th century date is to be suspected. Cf. 12:506, and the 1675 inscription 12:507.

ap The inscription commemorates the election of Pope Gelasius II, who fled from Rome after his election in 1118 and died elsewhere; in the meantime, Emperor Henry V placed an antipope in the office. Such a narration alone means that the inscription must date later than the events, and the style of the inscription connects it to a clearly later date; cf. 12:506.

aq Kajanto thinks of this as "not (...) as clear a *titulus honorarius* as the one above [p. 93 = 13:114]," while in my opinion Iulius II's name should be read as an ablative absolute.

ar Cf. comment on 13:11.

as Cf. comment on 13:11.

at = 13:429; not indicated by Forcella.

au Forcella's reading is in some parts erroneous, including the date. I read from the photo: "above: *(Iesu)s.* || below: *Salvatoris Laterani piae | hospitalitati et mans|urae bonarum artium | achademiae Ste(phanus) Nardi|₅nus car(dinalis) Mediol(anus) has ae|des suo aere positas vivens* | *dono dedit an‹n›o salutis* MCCCCLXXX." (Forcella: supra 1 omit.; infra 1 LATERAN, 4 ACCADEMIAE, 6 MCCCCLXX^v^.).

av https://upload.wikimedia.org/wikipedia/commons/3/38/Parione_-_Stefano_Nardini_-_dona_alla_compagnia_del_Salvatore_le_case_al_Governo_Vecchio_1480_03250030.JPG (accessed 22 July 2020).

aw Forcella indicates the horizontal line incorrectly between LL. 6–7 instead of 5–6; cf. also (2).

ax Judging from Forcella's poor transcription of this later inscription on the upper margin of the stone, either he has not seen the stone himself, or it was less visible to him. The text is fully legible (inspection 1 April 2017): "*1640 Gioan Lupo.*"

ay 1:1179, a double publication not indicated by Forcella, which is presented as if based on an inspection. But 13:1042, based on the 18th century scholar Pierluigi Galletti, has the correct reading (1477, not 1411, confirmed by my examination, 1 May 2017). The text is inscribed on the architraves of both side entrances. However, Forcella only records the one on the right-hand side in both instances: at 1:1179 he claims the right-hand text reads over both entrances, and at 13:1042 he only refers to one text over one entrance. The complete reading is as follows: left: "*Six(t)us p(a)p(a) IIII fundavit 1477.*" right: "*Sixtus p(a)p(a) IIII pont(ifex) max(imus) 1477.*"

az = 8:501; double publication not indicated by Forcella.

APPENDIX 2

Other Publications regarding Inscriptions

For bibliographical information, please refer to Appendix 1.

As stated in the introduction to my article, I am well aware that the following catalogue does not include everything that has in fact been published – and the ones marked as "unpublished" may have been discussed in publications not seen by me thus far. In accordance with the purposes of Appendix 1, this should rather be seen as a possible stepping stone for more comprehensive studies in the future.

Publication	Place	Date	Other publications	Photos	Referred to on p.
Blennow 15–19 no. 1	SS. Salvatore ai Monti	13th c. 1/2?, original from 1046 (Blennow; 1046 Forcella)	= 9:806.	*MEC* 1 tab. 21 no. 1; Blennow fig. 1	131
Blennow 19–26 no. 2	S. Maria in Campitelli	1073	= 13:786.	Blennow figg. 2–4	115
Blennow 27–32 no. 3	S. Cecilia in Trastevere	1073		Blennow fig. 5	
Blennow 31–2	–	16th c.		Blennow fig. 6	
Blennow 33–35 no. 4	S. Cecilia in Trastevere	1080	= 2:56.	Blennow fig. 7	115
Blennow 35–39 no. 5	S. Maria in Cuppella (lost)	1090	= 11:766		115
Blennow 39–42 no. 6	S. Gregorio al Monte Celio (lost)	1107			
Blennow 41	– (lost)	14th–16th c.?			120, 131, 148

 | DOI:10.1163/9789004436251_011

Publication	Place	Date	Other publications	Photos	Referred to on p.
Blennow 43–48 no. 7	S. Lorenzo in Lucina	1112	= 5:341; Mondini, *KSRM* 3, 307 no. I. – Cf. Mondini 279 note 70.	*MEC* 1 tab. 22 no. 2; Blennow fig. 9; *KSRM* 3, 279 fig. 230	115, 141
Blennow 46 note 95	S. Lorenzo in Lucina	1112	Cf. Mondini, *KSRM* 3, 289–293.	*MEC* 1 tab. 23 no. 2; *KSRM* 3, 291 fig. 238	115, 141
Blennow 48–53 no. 8	S. Maria del Buon Consiglio	1113	= 11:704 = 4:1311[bis 545].	*MEC* 1 tab. 22 no. 4; Blennow fig. 10	109, 141
Blennow 53–54 no. 9	S. Maria in Cuppella (lost)	1113	= 11:767.		
Blennow 55–58 no. 10	S. Tommaso ai Cenci	1114		Blennow fig. 11	115
Blennow 59–62 no. 11	S. Agnese in Piazza Navona (lost)	1123	= 9:1007.		148
Blennow 62–64 no. 12	S. Spirito in Sassia (lost)	1123			114, 141
Blennow 65–67 no. 13	S. Maria in Cosmedin	1123	= 4:743; Giovenale 63 no. 9	*MEC* 1 tab. 24 no. 2; Giovenale tab. 20 b; Blennow fig. 12	
Blennow 68–72 no. 14	–	1123	= 4:742; Giovenale 63 no. 10	*MEC* 1 tab. 25 no. 1; Giovenale tab. 20 a; Blennow fig. 13	
Blennow 71 with note 143	S. Maria in Cosmedin (lost)	11th–16th c.?			
Blennow 72–76 no. 15	S. Cecilia a Monte Giordano (lost)	1123	= 12:539.		115

Publication	Place	Date	Other publications	Photos	Referred to on p.
Blennow 76–84 no. 16	S. Crisogono	1123	= 2:486. – Cf. Favreau 253	*MEC* 1 tab. 24 no. 1; Blennow fig. 14	125, 141
Blennow 84–88 no. 17	S. Crisogono	1127	Cf. Favreau 253.	*MEC* 1 tab. 24 no. 3; Blennow fig. 15; Favreau 254 fig. 51	115
Blennow 88–92 no. 18	S. Nicola in Carcere	1128	= 4:261.	*MEC* 1 tab. 24 no. 4; Blennow fig. 16	
Blennow 92–96 no. 19	S. Lorenzo in Lucina	1130	= 5:343; Mondini, *KSRM* 3, 307–308 no. II. – Cf. Mondini 268 note 28 & 280 note 75.	*MEC* 1 tab. 23 no. 6; Blennow fig. 17; *KSRM* 3, 281 fig. 231	
Blennow 96–100 no. 20	S. Nicola ai Cesarini (lost)	1132	= 4:714.		
Blennow 99	→ Kajanto, Nyberg, and Steinby no. 104				
Blennow 100–104 no. 21	S. Tommaso in Parione	1139	= 7:1091.	*MEC* 1 tab. 24 no. 6; Blennow fig. 18	143
Blennow 105–10 no. 22	SS. Michele Arcangelo e Magno	1141 (Silvagni, Blennow; 12th c. Forcella)	= 6:967.	*MEC* 1 tab. 35 no. 2; Blennow fig. 19	125
Blennow 110–112 no. 23	S. Giovanni e Paolo (lost)	1157	= 10:4. – Cf. Mondini, *KSRM* 3, 76–77		141
Blennow 112–116 no. 24	S. Maria in Aquiro (lost)	1179	= 2:1333.		115, 145, 148
Blennow 116–119 no. 25	S. Nicola dei Funari	1180	= 2:193.	Blennow fig. 20	148

Publication	Place	Date	Other publications	Photos	Referred to on p.
Blennow 120–123 no. 26	S. Giovanni a Porta Latina	1190 or later? (Blennow; 1190 Forcella)	= 11:297. – Cf. Claussen, *KSRM* 3, 136–139	*MEC* 1 tab. 27 no. 3; *KSRM* 3, 137 fig. 108; Blennow fig. 21	138, 148
Blennow 123–127 no. 27	SS. Salvatore delle Coppelle	1195 (Blennow; 1156 Forcella)	= 8:1156.	*MEC* 1 tab. 27 no. 4; Blennow fig. 22	
Blennow 127–134 no. 28	S. Eustachio	1196	= 2:1177. – Cf. Claussen, *KSRM* 1, 454	*MEC* 1 tab. 26 no. 3; Blennow fig. 23; *KSRM* 1, 454 fig. 363	115, 120, 145, 148
Blennow 134–139 no. 29	S. Lorenzo in Lucina	1196	= 5:344; Mondini, *KSRM* 3, 309 no. IV. – Cf. Mondini 268 note 29 & 280 note 81.	*MEC* 1 tab. 26 no. 4; Blennow fig. 24; *KSRM* 3, 282 fig. 233	115
Blennow 138	S. Lorenzo in Lucina (extant?)	1196			
Blennow 139–141 no. 30	S. Maria in Campitelli (lost)	1217			115, 125
Blennow 141–145 no. 31	S. Sebastiano fuori le Mura (lost)	1218			
Blennow 145–154 no. 32	S. Adriano al Foro Romano (lost)	1228	= 2:139. – Cf. Claussen, *KSRM* 1, 36 note 63		142, 152

Publication	Place	Date	Other publications	Photos	Referred to on p.
Blennow 154–157 no. 33	S. Eusebio	1238	= 10:640. – Cf. Claussen, *KSRM* 1, 445 note 3	Blennow fig. 25; *KSRM* 1, 445 fig. 357	148
Blennow 157–161 no. 34	S. Tommaso ai Cenci	1240		Blennow fig. 26	148
Blennow 161–164 no. 35	S. Maria della Consolazione (lost)	1241	= 8:779.		125, 148
Blennow 165–70 no. 36	SS. Quattro Coronati	1246	= 8:718.	Blennow fig. 27–28	148
Blennow 171–178 no. 37	S. Adriano al Foro Romano	1248	= 2:141. – Cf. Claussen, *KSRM* 1, 36 note 65	*MEC* 1 tab. 28 no. 4; Blennow figg. 29–32	141, 145, 148
Blennow 178–182 no. 38	S. Sabina	1248	= 7:591. – Cf. Kajanto, *Dating*, 58 note 53	*MEC* 1 tab. 28 no. 5; Kajanto, *Classical and Christian*, fig. 4; Blennow fig. 33	145, 148
Blennow 182–173 no. 39	S. Giovanni e Paolo (lost)	1256	= 10:5. – Cf. Mondini, *KSRM* 3, 77		148
Blennow 184–179 no. 40	S. Agnese fuori le Mura	1256	= 11:542. – Cf. Claussen, *KSRM* 1, 54–55 note 15 & 065	Blennow fig. 34	146, 148
Blennow 190–194 no. 41(1)	S. Costanza	1256		Blennow fig. 35	120, 146, 148
Blennow 190–194 no. 41(2)	–	1256		Blennow fig. 35	148

Publication	Place	Date	Other publications	Photos	Referred to on p.
Blennow 194–198 no. 42	SS. Martina e Luca	1256	= 7:838.	Blennow fig. 36	115, 148
Blennow 198(1)	SS. Martina e Luca (lost)	probably 1250–1260			
Blennow 198(2)	– (lost)	probably 1250–1260			
Blennow 199–202 no. 43	S. Sabina	1263	= 7:592. – Cf. Kajanto, *Dating*, 58 note 53	*MEC* 1 tab. 28 no. 7; Blennow fig. 37	148
†Blennow 203–204[a]	S. Agata alla Suburra	1024	= 10:565.		
Blennow 204–210	S. Matteo in Merulana (lost)	14th–16th c.?[b]	= 10:727.		131, 139, 146
Blennow 210–215(1)	S. Annunziata all'arco dei Pantani (copy, 1518)	1220/1?[c]		Blennow fig. 38	148
Blennow 210–215(2)	– (copy, 1518)	1394, cf. (1)		Blennow fig. 38	119, 121, 141, 148
Blennow 210–215(3)	–	1518, cf. (1)		Blennow fig. 38	145, 148
Blennow 215–217(1)	S. Maria ad Martyres = Pantheon (copy, 1588)[d]	1209 (Blennow; 1588 Forcella)	= 1:1132.	Blennow fig. 39	
†Blennow 215–217(2)	–	1588[e]	= 1:1132.	Blennow fig. 39	
Blennow 217–220	S. Bibiana (lost)	unknown, later than 1229 (Blennow)			

Publication	Place	Date	Other publications	Photos	Referred to on p.
Blennow 220–221	– (lost)	after 1224, perhaps 15th c.?[f]	= 11:229. – Cf. Claussen, *KSRM* 1, 179 note 5		
†Blennow 221–224	S. Sabina	c.1586?[g]	= 7:590.		
Blennow 224–225	S. Adriano al Foro Romano (lost)	1244 or 15th–16th c.?[h]	= 2:140; Claussen, *Magistri*, 236 no. 16. – Cf. *KSRM* 1, 36 with note 64.		
Blennow 226–228	SS. Marcellino e Pietro	1281–1287[i]	= 11:609.	Cecchelli & Persico fig. 6	120, 131, 146, 148
Blennow 228–229	S. Maria di Monterone(?) (lost)	1351	= 2:205.		
Carunchio 25	Villa Lante al Gianicolo	1527	Cf. Chastel, *The Sack of Rome*, 91.[j]	Carunchio fig. 10; Chastel fig. 54.	144
Claussen, *Magistri*, 5 note 4 no. 1	→ Giovenale 63 no. 6				
Claussen, *Magistri*, 10	→ 6:191				
Claussen, *Magistri*, 15–16	→ Claussen, *KSRM* 1, 437				
Claussen, *Magistri*, 16	→ 12:565				
Claussen, *Magistri*, 17	→ 4:146				
Claussen, *Magistri*, 18	→ 4:818				
Claussen, *Magistri*, 22–26	S. Giovanni in Laterano (lost)	12th c. 2/2 (Claussen; 1170 Kendall)	Cf. Kendall 265 no. 129.	*Magistri*, figg. 17 & 19	
Claussen, *Magistri*, 23	→ Kendall 265 no. 129(2)				
Claussen, *Magistri*, 27–28	S. Bartolomeo all'Isola (lost)	12th c. 2/2			

Publication	Place	Date	Other publications	Photos	Referred to on p.
Claussen, *Magistri*, 28–31 & 108–109	S. Paolo fuori le Mura	1180–1200		*Magistri*, figg. 26–28 & 30–1	
Claussen, *Magistri*, 39–40	S. Silvestro in Capite (extant?)	12th c. 2/2		*Magistri*, fig. 40	
Claussen, *Magistri*, 54–55	→ Claussen, *KSRM* 1, 68				
Claussen, *Magistri*, 58–59	→ 7:982				
Claussen, *Magistri*, 60	→ Claussen, *KSRM* 1, 112				
Claussen, *Magistri*, 60–63	→ 1:473				
Claussen, *Magistri*, 64	Vatican (lost)	c.1180–1200			
Claussen, *Magistri*, 73–75	→ Claussen, *KSRM* 1, 139				
Claussen, *Magistri*, 75–76	→ 12:176				
Claussen, *Magistri*, 80–81	S. Ambrogio della Massima (lost)	11th–14th c.?	= Claussen, *KSRM* 1, 66		
Claussen, *Magistri*, 91–94	→ 7:395				
Claussen, *Magistri*, 92	→ 7:396				
Claussen, *Magistri*, 95–96	→ 10:3				
Claussen, *Magistri*, 111–112	Vatican (lost)	1220–1230			
Claussen, *Magistri*, 112	→ Claussen, *KSRM* 1, 118				
Claussen, *Magistri*, 112–115	→ 2:646				
Claussen, *Magistri*, 115–117	S. Sabba	c.1220 or slightly earlier		*Magistri*, fig. 130	
Claussen, *Magistri*, 117–118	S. Croce in Gerusalemme (extant?)	c.1200 or slightly earlier		*KSRM* 1, 438 fig. 350	
Claussen, *Magistri*, 118–119	S. Pudenziana (lost)	c.1210	Cf. Claussen, *KSRM* 1, 107.		

Publication	Place	Date	Other publications	Photos	Referred to on p.
Claussen, *Magistri*, 126–132	S. Giovanni in Laterano	c.1230–1240 (Claussen; 1215–1232/6 Kendall)	= *KSRM* 2, 259–260; Kendall 265 no. 130	*Magistri*, fig. 146; *KSRM* 2, 262 fig. 156	
Claussen, *Magistri*, 133 & 171	S. Paolo fuori le Mura	c.1250–1400		*Magistri*, fig. 175	
Claussen, *Magistri*, 133 note 738	–	c.1210–1230		*Magistri*, figg. 172, 177 & 179	
Claussen, *Magistri*, 141	→ Mondini, *KSRM* 3, 348–349				
Claussen, *Magistri*, 147–148	→ 2:10				
Claussen, *Magistri*, 151–152	unknown origin, now in Palazzo Venezia	13th c. 1/2		*Magistri*, fig. 214	
Claussen, *Magistri*, 153–155	S. Maria de Cambiatoribus (lost)	mid 13th c.			
Claussen, *Magistri*, 158	S. Maria in Monticelli (lost)	1227			
Claussen, *Magistri*, 165–167	→ 4:747				
Claussen, *Magistri*, 170	→ 1:1097				
Claussen, *Magistri*, 170	S. Urbano a Campo Carleo (lost)	13th c. 2/2?			
Claussen, *Magistri*, 172–173	→ Claussen, *KSRM* 1, 61				
Claussen, *Magistri*, 173–174	→ 10:364				
Claussen, *Magistri*, 208–210	→ 8:304				

Publication	Place	Date	Other publications	Photos	Referred to on p.
Claussen, *Magistri*, 212–213 & 222	S. Giacomo alla Lungara (lost)	late 13th – early 14th c?	= *KSRM* 3, 13		
Claussen, *Magistri*, 213	→ 4:746				
Claussen, *Magistri*, 214–216	→ 5:1015				
Claussen, *Magistri*, 216	→ 8:13				
Claussen, *Magistri*, 221	S. Giorgio in Velabro (lost)	late 13th – early 14th c?			
Claussen, *Magistri*, 234–235	S. Stefano Rotondo (extant?)	c.12th–13th c.			
Claussen, *Magistri*, 236 no. 16	→ Blennow 224–225				
Claussen, *Magistri*, 237	S. Silvestro in Capite (lost)	11th–15th c.?			
D'Achille 72–73 note 12	→ 11:723				
D'Achille 74–75 no. 1	→ 8:27				
D'Achille 74–75 no. 2	→ 8:25				
D'Achille 76 note 21	S. Giovanni(?) (lost)	1365			144
D'Achille 76–77 no. 3	→ 8:28(1)				
D'Achille 88 no. 17	→ 5:1123				
De Rossi, *BCAM* 1872, 54–58	S. Giovanni e Paolo (found in 1871; later whereabouts not indicated)	1073–1085	Cf. Favreau 38.		135
De Rossi, *ICUR*[1] II, 422 no. 36	Vatican	1049–1054	Cf. Petrella 48 no. 24.		

Publication	Place	Date	Other publications	Photos	Referred to on p.
Favreau 33–39 no. 1	S. Giovanni e Paolo	1073–1085	= Marini & Mai 211–212 no. 1. Cf. Petrella 46 no. 22; Mondini, *KSRM* 3, 74–75(2).	*MEC* 1 tab. 20 no. 6; Favreau 34; *KSRM* 3, 73 fig.	135
Favreau 221	S. Maria in Trastevere	1130–1143			
Favreau 222 no. 78	S. Maria Maggiore	11th–13th c.?			
Galletti vol. 3 450–452 no. 2(1)	S. Gregorio al Monte Celio	1235/6	Cf. Petrella 42 no. 5.		141
Giovenale 63 no. 6	S. Maria in Cosmedin	11th c.	= Claussen, *Magistri*, 5 note 4 no. 1	Giovenale tab. 24 a–e	
Giovenale 63 no. 7	→ 4:744				
Giovenale 63 no. 8	–	1123		Giovenale tab. 18 a	
Giovenale 63 no. 9	→ Blennow 65–67 no. 13				
Giovenale 63 no. 10	→ Blennow 68–72 no. 14				
Giovenale 64 no. 12	→ 4:748				
Giovenale 64 no. 13	→ 4:749				
Giovenale 64 no. 14	→ 4:747				
Giovenale 64 no. 14bis	–	1295–1304			
Giovenale 64 no. 15	→ 4:746				
Giovenale 64 no. 17	–	15th c.	k	Giovenale tab. 19 b	

Publication	Place	Date	Other publications	Photos	Referred to on p.
†Giovenale 72 no. 69	S. Valentino in Cosmedin	902	= Gray 142–143 no. 141. – Cf. Petrella 51 no. 28.	*MEC* 1 tab. 16 no. 2	117, 125, 139
Giovenale 72 no. 70	S. Maria in Cosmedin	12th c.	l	*MEC* 1 tab. 41 no. 3	152
Giovenale 73 no. 71	–	1140	m	*MEC* 1 tab. 24 no. 7	
†Gray 142 no. 140	S. Sabina	c.920 (Gray, Silvagni)		*MEC* 1 tab. 39 no. 6; Gray tab. 24, 4	
Gray 142–143 no. 141	→ Giovenale 72 no. 69				
†Guillou no. 112	Grottaferrata	1131/2			136
Guillou no. 115	S. Giorgio in Velabro	10th c. 2/4		*MEC* 1 tab. 34 no. 4; Guillou figg. 113–114	
†Guillou no. 116	–	10th c.?		Guillou fig. 115	
Guillou no. 117	→ 10:540				
Hülsen 169	S. Agnese a Duo Forni? (extant?)	13th–14th c.			
Kajanto, Nyberg, and Steinby 169–170 no. 104 + Blennow 99	S. Nicola ai Cesarini	c.1132 (Blennow; 12th c. others)		*MEC* 1 tab. 40 no. 6; Kajanto, Nyberg, and Steinby tab. 59, 3	
†Kajanto, Nyberg, and Steinby 180 no. 111	–	c.1000–1450?	Too fragmentary.	Kajanto, Nyberg, and Steinby tab. 60, 5	

Publication	Place	Date	Other publications	Photos	Referred to on p.
Kajanto, *Papal Epigraphy*, 60, (1)	Vatican (lost)	1195	= 6:7.		
Kajanto, *Papal Epigraphy*, 60, (2)	→ 6:12				
Kajanto, *Papal Epigraphy*, 60, (3)	– (lost)	1437	= 6:47.		
Kajanto, *Papal Epigraphy*, 60–61	Acqua Vergine (lost)	1453	= 13:141.		120
Kajanto, *Papal Epigraphy*, 61, (1)	S. Stefano Rotondo	1453	= 8:538.		
Kajanto, *Papal Epigraphy*, 81	Capitolium	1471–1484		Kajanto fig. 12	
Kajanto, *Papal Epigraphy*, 82 no. 1	Ponte Sisto (lost)	1473			120
Kajanto, *Papal Epigraphy*, 82 no. 2	–	1475	= 13:92.	Kajanto fig. 9; Petrucci tab. 13	
Kajanto, *Papal Epigraphy*, 82 no. 3	–	1475	= 13:93.	Kajanto fig. 10	
Kajanto, *Papal Epigraphy*, 83 with note 63	Via Sistina (lost)	1480	= 13:110.		
Kajanto, *Papal Epigraphy*, 83, (2)	→ 13:111				
Kajanto, *Papal Epigraphy*, 84	Fontana di Trevi (lost)	1471–1484			
Kajanto, *Papal Epigraphy*, 84–85	Capitolium (lost)	1471–1484			
Kajanto, *Papal Epigraphy*, 85	–	1471	= 1:16.	Kajanto fig. 8	141
Kajanto, *Papal Epigraphy*, 86–87	Piazza di Macel de' Corvi (lost)	1484–1492 (prosopography; Forcella 15th c.)	= 13:112.		

Publication	Place	Date	Other publications	Photos	Referred to on p.
Kajanto, *Papal Epigraphy*, 88	→ 6:103				
Kajanto, *Papal Epigraphy*, 88–89	→ 13:9				
Kajanto, *Papal Epigraphy*, 93	Via del Banco di S. Spirito (lost?)	1512	= 13:114.	Kajanto fig. 14	
Kajanto, *Papal Epigraphy*, 94	→ 13:11				
Kajanto, *Papal Epigraphy*, 94 note 13, (1)	→ 13:12				
Kajanto, *Papal Epigraphy*, 94 note 13, (2)	→ 13:13				
Kajanto, *Papal Epigraphy*, 94 note 13, (3)	→ 13:14				
Kajanto, *Papal Epigraphy*, 95, (1)	Piazza Giudia	1508	= 13:113.		
Kajanto, *Papal Epigraphy*, 95–96	Vatican (extant?)	1504	= 6:122.		150
Kajanto, *Papal Epigraphy*, 98	S. Maria ad Martyres = Pantheon	1513–1521	= 1:1121.	Kajanto fig. 17	
Kajanto, *Papal Epigraphy*, 98–99	Capitolium	1513–1521	= 1:40(1).	Kajanto fig. 15–16	122, 135, 152
Kendall 265 no. 129(2)	S. Giovanni in Laterano	12th c. (Silvagni, Claussen; 1170 Kendall)	= Claussen, *Magistri*, 23; *KSRM* 2, 84. – Cf. *KSRM* 2 71 & 75.	*MEC* 1 tab. 41 no. 5; *KSRM* 2, 76 figg. 24–25	
Kendall 265 no. 130	→ Claussen, *Magistri*, 126–132				
Kendall 265–266 no. 131	→ Claussen, *KSRM* 3, 27–29				
Kendall 267–268 no. 133	→ 10:2				

Publication	Place	Date	Other publications	Photos	Referred to on p.
Claussen, *KSRM* 1, 25 note 16	S. Adriano al Foro Romano	11th c.			
Claussen, *KSRM* 1, 58	S. Agnese fuori le Mura (lost)	c.1050–1300			
Claussen, *KSRM* 1, 61	–	c.1250–1260	= Claussen, *Magistri*, 172–173	*KSRM* 1, 56 fig. 40	
Claussen, *KSRM* 1, 66	→ Claussen, *Magistri*, 80–81				
Claussen, *KSRM* 1, 68	S. Anastasia (lost)	1134–1139	= Claussen, *Magistri*, 54–55		
Claussen, *KSRM* 1, 75	–	c.1100–1250		*KSRM* 1, 75 fig. 49	
Claussen, *KSRM* 1, 80(1)	S. Angelo in Pescheria	1291			153
Claussen, *KSRM* 1, 80(2)	–	12th c. 2/2			153
Claussen, *KSRM* 1, 83	S. Antonio Abbate	1259	= 11:246.	*KSRM* 1, 84 fig. 55	
Claussen, *KSRM* 1, 112	SS. XII Apostoli (lost)	1180–1200	= Claussen, *Magistri*, 60		
Claussen, *KSRM* 1, 114	→ 2:662				
Claussen, *KSRM* 1, 116	→ 2:646				
Claussen, *KSRM* 1, 118	– (lost?)	1220	= Claussen, *Magistri*, 112		
Claussen, *KSRM* 1, 121 note 3	→ 6:1072				
Claussen, *KSRM* 1, 121 note 2	S. Balbina	1256–1268		*KSRM* 1, 121 fig. 79	
Claussen, *KSRM* 1, 135–136	S. Bartolomeo all'Isola	1113	= 4:1286. – Cf. Claussen, *Magistri*, 27 note 153	*MEC* 1 tab. 23 no. 3; *KSRM* 1, 146–147 fig. 102	115–116

Publication	Place	Date	Other publications	Photos	Referred to on p.
Claussen, *KSRM* 1, 139	–	c.1180	= Claussen, *Magistri*, 73–75	*Magistri*, figg. 86–7; *KSRM* 1, 164 fig. 112	110
Claussen, *KSRM* 1, 139–140 with notes 30–31	– (lost)	1201–1204 (Claussen; 12th c. Forcella)	= 4:1287.[n]		129, 137
Claussen, *KSRM* 1, 178, 1	→ 9:818				
Claussen, *KSRM* 1, 178, 2	S. Biagio in Cantosecuto = della Pagnotta (lost)	1072			
Claussen, *KSRM* 1, 179 note 6	S. Bibiana (lost)	c.1227–1300		*KSRM* 1, 179 fig. 124 (painting)	
Claussen, *KSRM* 1, 180 note 8	→ 11:230				
Claussen, *KSRM* 1, 226	S. Cecilia in Campo Marzio	12th–14th c.?[o]		*KSRM* 1, 224 fig. 159	120
Claussen, *KSRM* 1, 262	→ 2:59				
Claussen, *KSRM* 1, 303 note 17	S. Clemente	1104–1125? (Claussen; 1127 Silvagni)		*MEC* 1 tab. 27 no. 1	136
Claussen, *KSRM* 1, 345	→ 4:1245				
Claussen, *KSRM* 1, 348	→ 10:540				
Claussen, *KSRM* 1, 349 note 6	SS. Cosma e Damiano in Trastevere	1238			153
Claussen, *KSRM* 1, 370	→ 4:146				
Claussen, *KSRM* 1, 437	S. Croce in Gerusalemme (lost)	1145–1148 (Claussen; 1148 Forcella)	= 8:499; Claussen, *Magistri*, 15–16		
Claussen, *KSRM* 1, 460	→ 2:1180				

Publication	Place	Date	Other publications	Photos	Referred to on p.
Claussen, *KSRM* 1, 478	→ 2:10				
Claussen, *KSRM* 2, 19 no. 3	S. Giovanni in Laterano	c.1180–1200			
Claussen, *KSRM* 2, 20 no. 7	– (lost)	1237			
Claussen, *KSRM* 2, 65 note 220	→ 8:54				
Claussen, *KSRM* 2, 84 (cf. 71 and 75)	→ Kendall 265 no. 129(2)				
Claussen, *KSRM* 2, 90 note 367	– (lost)	1216–1227			153
Claussen, *KSRM* 2, 174 note 731, 1	→ 8:29				
Claussen, *KSRM* 2, 174 note 731, 2	– (lost)	1345			144
Claussen, KSRM 2, 174 note 731, 3	→ 8:27				
Claussen, *KSRM* 2, 191–192 note 842(1)	→ 8:30(1)				
Claussen, *KSRM* 2, 191–192 note 842(2)	→ 8:30(2)				
Claussen, *KSRM* 2, 202 note 893	– (lost)	13th c. 1/2		*KSRM* 2, 200 fig. 107 (drawing)	
Claussen, *KSRM* 2, 259–260	→ Claussen, *Magistri*, 126–132				
Claussen, *KSRM* 2, 262–265	→ 8:1192				
Claussen, *KSRM* 2, 318	–	1196	= 8:7.	*KSRM* 2, 318 fig. 213	135
Claussen, *KSRM* 2, 341	–	1291	= 8:14. – Cf. *KSRM* 2, 56 note 166 & 106–109	*KSRM* 2, 109 fig. 44	116, 126, 148, 152

Publication	Place	Date	Other publications	Photos	Referred to on p.
Claussen, *KSRM* 2, 342–343	–	1291	= 8:16.	*KSRM* 2, 139 fig. 68	152
Claussen, *KSRM* 2, 343–345	–	1291	= 8:15.	*KSRM* 2, 138 fig. 67	152
†Senekovic, *KSRM* 2, 387 note 145P	S. Giovanni in Laterano	1518–1523		*KSRM* 2, 319 fig. 214	
Claussen, *KSRM* 3, 12	→ 6:1060				
Claussen, *KSRM* 3, 13	→ Claussen, *Magistri*, 212–213 & 222				
Claussen, *KSRM* 3, 16 note 16	S. Giorgio in Velabro	1259	= 11:597. – Cf. Petrella 51 no. 27	*KSRM* 3, 19 fig. 3	141, 146
Claussen, *KSRM* 3, 27–29	–	13th c. 1/2 (Claussen, Kendall; 13th c. Forcella, Silvagni)	= 11:596; Kendall 265–266 no. 131	*MEC* 1 tab. 41 no. 7; *KSRM* 3, 28 fig. 10 & 36–38 figg. 20–22	
Claussen, *KSRM* 3, 65	S. Giovanni in Oleo	12th c.		*KSRM* 3, 68 fig. 46	
Mondini, *KSRM* 3, 76 with note 46	→ 10:1				
Mondini, *KSRM* 3, 90–91	→ 10:2				
Mondini, *KSRM* 3, 113–114	→ 10:3				
Senekovic, *KSRM* 3, 190 note 25	S. Gregorio al Monte Celio	1480			
Senekovic, *KSRM* 3, 190 note 26	– (lost)	1471–1484			
Mondini, *KSRM* 3, 283	S. Lorenzo in Lucina	12th c.?		*KSRM* 3, 285 fig. 235	
Mondini, *KSRM* 3, 307 no. I	→ Blennow 43–48 no. 7				
Mondini, *KSRM* 3, 307–308 no. II	→ Blennow 92–96 no. 19				

Publication	Place	Date	Other publications	Photos	Referred to on p.
Mondini, *KSRM* 3, 308 no. III	–	1112 or shortly after (1112 Forcella)	= 5:342. – Cf. Mondini 267 note 27 & 280 note 77	*MEC* 1 tab. 22 no. 3; *KSRM* 3, 281 fig. 232	123, 152
Mondini, *KSRM* 3, 309 no. IV	→ Blennow 134–139 no. 29				
Mondini, *KSRM* 3, 337–339 & 442	S. Lorenzo fuori le Mura	1194–1198			129
Mondini, *KSRM* 3, 339[q]	–	1198–1216		*KSRM* 3, 341 fig. 278	125
Mondini, *KSRM* 3, 348–349	–	1254	= 12 508; Claussen, *Magistri*, 141. – Cf. Mondini, *KSRM* 3, 426.	*KSRM* 3, 348 fig. 286	117
Mondini, *KSRM* 3, 350 with note 131	– (lost)	c.1300–1320			
Mondini, *KSRM* 3, 351 note 135	→ 12:575				
Mondini, *KSRM* 3, 411	→ 12:565				
Mondini, *KSRM* 3, 522, 1	–	1024–1032		*MEC* 1 tab. 18 no. 7; *KSRM* 3, 332 fig. 264	
Mondini, *KSRM* 3, 522, 2	–	1200	Cf. Mondini 349.	*KSRM* 3, 350 fig. 288	
Mondini, *KSRM* 3, 531 note 20	→ 5:1120				
Senekovic, *KSRM* 3, 548 no. I[r]	S. Lucia della Tinta (lost)	1122 (Senekovic; 1002 Forcella)	= 11:374. – Cf. Senekovic 543–544		
Marini & Mai 17 no. 1	S. Giovanni e Paolo	early 11th c.? (Gray; 984 Silvagni)	Cf. Gray 147 no. 149.	*MEC* 1 tab. 17 no. 3; ILCV tab. 40c	

Publication	Place	Date	Other publications	Photos	Referred to on p.
†Marini & Mai 17 no. 2	S. Adriano al Foro Romano[s]	994	Cf. Gray 146.		125, 139
Marini & Mai 38 no. 1	S. Sabina	1145–1153 (Silvagni)		*MEC* 1 tab. 25 no. 3	
Marini & Mai 46 no. 1	S. Cecilia in Trastevere	1185–1187		*MEC* 1 tab. 41 no. 4	
†Marini & Mai 214 no. 2	S. Paolo fuori le Mura (lost)	946–955	Cf. Petrella 46 no. 20.		
†Marini & Mai 215 no. 1	S. Paolo fuori le Mura	probably 928–929 (Gray; 10th century Silvagni; Leo IV's reign, others)	Cf. Giovanni Battista de Rossi, *ICUR*[1] II, 423 no. 40; Gray 143–144 no. 142; Filippi 43 no. SP 673	*MEC* 1 tab. 38 no. 5; Gray tab. 19, 2; Filippi fig. 154	135
Marini & Mai 219 no. 1	S. Nicola in Carcere	1088–1099	Cf. Petrella 45 no. 16.	*MEC* 1 tab. 21 no. 6	146, 147
Marini & Mai 223 no. 1	S. Sabba	1073–1085		*MEC* 1 tab. 20 no. 5	
†unpublished (Filippi 35 no. 50793)	S. Paolo fuori le Mura	"medieval"	Too fragmentary.	Filippi fig. 68	
†unpublished (Filippi 41 no. 51443)	–	"medieval"	Too fragmentary.	Filippi tab. 154	
unpublished (Filippi 41 no. 56169)	–	12th c.		*MEC* 1 tab. 41 no. 1	
†unpublished (Filippi 41 nos. 51434+51440)	–	before 1300	Too fragmentary.	Filippi tab. 94	120
†unpublished (Filippi 43 no. SP 649)	–	c.1000–1450	Too fragmentary.	Filippi tab. 146	
unpublished (Filippi 43 no. SP 674)	–	before 1300		Filippi tab. 94	

Publication	Place	Date	Other publications	Photos	Referred to on p.
unpublished (Filippi 43 no. SP 675)	–	before 1300		Filippi tab. 154	
†unpublished (Filippi 56 no. SP 1782)	–	"medieval"	Too fragmentary.	Filippi tab. 192	
†unpublished (Filippi 56 no. SP 1787)	–	"medieval"	Too fragmentary.	Filippi tab. 192	
†unpublished (Ravenna, Museo Archivescovile, inv. no. 32)[t]	(unknown)	1300			132, 153
†unpublished	SS. Cosma e Damiano(?)	984	Cf. LL. 1–2 transcribed in Gray 146 no. 148.	*MEC* 1 tab. 17 no. 2	139
†unpublished[u]	SS. Vincenzo ed Anastasio alle tre Fontane	12th–14th c.?	Too fragmentary.		
unpublished	S. Clemente	11th c.		*MEC* 1 tab. 35 no. 5	
unpublished[v]	S. Maria dell'Orto	1495			
unpublished	S. Maria in Aventino (extant?)	late 12th c. (Silvagni; 8th c. Diehl)		*MEC* 1 tab. 40 no. 7; Diehl tab. 38a	137
unpublished	S. Maria in Trastevere	1291		Petrucci tab. 9	
unpublished	S. Maria in Trivio	12th c.		*MEC* 1 tab. 40 no. 4	
unpublished	S. Silvestro in Capite	1060		*MEC* 1 tab. 19 no. 4	153
unpublished	SS. Cosma e Damiano in Pigna	1172		*MEC* 1 tab. 35 no. 3	

a Not an inscription; cf. Blennow.

b A forgery from the indicated period. It is not fully certain if this has ever existed as an actual inscription (cf. Blennow).

c As Blennow puts forth, the two first parts of the extant inscription, made in 1518, derive from earlier sources: c. 1220/1 (quite possibly an inscription, but could have been a document), and 1394 (certainly an inscription). The formulae are compatible with the respective dates.

d This first part (LL. 1–12) of the inscription, erected in 1588, seems to be a fairly accurate copy of a medieval original (cf. Blennow).

e This second part refers to the action taken in 1588.

f Blennow: "The title pontifex maximus is not found in the inscriptions from the period, and neither the formula anno post Christum natum." In addition to these features, the abbreviation style (as recorded in the transcription) strongly suggests a Renaissance date.

g 1238 Forcella. Blennow: "[The inscription] was apparently, judging from the sixteenth-century palaeography, executed several centuries after the thirteenth century. It was most probably made in connection with the rebuilding made by pope Sixtus V in the late sixteenth century, at which time the high altar was renewed (...). The inscription is modelled on the style of a papal letter of indulgence (...). The dating in the end of the text is of the short type used in the letters (...). Lines 8–14 are echoed in a papal indulgence document from 1247 (...). Thus, either a document from the time must have been used for creating the inscription, or the text was composed in order to look like such a document."

h The date is problematic. Forcella, based on an earlier description of the lettering and the adjoining insignia, and the formulation "*magister Iulianus refecit hoc opus,*" favours a 15th century date, followed by Blennow, who adds that "the formulations (...) are not found in our period" Claussen, *Magistri*, questions this (for an erroneous reason, since the remake does not refer to the main altar, but to an altar of St Agnes), but more recently (*KSRM* 1) accepts the "*refecit*" formulation to refer to a later "*Erneuerung oder Wiederherstellung.*"

i Blennow, Cecchelli and Persico; 1256 Forcella. Blennow: "The expressions (...) are not found in the inscriptions of our period. (...) The last two lines reveal that the inscription in fact was executed some decades after the consecration in 1256. (...) Thus the inscription should be dated to 1281–1287 (...). This is also suggested by the language." Note that 1:1098 dating from 1270 (S. Maria ad Martyres = Pantheon), shares the rare expression "*anno nativitatis*" (see my article, 120 graph 4.4, and note 48).

j Chastel's reading is slightly inaccurate, and the photo is inferior to Carunchio's.

k A label in Greek.

l Silvagni describes this and the following inscription, Giovenale no. 71, jointly as "*Charta publica de dono populi Tiburtini ecclesiae S. Angeli*" – the former as the upper part of an uncertain date, and the latter as the lower and dated part. The inscriptions seem to be related to the same event but are to be considered as separate: they differ in the style of the language and in paleography.

m Cf. above Giovenale no. 70.

n L. 3: Forcella omits the indication that the verse was recorded as incomplete (as the metre requires).

o The object is placed in a niche (*KSRM* 1, 224 fig. 159), with only one inscribed side visible, but according to a 1604 writer, the date *MCXXXI* is recorded on another side which cannot be seen today. Claussen calls this date into question since the visual features point towards the following century, and "[*m*]*an muss für die Regierungszeit des Gegenpapstes Anaklet II (1130–38) immer wieder mit nachträglich „bereinigten" Inschriften rechnen.*" For some reason, Claussen transcribes HAEC and CAECILIA where the photo shows only E.

- p Commemoration of the deceased Iulianus de Dathis.
- q Judging from the photo (p. 341 fig. 278), many indications of lacunae have been omitted in Mondini's transcription. For the most part, this does not affect the reading, but on LL. 2–3 CELESTINI III ET INNOCE|NTII ROMANORVM, the latter words should be read *Innoce|nt[i* I]ỊI *Rom[a]ṇor(um)*: first, the abbreviation of *(um)* has been omitted, and second, in the lacuna the I after the T was likely smaller and written under the preceding T, as at the end of the same line or in *Lau̯rentii* on L. 5. The rest of the lacuna with the succeeding surviving letters accommodate the numeral *III* easily. The ordinal number is to be expected, especially since Celestine III has it.
- r Reconstructing this lost inscription is difficult because of considerable variation in the surviving transcriptions; cf. Senekovic, *KSRM* 3, 543–544 and 548 no. I.
- s A surviving fragment is on display in the museum of Crypta Balbi.
- t This church bell is not included in the material, because I only came across it at a later stage. For further details, see my article note 92. Thus far, I am unaware of a proper publication of this object.
- u In the southern wall, bordering the path past SS. Vincenzo ed Anastasio to S. Paolo alle Tre Fontane (inspection 9 May 2017). A medieval building inscription: "[- - -]+*id(us) fec(it) h(oc) op(us) fieri et L*[- - -] | - - - - -". The first letters belong most likely to a personal name ending with -idus.
- v In the façade, above the right-hand side entrance (inspection 30 April 2017).

The following epitaphs have been referred to on pages indicated in the parentheses: 1:1252; 1:1280–1281; 1:1655–1657; 2:268; 2:460; 2:678 (131); 2:1052 (153); 2:1344; 3:29; 3:533; 3:558; 3:791–792; 3:1066; 4:180; 4:932; 4:1247; 5:73–74; 5:820; 5:824 (131); 6:32 (139); 6:55 (131); 6:59 (132); 6:110; 7:606 (131); 8:1 (112, 148, 152); 8:926 (131); 9:743 (134); 10:120 (153); 11:286 (131); 12:569 = Garms et al., *Mittelalterliche Grabmäler*, 96–97 no. xxi, 2 (70, 131, 133); 13:1265 (131); Kajanto, *Papal Epigraphy*, 20–21 (152).

APPENDIX 3

Topographic Summary of Inscriptions

For bibliographical information, please refer to Appendix 1.

The following two tables summarise the provenance of the inscriptions in Appendices 1 and 2: the first one includes all churches and the Vatican, and the second includes all other buildings and places. Dash "/" separates multiple lemmata in one publication or Forcella's volume, and "||" separates publications and Forcella's volumes.

Churches	
Adriano al Foro Romano	2: 144–145 || Blennow 145–154 no. 32 / 171–178 no. 37 / 224–225 || *KSRM* 1, 25 note 16 || †Marini & Mai 17 no. 2
Agata alla Suburra	10: 564 / 568–570 || †Blennow 203–204
Agnese a Duo Forni?	Hülsen 169
Agnese fuori le Mura	11: 543–545 || Blennow 184–179 no. 40 || *KSRM* 1, 58 / 61
Agnese in Piazza Navona	Blennow 59–62 no. 11
Agostino	5: 223 / 239
Ambrogio della Massima	9: 636 || Claussen, *Magistri*, 80–81
Anastasia	10: 72 || *KSRM* 1, 68 / 75
Andrea a Ponte Molle	12: 245
Andrea della Valle	8: 1203
Angelo in Pescheria	4: 224 || *KSRM* 1, 80, 1 / 80, 2
Annunziata all'Arco dei Pantani	Blennow 210–215(1)–(3)
Antonio Abbate	11: 248 || *KSRM* 1, 83
Balbina	6: 1072 || *KSRM* 1, 121 note 2
Barbara dei Librai	7: 788
Bartolomeo all'Isola	Claussen, *Magistri*, 27–28 || *KSRM* 1, 135–136 / 139 / 139–140 with notes 30–31
Benedetto in Piscinula	10: 152
Biagio al Campidoglio	2: 188 / 191
Biagio in Cantosecuto = della Pagnotta	9: 818 || *KSRM* 1, 178, 2

 | DOI:10.1163/9789004436251_012

Biagio della Fossa	9: 843
Bibiana	11: 228 / 230 || Blennow 217–220 / 220–221 || *KSRM* 1, 179 note 6
Bonifacio ed Alessio	7: †735 || 13: 1210
Brigida a Piazza Farnese	9: 622–624
Cecilia in Campo Marzio	*KSRM* 1, 226
Cecilia a Monte Giordano	Blennow 72–76 no. 15
Cecilia in Trastevere	2: 59 / †62 / 63 / 82 || Blennow 27–32 no. 3 / 31–2 / 33–35 no. 4 || Marini & Mai 46 no. 1
Clemente	4: 1244–1245 || 13: 1029 || *KSRM* 1, 303 note 17 || unpublished
Cosma e Damiano dei Barbieri	6: 86 || 13: 586
Cosma e Damiano al Foro Romano	4: 146
Cosma e Damiano in Pigna	unpublished
Cosma e Damiano in Trastevere	10: 540 / 542–543 || *KSRM* 1, 349 note 6
Cosma e Damiano(?)	†unpublished
Costanza	Blennow 190–194 no. 41(1) / 190–194 no. 41(2)
Crisogono	2: 487 / 1591 || Blennow 76–84 no. 16 / 84–88 no. 17
Croce in Gerusalemme	8: 498 / 500 / 502 / 504(1)–(3) / 505 || Claussen, *Magistri*, 117–118 || *KSRM* 1, 437
Eligio dei Ferrari	11: 449
Eusebio	Blennow 154–157 no. 33
Eustachio	2: 1178–1180 / 1186 / 1204 / 1230 || 13: 432 || Blennow 127–134 no. 28
Francesco delle Stimate	4: 1144
Giacomo alla Lungara	6: 1060–1061 / 1067 || Claussen, *Magistri*, 212–213 & 222
Giacomo degli Spagnoli	3: 494 / 507 / 538 / 545 || 13: 428
Giorgio in Velabro	Claussen, *Magistri*, 221 || Guillou no. 115 / †no. 116 || *KSRM* 3, 16 note 16 / 27–29
Giovanni in Ayno	9: 981
Giovanni Decollato	7: 121
Giovanni dei Fiorentini	7: 19 || 13: 144–145
Giovanni dei Genovesi	7: 327

Giovanni in Laterano	8: 3–6 / 12–13 / 17–18 / 25–29 / 30(1)–(2) / 31 / 42 / 44–45 / 54 / 1192–1193 || Claussen, *Magistri*, 22–26 / 126–132 || Kendall 265 no. 129(2) || *KSRM* 2, 19 no. 3 / 20 no. 7 / 90 note 367 / 174 note 731, 2 / 202 note 893 / 318 / 341 / 342–343 / 343–345 / †p. 387 note 145
Giovanni della Malva	9: 712–713
Giovanni in Oleo	*KSRM* 3 65
Giovanni della Pigna	9: 969 / 975
Giovanni a Porta Latina	Blennow 120–123 no. 26
Giovanni(?)	11: 298 || D'Achille 76 n. 21
Giovanni e Paolo	10: 1–3 / 7 || Blennow 110–112 no. 23 / 182–173 no. 39 || De Rossi, *BCAM* 1782, 54–58 || Favreau 33–39 no. 1 || Marini & Mai 17 no. 1
Girolamo della Carità	4: †622 / 623 || 13: †809–†815 / 853
Gregorio al Monte Celio	2: 258 / 289 || Blennow 39–42 no. 6 / 41 || Galletti vol. 3 450–452 no. 2(1) || *KSRM* 3, 190 note 25 / 190 note 26
Lorenzo in Damaso	13: 588 / 616
Lorenzo fuori le Mura	12: 565 / 570 / 575 || *KSRM* 3, 337–339 & 442 / 339 / 348–349 / 350 with note 131 / 522, 1 / 522, 2
Lorenzo in Lucina	5: 345 / 347 || Blennow 43–48 no. 7 / 46 note 95 / 92–96 no. 19 / 134–139 no. 29 / 138 || *KSRM* 3, 283 / 308 no. III
Lorenzo ai Monti	5: 433–437
Lorenzo in Panisperna	5: 1120 / 1123
Lorenzolo	11: 723
Lucia del Gonfalone	7: 860
Lucia della Tinta	11: 375 || *KSRM* 3, 548 no. I
Marcellino e Pietro	11: 611 || Blennow 226–228
Marco	4: 818 / 821
Margherita	13: 1328
Maria dell'Anima	3: 1071
Maria in Aquiro	2: 1334 || Blennow 112–116 no. 24
Maria in Aracoeli	1: 415 / 453 / 456 / 473–474 / 497 / 2078(1) / †2078(2)
Maria in Aventino	unpublished

Maria del Buon Consiglio	11: 705 || Blennow 48–53 no. 8
Maria de Cambiatoribus	Claussen, *Magistri*, 153–155
Maria in Campitelli	5: 1015 || Blennow 19–26 no. 2 / 139–141 no. 30
Maria in Campo Marzo	10: 867–868 / 870
Maria in Campo Santo	13: 520
Maria della Consolazione	8: 783 / 1206 / 1211 / 1230–1234 || Blennow 161–164 no. 35
Maria in Cosmedin	4: 744 / 746–749 || Blennow 65–67 no. 13 / 68–72 no. 14 / 71 with note 143 || Giovenale 63 no. 6 / 63 no. 8 / 64 no. 17 / 64 no. 14bis / 72 no. 70 / 73 no. 71
Maria in Cuppella	Blennow 35–39 no. 5 / 53–54 no. 9
Maria in Domnica	8: 485–486 || 13: 1
Maria Maggiore	11: 1 / 4–5 / 7–8 / 18 / 44 || Favreau 222 no. 78
Maria ad Martyres	1: 1097–1098 / 1104 / 1114 || Blennow 215–217(1) / †pp. 215–217(2) || Kajanto, *Papal Epigraphy*, 98
Maria di Monterone(?)	Blennow 228–229
Maria in Monticelli	5: 1353–1354 || Claussen, *Magistri*, 158
Maria Nuova	2: 5 / 10
Maria dell'Orto	unpublished
Maria della Pace	5: 1296
Maria del Popolo	1: 1175 / 1196–1197 / 1201 / 1209 / 1224 / 1237 / 1254 / 1320 || 13: 1042 / 1279
Maria Porta Paradisi	12: 141
Maria in Publicolis	4: 1114 || 13: 1138
Maria del Sole	12: 442 / †p. 311
Maria sopra Minerva	1: 1567 / 1629 / 1641 / 1690–1691 || 2: 1056 || 13: 425 / 433 / 903 / 905
Maria Traspontina	6: 1093 || 13: 422–423
Maria in Trastevere	2: 1035(1) / †1035(2) / 1037 / 1054 / 1072 / 1080 || Favreau 221 || unpublished
Maria in Trivio	unpublished
Maria Maddalena	8: 1001
Martina e Luca	Blennow 194–198 no. 42 / 198(1) / 198(2)
Martinello al Monte di Pietà	10: 360 / 362
Matteo in Merulana	10: 728 || Blennow 204–210
Michele Arcangelo e Magno	Blennow 105–10 no. 22

Nicola in Carcere	4: †260 / 263 / 280 || Blennow 88–92 no. 18 || Marini & Mai 219 no. 1
Nicola ai Cesarini	Blennow 96–100 no. 20 || Kajanto, Nyberg, and Steinby 169–170 no. 104 + Blennow 99 / †p. 180 no. 111
Nicola dei Funari	Blennow 116–119 no. 25
Nicola dei Prefetti	10: 364
Onofrio	5: 816 / 821
Pancrazio	11: 572 / †573
Pantaleo	4: 1314 / 1316
Paolo fuori le Mura	12: 2 / 4–6 / 11–13 / 17 || Claussen, *Magistri*, 28–31 & 108–109 / 133 & 171 / 133 note 738 || †Marini & Mai 214 no. 2 / †p. 215 no. 1 || †unpublished (Filippi 35 no. 50793) / † (41 nos. 51434+51440) / † (41 no. 51443) / (41 no. 56169) / † (43 no. SP 649) / (43 no. SP 674) / (43 no. SP 675) / † (56 no. SP 1782) / † (56 no. SP 1787)
Paolo alla Regola	4: 1265
Pietro in Montorio	5: 686 / 696 || 13: 1142
Prassede	2: 1490–1492 / 1494–1495
Prisca	11: 311–313
Pudenziana	11: 262 || Claussen, *Magistri*, 118–119
Quattro Coronati	8: 717(1)–(2) || Blennow 165–70 no. 36
Quirico e Giulitta	8: 736–738
Sabba	12: 176–178 || Claussen, *Magistri*, 115–117 || Marini & Mai 223 no. 1
Sabina	7: 622–624 || Blennow 178–182 no. 38 / 199–202 no. 43 / †pp. 221–224 || †Gray 142 no. 140 || Marini & Mai 38 no. 1
Salvatore delle Coppelle	Blennow 123–127 no. 27
Salvatore in Lauro	7: 171
Salvatore ai Monti	Blennow 15–19 no. 1
Salvatore in Pensilis	3:721
Salvatore al Ponte Senatorio	9: 531
Salvatore a Porta Paolo	12: 456–457
Salvatore in Primicerio	11: 321
Salvatore alla Scala Santa	8: 301–305
Sebastiano fuori le Mura	12: 186 || Blennow 141–145 no. 31
Sebastiano al Palatino	12: †505–†506
Sergio e Bacco	9: 701

Silvestro in Capite	9: 149 || Claussen, *Magistri*, 39–40 / 237 || unpublished
Silvestro al Quirinale	4: 89
Silvestro e Dorotea	9: 739–740
Silvestro e Martino ai Monti	4: 2 / 10
Simeone Profeta	2: 581
Sisto	10: 901 / 904–905
Spirito in Sassia	6: 1173 / 1183 / 1214 || Blennow 62–64 no. 12
Stanislao dei Polacchi	3: 722
Stefano del Cacco	7: 982
Stefano Rotondo	8: 539 || Claussen, *Magistri*, 234–235 || Kajanto, *Papal Epigraphy*, 61, (1)
Stefano del Trullo	2: 1485
Teodoro	10: 448
Tommaso di Cantorbery	7: 335
Tommaso ai Cenci	Blennow 55–58 no. 10 / 157–161 no. 34
Tommaso in Formis	7: 395–398
Tommaso in Parione	Blennow 100–104 no. 21
Urbano a Campo Carleo	9: 999 || Claussen, *Magistri*, 170
Valentino in Cosmedin	†Giovenale 72 no. 69
Venanzio ed Ansovino	7: 237 / 245
Vincenzo ed Anastasio alle tre Fontane	12: †444 / 445 || †unpublished
Vitale	11: 356–357
Vito e Modesto	11: 285
XII Apostoli	2: 645–646 / 662 / 664 || *KSRM* 1, 112 / 118
Vatican	6: 4–6 / †8–†9 / 12 / 14 / 17 / 21–22 / 24–26 / 46 / 48 / †50 / 51 / 53 / 56–58 / 63 / 67[bis 40] / †69 / 75–76 / 88 / 90 / 92 / 97 / 103 / 105–106 / 115 / 118 / 124–126 / 129–132 / 135–147 / 149–157 / 165–166 / 168 / 191 / 1647 /p. 4 /p. 5 || 13: 331 || Claussen, *Magistri*, 64 / 111–112 || De Rossi, *ICUR*^{1} II, 422 no. 36 || Kajanto, *Papal Epigraphy*, 60, (1) / 60, (3) / 95–66

Other Buildings and Places

Acqua Vergine	Kajanto, *Papal Epigraphy*, 60–61
Arch of Constantine	13: 325

Villa Lante al Gianicolo unknown origin	Carunchio 25 Claussen, *Magistri*, 151–152 || †unpublished (Ravenna, Museo Archivescovile, inv. no. 32)

APPENDIX 4

Calendrical Contrivances Used in Rome between the 10th and 16th Centuries

Calendrical contrivances with average datings used in Marko Halonen's article. For datings, please refer to the catalogues of the libraries in question.

Signum	Location	Average year
Arch.Cap.S.Pietro.A9	Città del Vaticano, Biblioteca Apostolica Vaticana (BAV)	1339
Arch.Cap.S.Pietro.A.9_2	BAV	1339
Stamp.Arch.Cap.S.Pietro.A.41	BAV	1496
Arch.Cap.S.Pietro.B.65	BAV	1427
Arch.Cap.S.Pietro.B65_2	BAV	1426
Arch.Cap.S.Pietro.B.67	BAV	1425
Arch.Cap.S.Pietro.B.67_2	BAV	1425
Arch.Cap.S.Pietro.B.69	BAV	1463
Arch.Cap.S.Pietro.B.71	BAV	1350
Arch.Cap.S.Pietro.B.72	BAV	1475
Arch.Cap.S.Pietro.B.72_2	BAV	1475
Arch.Cap.S.Pietro.B.79	BAV	1163
Arch.Cap.S.Pietro.B.79_2	BAV	1257
Arch.Cap.S.Pietro.B.81	BAV	1467
Arch.Cap.S.Pietro.B.84	BAV	1350
Arch.Cap.S.Pietro.B.84_2	BAV	1350
Arch.Cap.S.Pietro.B85	BAV	1400
Arch.Cap.S.Pietro.B.85_2	BAV	1400
Arch.Cap.S.Pietro.B.87	BAV	1300

 | DOI:10.1163/9789004436251_013

Signum	Location	Average year
Arch.Cap.S.Pietro.B.87_2	BAV	1475
Arch.Cap.S.Pietro.B.88	BAV	1463
Arch.Cap.S.Pietro.B88_2	BAV	1463
Arch.Cap.S.Pietro.D.156	BAV	1150
Arch.Cap.S.Pietro.D.156_2	BAV	1150
Arch.Cap.S.Pietro.E4	BAV	1450
Arch.Cap.S.Pietro.E6	BAV	1250
Arch.Cap.S.Pietro.E6_2	BAV	1400
Arch.Cap.S.Pietro.E7	BAV	1350
Arch.Cap.S.Pietro.E7_2	BAV	1350
Arch.Cap.S.Pietro.E.14	BAV	1263
Arch.Cap.S.Pietro.E.14_2	BAV	1263
Arch.Cap.S.Pietro.E.15	BAV	1400
Arch.Cap.S.Pietro.E.15_2	BAV	1400
Arch.Cap.S.Pietro.F27	BAV	1350
Arch.Cap.S.Pietro.F.27_2	BAV	1350
Arch.Cap.S.Pietro.H.28	BAV	1071
Arch.Cap.S.Pietro.H.56	BAV	1150
Arch.Cap.S.Pietro.H.56_2	BAV	1250
Arch.Cap.S.Pietro.H.56_3	BAV	1450
Arch.Cap.S.Pietro.H.57	BAV	1405
Arch.Cap.S.Pietro.H.57_2	BAV	1405
Arch.Cap.S.Pietro.H.57_3	BAV	1498
Arch.Cap.S.Pietro.H.58	BAV	1100
Barb.lat.421	BAV	1000
Barb.lat.563	BAV	1433
Barb.lat.563_2	BAV	1433

Signum	Location	Average year
Barb.lat.598 (XIII7)	BAV	1450
Barb.lat.598 (XIII7)_2	BAV	1450
Barb.lat.609	BAV	1548
Barb.lat.646/Vat.lat.4885	BAV	1111
Barb.lat.646/Vat.lat.4885_2	BAV	1111
Chig.C.V.138	BAV	1157
Chig.C.V.138_2	BAV	1157
Chig.C.VI.174	BAV	1300
Inc.IV.10	BAV	1488
Ott.lat.3	BAV	1200
Ott.lat.3_2	BAV	1200
Ott.lat.38	BAV	1059
Ott.lat.38_2	BAV	1059
Ott.lat. 356	BAV	1275
Ott.lat. 356_2	BAV	1322
Ott.lat. 356_3	BAV	1322
Ott.lat. 356_4	BAV	1400
S.Maria.Magg.91	BAV	1350
S.Maria.Magg.97	BAV	1300
S.Maria.Magg.97_2	BAV	1300
S.Maria.Magg.99	BAV	1450
S.Maria.Magg.105 (INC)	BAV	1476
S.Maria.Magg.105 (INC)_2	BAV	1476
S.Maria.Magg.106	BAV	1350
S.Maria.Magg.119	BAV	1480
S.Maria.Magg.120	BAV	1478
Vat.lat.378.	BAV	950

Signum	Location	Average year
Vat.lat.378._2	BAV	1050
Vat.lat.4406	BAV	1200
Vat.lat.4406_2	BAV	1200
Vat.lat.4885/(Barb.lat.646)	BAV	1100
Vat.lat.6827 i	BAV	1417
Vat.lat.6827 i_2	BAV	1350
Vat.lat.6827 ii	BAV	1450
Vat.lat.6827 ii_2	BAV	1450
Vat.lat.7209	BAV	1456
Vat.lat.7658	BAV	1250
Vat.lat.7658_2	BAV	1335
Vat.lat.7722	BAV	1350
Vat.lat.7722_2	BAV	1450
Vat.lat.7792	BAV	1483
Vat.lat.9490	BAV	1469
Vat.lat.12986	BAV	1320
Vat.lat.12986_2	BAV	1391
Vat.lat.12986_3	BAV	1392
Vat.lat.12988	BAV	1342
Vat.lat.12988	BAV	1450
Vat.lat.14701	BAV	1450
MSS.24	Roma, Biblioteca Angelica = BA	1415
MSS.24_2	Roma, Biblioteca Angelica	1415
MSS.1098	Roma, Biblioteca Angelica	1475
MSS.1909	Roma, Biblioteca Angelica	1450
MSS.1909_2	Roma, Biblioteca Angelica	1450
Rari.I.2.7	Roma, Biblioteca Angelica	1582
Z.14.10/1	Roma, Biblioteca Angelica	1551

Signum	Location	Average year
F.85	Roma, Biblioteca Vallicelliana	1188
Latin.826	Paris, Bibliothèque Nationale de France (BNF)	1313
Latin.826_2	BNF	1313
Latin.755	BNF	1250
Latin.755_2	BNF	1296
Add.MS.30034	BL	1381
Add.MS.30034_2	BL	1381
Inc.40.4113	Nürnberg, Germanisches Nationalmuseum	1496
BSB.Ink.M.459	München, Bayerische Staatsbibliothek = BSB	1482
BSB.Ink.M.471	BSB	1488

Index of Persons

Index of Places

www.ingramcontent.com/pod-product-compliance
Lightning Source LLC
Chambersburg PA
CBHW071248180926
51613CB00002B/173
9789004436244